A Guide to the National Parks

A Guide
TO THE
National Parks

THEIR LANDSCAPE AND GEOLOGY

William H. Matthews III

FOREWORD BY PAUL TILDEN

Doubleday / Natural History Press
GARDEN CITY, NEW YORK

This book is a revised and updated version of *A Guide to the National Parks,* Volumes I and II, published by the Natural History Press in 1968
Natural History Press Paperback Edition: 1973
The following five National Parks maps were provided by *National Parks Magazine:* Grand Canyon, Mount McKinley, Mount Rainier, Petrified Forest, Yellowstone.

ISBN: 0-385-06298-2
Library of Congress Catalog Card Number 72-89824

TO THE MEN AND WOMEN IN GREEN—
THE UNIFORMED PERSONNEL
OF THE NATIONAL PARK SERVICE

Contents

ABOUT THE AUTHOR ix

FOREWORD xi

PREFACE TO THE REVISED EDITION xvi

ABOUT THIS BOOK xvii

1 *The National Park Story* 1

2 *Science and Scenery* 11

3 *The Parks Are Yours—Enjoy and Protect Them* 46

4 *Acadia National Park, Maine* 63

5 *Big Bend National Park, Texas* 75

6 *Bryce Canyon National Park, Utah* 85

7 *Canyonlands National Park, Utah* 91

8 *Carlsbad Caverns National Park, New Mexico* 103

9 *Crater Lake National Park, Oregon* 117

10 *Everglades National Park, Florida* 129

11 *Glacier National Park, Montana* 137

12 *Grand Canyon National Park, Arizona* 155

13 *Grand Teton National Park, Wyoming* 169

14 *Great Smoky Mountains National Park,*
North Carolina-Tennessee 181

15 *Guadalupe Mountains National Park, Texas* 193

16 *Haleakala National Park, Hawaii* 199

17 *Hawaii Volcanoes National Park, Hawaii* 205

18 *Hot Springs National Park, Arkansas* 213

19 *Isle Royale National Park, Michigan* 221

20 *Lassen Volcanic National Park, California* 233

21 *Mammoth Cave National Park, Kentucky* 243

22 Mesa Verde National Park, Colorado   257
23 Mount McKinley National Park, Alaska   271
24 Mount Rainier National Park, Washington   285
25 North Cascades National Park, Washington   307
26 Olympic National Park, Washington   315
27 Petrified Forest National Park, Arizona   331
28 Platt National Park, Oklahoma   347
29 Redwood National Park, California   353
30 Rocky Mountain National Park, Colorado   361
31 Sequoia-Kings Canyon National Parks, California   377
32 Shenandoah National Park, Virginia   393
33 Virgin Islands National Park, St. John   405
34 Voyageurs National Park, Minnesota   415
35 Wind Cave National Park, South Dakota   421
36 Yellowstone National Park, Wyoming   429
37 Yosemite National Park, California   453
38 Zion National Park, Utah   475

GLOSSARY   488
SELECTED READINGS   507
INDEX   513

William H. Matthews III is Professor of Geology at Lamar University in Beaumont, Texas, and has served as Director of Education for the American Geological Institute in Washington, D.C., and as Visiting Scientist for the Texas Academy of Science. He is active in a number of educational and industrial organizations, the Earth Science Curriculum Project at the University of Colorado, Encyclopaedia Britannica Films, the Texas State Parks Commission, the Council on Education in the Geological Sciences, and the National Science Foundation.

He has written numerous articles for professional and educational periodicals and many books, among them *Texas Fossils, Fossils: An Introduction to Prehistoric Life, Wonders of the Dinosaur World, Exploring the World of Fossils, The Story of the Earth, Wonders of Fossils, Geology Made Simple, Invitation to Geology,* and *Introducing the Earth.* In 1965 he was awarded the Neil Miner Award of the National Association of Geology Teachers "for exceptional contributions to the stimulation of interest in the Earth Sciences."

Foreword

As a youngster growing up in the mountain country of northern New England I often used to hear older folks say, when the dark clouds gathered, that "rain on the hill meant water at the mill." And since all the little woodworking plants and sawmills and woolen factories I knew about were perched athwart leaky wooden dams on valley streams, the proposition seemed reasonable. It demanded nothing more than a firm belief in the proposition that water runs downhill.

When I later became involved in the American conservation movement, and especially in its preservation facet, it soon became apparent that there is no escape from the truisms, even those specialized by avocation; good words that can be painted or sprayed over ideas like varnish to enhance their sales appeal. One phrase of this sort, of recent coinage, is current in the conservation world, and especially in that part of Government that is responsible for the administration and protection of the world's greatest system of National Parks. The easy expression is, that *parks are for people*.

And who would deny it? Parks really are for people. Now and again one may hear a quiet voice asking whether an especially fine example of remaining primitive America, with its plants, animals and geologic story, might not be protected by the public for nothing more than its own sake; but the thought becomes lost in the daily shuffle of practicality. The truism has merit. Parks *are* for people. They buy them, bear the cost of operation, and use them today in numbers that have become a little frightening to their administrators, the devoted folk of the National Park Service. If we may assume that it took the author of this volume perhaps a year to write his text, then during that time some forty million Americans had descended on their thirty-three National Parks.

A National Park is easily bounded, described, and measured; but its visitors and their motives have, over the years, remained largely unexplored and unclassified. Exactly what were these millions looking for in the Parks? There has never been a precise answer to this question, nor, probably, will there ever be one, although the National Park Service has been groping more or less blindly for an answer ever since it came into jurisdiction of the Park system back in 1916. But perhaps I might be allowed to speculate, at least, with a number of years' unofficial work in Park matters as license.

Without doubt a small number of visitors arrive in the National Parks every year through sheer accident. One may easily make a wrong turn in the road or be curious to learn where the road leads, and unwittingly become a figure in the dry columns of Park use that are published by the Park Service in Washington every month.

Another small fraction of visitation includes people who arrive in the National Parks with a definite, single purpose in mind—professional folk embarking on a mission of investigation into some field of the natural sciences; camera enthusiasts taking advantage of grand, unspoiled country as background for wildlife pictures; wilderness hikers who would shun the company of their fellow man for a while.

But by far the greater part of the forty million were plain, ordinary Americans fleeing the dullness and the routine of ordinary lives; a majority of them nowadays probably are from the city and the suburb, taking themselves and their children into the clean out-of-doors. They camp, swim, fish, hike, paddle a canoe, and generally use their great system of national havens for exactly those purposes for which they were established—for re-creation. They flee the smog, the pollution, the city traffic; the urgent telephone calls, the conferences, and the ideas that must be fitted into some important frame of reference. If there are individual problems in the Parks, they are of the pleasant or the temporary sort. How does an unskilled hand produce super-sized sticks from a large and half-green log with nothing but a light hatchet? Who forgot to fill the tank of the gasoline stove when there was plenty of light and no mosquitoes?

If one could sound out this vast unclassified group, I doubt that many of its individuals would admit that their National Park visit was motivated by an urge for education; nor, likely, would many be found who were hostile to the notion. And yet, ever since the National Park system was established, both the National Park Service and its private following of Park protectionists have been suggesting the educational opportunities that are inherent in the Parks—thinking of education in its very broadest sense. A national nature reserve is authorized by Congress "for the benefit, inspiration, education, recreational use and enjoyment of the public," say the prologues to some of our park laws. Both Park Service and conservationists like to think of the Parks, over and above their recreational purposes, as units of a national educational institution which offers informal classes in the architecture and history of the earth; an institution that enforces no hours, grants no degrees, and posts no entrance requirements, other than a certain minimum level of curiosity about the world of nature. How best, then, to encourage the public to enroll?

The National Park Service tackled this question a good many years ago. Only twelve years after the Service was established a distinguished committee of conservationists and scientists was at work molding an approach to informal education for the public in the Parks. The names of Bryant, Bumpus, Kellogg, Merriam, and Oastler will be remembered as long as the American Park system exists for purposes other than purely physical recreation; for the report that was turned in then to the Secretary of the Interior constituted a great break in the course of Park Service thinking. Whatever the views of individuals may have been, the official aim of park use had been, up to then, almost purely recreation in the physical sense—sightseeing, if one prefers the term. But from this study and report eventually evolved the Park-wide program of pleasant education—more recently and perhaps more accurately described as interpretation—which one finds in the Parks today, with its illustrated lectures, self-guiding trails, dioramas, pamphlets, booklets, and such other devices as ingenuity can create. All of these are aimed toward bringing the National Park visitor into a little closer contact with Mother Earth, and into a closer kinship with her lesser inhabitants.

But there have been many private as well as public practitioners of the craft of good park interpretation. Conservationists—especially the Park enthusiasts among them—have constantly sought the printed word to encourage a public attitude toward the National Park system, which looks beyond mere physical recreation. And increasingly, in recent years, the conservationists have found powerful and literate allies in the world of science.

For our scientists are finding that they need Parks too. Quite aside from esthetic considerations, men of science find the National Park system a great natural laboratory in which man and his developments have produced a minimum of distortion. In a Park the geologist finds intact the nation's finest examples of earth-making and earth-destruction. In a Park the biologist finds his plants and animals living in nearly primitive relationship, free of the imbalances impressed upon them elsewhere by man's search for food and shelter. In the National Park the hydrologist finds the free-flowing streams that he must have as benchmarks for measuring the subtle changes wrought by man on the rivers and river basins of the outside world. When natural phenomena must be measured, reconstructed, or counted, science turns increasingly to the National Park system as a research laboratory in which disturbing factors are at a bare minimum. And in carrying out these works, many a scientist has fallen in love with the Parks, as has the author, obviously, of this book; and from the love has come the desire to share with the Park-going public—which can, in truth, be only dimly identified—some of the beauties of specialized knowledge.

Parks really are for people. Parks are for all kinds of people —the scholar, the lonely hiker, the hopeful cameraman, the patient camper, the myriad other Americans who arrive in

their great natural preserves for myriad reasons. It is partly
through interpretive works such as this that some of these ar-
rivals—many, I hope—will be induced to enroll in the great
university of the Parks, both for their own deeper satisfaction
and for the ultimate benefit of the National Park idea.

PAUL M. TILDEN

Assistant to the President,
National Parks Association, and Editor,
National Parks *Magazine*

Preface to the Revised Edition

This version of *A Guide to the National Parks* contains the original two-volume set in a single paperbound edition. Hopefully, the new edition will not only be more convenient for the traveler, its lower price should make it more readily available to visitors to the National Parks.

The major changes have been within the activities available at the various parks. The changes—most of which were made at the suggestion of the National Park Service—relate to facilities and activities that have been, and will continue to be, subject to change. A new section, "Tips for Tourists," has been added for each Park. Here will be found information relating to highlights of all special activities that are available in the respective Parks. As always, however, it is advisable to check at an entrance station or visitor center in order to determine what activities and programs are available.

Another major change in the revised edition is the elimination of a number of photographs. This has been necessary in order to reduce the size of the book and make it available at lower cost. But the basic story remains: the fascinating geologic story behind the scenery in the National Parks. Hopefully, this edition, like its predecessor, will help the visitor understand and appreciate what he has seen in the National Parks.

WILLIAM H. MATTHEWS III

July 1, 1972

About This Book

Someone has called our National Parks "outdoor geological laboratories." This describes them well, for few places provide better opportunity to see how geologic processes have sculptured the landscape. Indeed, most National Parks have been established because of their outstanding geologic significance and to preserve the natural features that can be seen there.

Yet most visitors take the Parks' scenery for granted. They do not stop to consider the natural forces that produced the spectacular landscape around them. But the geologist views the scenery quite differently: he sees a landscape that owes its origin to the rocks that underlie it and to the geologic tools that Nature has used to shape the land.

This book tells the geologic story of the National Parks. It explains, in non-technical language, the geologic phenomena that produced the natural features for which each Park is noted. As you read this geologic story, you will learn much about the processes and materials of geology and about the agents that create landscape features. A better understanding of these phenomena will lead to an increased appreciation of the Parks from both a scientific and an aesthetic point of view. It is hoped, moreover, that this book will inspire the reader to visit our National Parks and to recognize and understand the meaning behind their magnificent scenery.

Because many of the National Parks owe their existence to a variety of geologic processes, there is occasional repetition from one chapter to another. This has been done in order that the reader may treat each Park as a separate unit and not have to carry all the facts in mind. In this way, cross-referencing has been kept to a minimum.

Although emphasis is placed on the geology and natural history of the National Parks, additional information is also provided. There is a "thumbnail sketch" of each Park available for quick reference to factual information that will help in planning your visit. Included also is a map of each area that is discussed.

It should be noted that some of the information provided about each Park is subject to change. Many of the National Parks are in a state of continual expansion as more and better facilities become available each year. For this reason the reader should check the latest informational brochure and schedule of interpretive services for the particular Park concerned. This information is normally made available as the visitor enters the Park, or it may be obtained free of charge by mail. (The mailing addresses of the National Parks will be found at the end of each chapter.)

In the back of the book is a Glossary of geologic terms that are commonly used in discussing the geology of the Parks. Use it freely and frequently.

ACKNOWLEDGMENTS

Despite the many years of travel and research that have gone into this book, it could not have been written without the help of many individuals and organizations. More specifically, it could never have been produced without the generous and enthusiastic co-operation of the National Park Service. In order to insure accuracy and the best possible coverage, each chapter has been reviewed by the Superintendent and/or other personnel of the Park concerned. Park personnel have also provided much basic information about their respective areas and made available many of the photographs that appear in the book. For this—and a host of other courtesies—the author gratefully acknowledges the help of the following men and women of the National Park Service: Arthur C. Allen, Jack K. Anderson, William Arnold, Robert J. Badaracco, Robert L. Barrel, Merrill D. Beal, Glen T. Bean, Ernest J. Borgman, Dick Boyer, Myrl Brooks, Perry E. Brown, W. W. Bryant, Leslie W. Cammack, Bernard T. Campbell, Robert W. Carpenter, Robert J. Carr, James W. Carrico, Ernst Christensen, Willard W. Danielson, Willard Dilley, Jack B. Dodd, William W. Dunmire, Henry P. During, Francis H. Elmore, Douglas B. Evans, Paul G. Favour, Jr., George W. Fry, Bennett T. Gale, Charles J. Gebler, Frank R. Givens, John M. Good, Russell K. Grater, Neal G. Guse, Lawrence C. Hadley, Allen R. Hagood, Louis W. Hallock, Dwight L. Hamilton, Allyn F.

Hanks, Bryan Harry, Arthur J. Hayes, Larry E. Henderson, Robert C. Heyder, R. Taylor Hoskins, Warren D. Hotchkiss, Jerry B. House, Douglass H. Hubbard, Harold A. Hubler, Richard Hughes, Charles E. Humberger, Carl E. Jepson, C. E. Johnson, Fred T. Johnston, William R. Jones, Stanley C. Joseph, Louis G. Kirk, Joseph Kuleza, Rickard L. Lake, Jess H. Lombard, Paul F. McCrary, Leonard W. McKenzie, Mrs. June Maguire, George D. Marler, Alan Mebane, Paul McG. Miller, R. B. Moore, Robert C. Morris, Frank R. Oberhansley, George Olin, Luther T. Peterson, Jr., Mrs. Jean M. Pinkley, Richard S. Raynor, H. V. Reeves, Jr., George B. Robinson, Edwin L. Rothfuss, Dick Russell, John A. Rutter, Peter G. Sanchez, James W. Schaak, E. Ray Schaffner, Henry G. Schmidt, William A. Schnettler, Franklin G. Smith, Donald M. Spalding, John A. Stephens, Howard B. Stricklin, Frank E. Sylvester, Lynn R. Thompson, William G. Truesdell, John Tyers, Robert F. Upton, Philip F. Van Cleve, W. Verde Watson, John Wagoner, Harry W. Wills, Bates E. Wilson, Andrew C. Wolfe, W. Ward Yeager, and Mrs. Gale Koschmann Zimmer.

Other members of the National Park Service who assisted in providing information and materials include Elizabeth H. Coiner, science writer, and Carlos S. Whiting, former Acting Chief of Information. In addition, the author is especially indebted to Robert H. Rose, former National Park Service Research Geologist; Francis X. Kelly, former Chief of Press Relations; and John Vosburgh, Chief, Branch of Features, for their invaluable assistance throughout the entire preparation of the manuscript.

For the foreword to this book, the author is indebted to Paul M. Tilden, editor of *National Parks* magazine. Mr. Tilden also supplied certain of the maps and other helpful information.

Thanks are also due to the following persons who supplied information or checked portions of the manuscript for technical accuracy or adequacy of content: Dr. John Eliot Allen and Dr. Robert O. Van Atta, Portland State College; Fred Amos and David E. Jensen, Ward's Natural Science Establishment; Dr. Saul Aronow and Dr. H. E. Eveland, Lamar University; Dr. Donald M. Baird, University of Ottowa; Dr. Robert E. Boyer, The University of Texas; Thomas S. Childs, Jr., Natural History Press; Dr. D. R. Crandell, U. S. Geological Survey; Barbara De Groot; K. C. Den Dooven, *Western Gateways* magazine; Dr. Thomas W. Donnelly, Rice University; Dr. Erling Dorf, Princeton University; the late Dr. James L. Dyson, Lafayette College; Dr. Peter T. Flawn, Bureau of Economic Geology of The University of Texas; Dr. Robert L. Heller, University of Minnesota, Duluth; Dr. Jack L. Hough, University of Michigan; Donald Hummel, Glacier Park, Inc.; Dr. Daniel J. Jones, University of Utah; Dr. Robert E. Kelly, Michigan Geological Survey; Dr.

Preston McGrain, Kentucky Geological Survey; Maureen Mahon; Dr. Gordon B. Oakeshott, California Division of Mines and Geology; Charles J. Ott, McKinley Park, Alaska; Dr. John S. Shelton, Claremont, California; Dr. Webster F. Stickney, Maine Department of Economic Development; Dr. Howel Williams, University of California at Berkeley; and Ginny Wood, Camp Denali, McKinley Park, Alaska.

Special thanks are due my son, Jim, who assisted me with photography and a host of other details. Finally, I wish to thank my wife, Jennie, for help and encouragement in every phase of the project. She accompanied me to the Parks, assisted in basic research, prepared the index, and critically reviewed the entire manuscript. Her many suggestions contributed greatly to the final presentation.

W.H.M. III

A Guide to the
National Parks

1. The National Park Story

Most Americans are familiar with and justly proud of their country's National Parks. Illustrations in their science and geography books acquaint school children with the breath-taking scenery of Yosemite, the subterranean beauty of Carlsbad Caverns, and the famed bears and geysers of Yellowstone. And, at some time during their youth, the more fortunate of these youngsters will accompany their parents on vacation trips to one or more of these spectacular areas. Still later, many of them will return to allow their own children to enjoy the scenic and scientific wonders of the National Parks.

But for all of this, how much do we *really* know about our National Park system? When, for example, was it formed and what was our first National Park? Who administers the Parks? What is the difference between a National Park and a National Monument? How are National Parks established? The answers to these frequently asked questions can considerably heighten the visitor's appreciation and understanding of the areas administered by the National Park Service.

How It All Began

Although the fabled Yellowstone area was not really explored until 1870, rumors of a fantastic natural wonderland in the "Yellow Rock" country south of the Montana Territory had been circulating for years. Unfortunately the men who originated these reports were not considered reliable. It was no secret that John Colter and Jim Bridger, like so many of their fellow frontiersmen, were well known for their ability to tell a good story, often at the expense of the facts. So it is not surprising that their tales of cliffs of black glass, smoking hills, and hot springs which spewed torrents of boiling water were treated with skepticism by even the most gullible listener. Indeed, some of Jim Bridger's descriptions of Yellowstone's scenic wonders, suggests that his ability as an outdoorsman was probably exceeded only by his incredibly vivid imagination:

There exists in the Park country a mountain which was once cursed by a great medicine man of the Crow nation. Everything upon the mountain at the time of this dire event became instantly petrified and has remained so ever since. All forms of life are standing about in stone where they were suddenly caught by the petrifying influences, even as the inhabitants of ancient Pompeii were surprised by the ashes of Vesuvius. Sage brush, grass, prairie fowl, antelope, elk, and bears may there be seen as perfect as in actual life. Even flowers are blooming in colors of crystal, and birds soar with wings spread in motionless flight, while the air floats with music and perfumes siliceous, and the sun and the moon shine with petrified light!*

Fortunately not all of the early frontiersmen painted their accounts of Yellowstone with such a broad brush, and their more factual reports stimulated further interest in this unusual area. Finally a group of Montanans under the leadership of General Henry D. Washburn, Surveyor General of the Montana Territory, decided to explore the legendary "Yellow Rock" country, and on the night of September 19, after spending several awe-filled weeks exploring the astonishing Yellowstone area, the party gathered around a campfire. Now that the "tall tales" of John Colter and Jim Bridger had been found not to be completely within the realm of fancy, what should be done about it? Would they, like Bridger and Colter, be ridiculed when they reported the incredible natural phenomena that they had seen?

The problem was finally solved by Judge Cornelius Hedges.

* From *The Yellowstone National Park,* by Hiram Martin Chittenden, edited and with an introduction by Richard A. Bartlett. Copyright © 1964 by the University of Oklahoma Press.

This country, he said, should be owned by the government—it should be set aside as a great national park for all of the people to use and enjoy. Thus, Judge Hedges by virtue of his vision and imagination, sparked a chain of events that were to eventually lead to the establishment of the world's first National Park.

Upon their return to civilization the members of the Washburn Expedition immediately started to promote the National Park idea. Judge Hedges wrote a series of newspaper articles extolling the marvels of the Yellowstone region and Lieutenant Doane, who had commanded the expedition's army escort, wrote an official report of the trip which was publicized in a number of newspapers and magazines. But by far the most active and fervent missionary for the National Park project was Nathaniel Langford, who had kept the official record of the trip, and who now traveled widely giving public lectures and wrote magazine articles. More important, he got in touch with high government officials and he impressed upon them the unique character of the Yellowstone country and the importance of preserving it for posterity.

Fortunately the work of Langford and his friends was not in vain, and in 1871 a scientific party under the direction of the United States Geologist Dr. Ferdinand V. Hayden spent several months exploring the Yellowstone region. In addition to scientific personnel, there were two painters and a photographer in the group. They hoped to bring back graphic, indisputable proof of the natural wonders of this amazing area.

At the completion of their expedition, members of the Hayden party prepared a detailed geologic and descriptive report of Yellowstone. Their findings, published as an official government document, clearly substantiated the authenticity of the earlier reports. In fact, one member of the Hayden group suggested that Langford had been far too conservative in his eulogy of the "Yellow Rock" country.

Finally on December 18, 1871, Judge Hedge's vision of a "national park" took its first step toward reality: William Clagget, Delegate to Congress from the Montana Territory, introduced a Yellowstone Park Bill into both Houses of Congress,

and on March 1—only seventy-three days after it had been introduced—the bill was signed by President Grant, thus officially establishing Yellowstone National Park.

As a fitting end to the story, "National Park" Langford,
he was now called, was made Superintendent of the newly created Park. But at times he must have considered his appoint-
ment a rather dubious honor, for he served for five years without pay and without a Park appropriation. Eventually, in
1878, Congress appropriated $10,000 for the preservation and
development of Yellowstone. The Second Superintendent,
P. W. Norris, used this money to build roads and rough shelters
and to hire additional personnel to patrol and develop the
Park.

Recognizing the wisdom of earlier legislative action and
heeding the pleas of conservationists, Congress in 1890 established General Grant (now a part of Kings Canyon National
Park), Sequoia, and Yosemite National Parks in California.
A fifth Park, Mount Rainier in Washington, was added to the
system in 1899.

As time passed public interest in the Parks movement grew
and Congress took further steps to protect the nation's areas of
historic and scientific significance. One of the most important of
these moves was the passage of the Antiquities Act of June 8,
1906. Introduced by Representative John F. Lacey of Iowa,
this measure was especially drafted to halt vandalism of the
archeological ruins of the Southwest. In reality it did much
more—it gave the President of the United States power to establish National Monuments by presidential proclamation.

Ten years later Congress passed the National Parks Act providing for establishment of the National Park Service as a separate bureau of the Department of the Interior. Its first Director
was Stephen T. Mather, a wealthy Chicago businessman who
had served as Assistant to the Secretary of the Interior since
1915. A remarkably dedicated and energetic worker, Mather
not only gave of himself, he spent much of his personal fortune
on the development and promotion of the National Parks. He
also did much to further the educational and interpretive work
in the Parks and to improve the various concession systems

which were then in operation. There is no doubt that the National Park system would probably never have reached its present status had it not been for his tremendous drive and devotion.

The National Park Service

From Yellowstone's approximately two million acres has grown the world's first and finest system of National Parks. As of January 1, 1972, 285 areas in fifty states, the Virgin Islands, Puerto Rico, the District of Columbia, and the Panama Canal Zone are administered by the National Park Service. These federally designated areas encompass more than thirty million acres and include (in addition to the National Parks) National Monuments, National Historical Parks, National Seashores, National Cemeteries, National Memorials, and numerous other areas of scenic, scientific, or historic interest.

Creation and Authority of the Service. As noted earlier, the National Park Service was established in the Department of the Interior by the act of August 25, 1916. Subsequent legislation, executive orders, and proclamations have added greatly to the National Park system and has considerably expanded the activities of the Service.

Uniformed Personnel. The general operation of areas administered by the National Park Service is under the direct supervision of a Superintendent, who may be assisted by Park Rangers, Naturalists, Historians, and Archeologists.

The major functions of the Ranger are those of conservation and protection. Thus, his duties include enforcing park regulations, giving out public information, supervising entrance stations, and protecting the area's natural features. But when necessary he may be called upon to fight fires, find lost persons, and to perform various rescue operations—often at great personal risk. Most Parks also have an official Park Naturalist. These persons, the majority of whom have college training in geology, biology, forestry, or some other natural science, are concerned primarily with research and the interpretation of the area's natural features. By means of nature walks, illustrated lectures, and museum displays, the Park Naturalist helps the visitor to understand better the natural history or scientific significance of each unit. Certain of the National Historical Parks and Historic Sites have Park Historians in attendance. Well versed in the history and folklore of the units committed to their charge, they answer questions and give lectures on these topics.

The "Men and Women in Green"—as uniformed Park personnel are commonly called—are placed in the Parks to assist and protect you. They gladly answer questions, direct you to a parking area, or assign you a campsite. They will try to find you if you should become lost, give a campfire talk on the geology or natural history of the area, or lead a nature walk. They also warn about fire hazards, enforce traffic regulations, and ask hikers to stay on the trail rather than take that tempting short cut. They make sure that no one touches the cave formations in Carlsbad Caverns, and they will request that you not take home even a "tiny sample" of the Petrified Forest. And at Yellowstone they will also remind you not to feed those "friendly" bears.

The Rangers and Naturalists are interested in each and every visitor and want him to enjoy his visit to the utmost, yet they are also pledged to protect and conserve the areas for which they are responsible. By enforcing the few, simple, and reasonable Park regulations, these dedicated conservationists are doing what must be done if these areas are to be preserved for future generations of Americans.

Tourists, who normally expect to see Rangers at National Parks and Monuments, are sometimes startled to find them on duty at the Statue of Liberty in New York or George Washington Carver's birthplace in Missouri. Most people do not realize that in addition to caring for the nation's scenic wonders, the National Park Service is also responsible for the protection of many of our more famous historic sites, monuments, and battlefields.

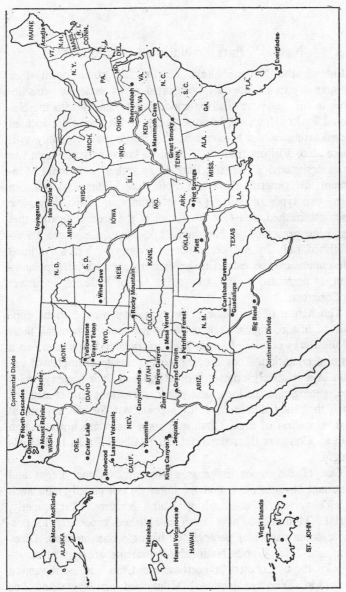

Map of United States showing locations of National Parks.

National Park or National Monument?

Although the National Park Service has some twenty-four different types of federally designated areas under its jurisdiction, in this book we shall be concerned primarily with the National Parks. However, because many of the Parks are located near certain of the Monuments the reader will probably visit these also. Visitors frequently want to know why one area has been designated a National Park and another a National Monument. In general, there are two basic differences between these two types of units: (1) the manner in which they have been established, and (2) the nature of the area or feature that has been preserved. A National Park, for example, can be established only by a specific act of Congress. Yet a National Monument can be established either by presidential proclamation, as permitted by the Antiquities Act of 1906, or by an act of Congress.

The nature of the place to be preserved must also be considered in the original designation of the area. In general, most of the Parks consist of larger areas of outstanding scenery in a natural condition. They are, moreover, classic examples of nature in a relatively undisturbed setting and their scenery, wildlife, geologic formations, and other attractions are considered such that they should be forever preserved for the enjoyment and education of man. Obviously such areas are not common and so Congress does not establish National Parks indiscriminately.

One of the major differences between National Parks and National Monuments is that the latter do not usually offer such a wide variety of outstanding natural features. Consequently, certain Monuments have been established especially to preserve some particular geologic or biologic phenomenon. Dinosaur and Muir Woods National Monuments are typical examples of these. Located in northeastern Utah and northwestern Colorado, Dinosaur National Monument displays many fine dinosaur remains in a relatively small area, while Muir Woods

National Monument, in California, is famous for its virgin stand of Pacific Coast redwood. But most National Parks (Yellowstone, for example) exhibit a far wider array of natural and scenic attractions than do these two Monuments. Many of the National Monuments are dedicated to the preservation of historical and archeological sites. However, only one National Park—Mesa Verde in southwestern Colorado—has been established exclusively for the preservation of an area of this type.

Although most of the Parks are large, size is not necessarily a criterion. Yellowstone, the largest of the National Parks, sprawls over 2,221,772 acres and parts of three states, Wyoming, Montana, and Idaho. Most of the other National Parks also number tens of thousands of acres; but, the smallest, Platt National Park in south-central Oklahoma, is confined to less than 912 acres. Conversely, Alaska's Katmai National Monument encompassing 2,792,137 acres, and Glacier Bay National Monument, Alaska, which covers 2,803,840 acres, are both larger than our largest National Park.

At times more practical considerations have entered into the matter of designating National Monuments. Because of existing laws, an area can often obtain protected status sooner if it is designated a National Monument rather than a National Park.

With so many factors to be considered, it is not surprising that there has been occasional confusion as to whether an area should be established as a Park or a Monument. In some respects, the two may be quite similar; in others, totally different. The confusion surrounding what a National Monument "should" be has been summarized by one writer as follows: "A National Monument is a piece of land containing from 1 to 1,000,000 acres, either flat or rough, timbered or bare . . . But the most clearly outstanding character of the National Monument is its complete inconsistency."

Occasionally a National Monument will be elevated to the status of a Park. The most recent example of this took place in 1962 when the Petrified Forest, a well-known area in eastern Arizona which had been a National Monument since 1906, was officially designated Petrified Forest National Park.

Although the President has the authority to establish National Monuments by proclamation, he cannot abolish them. The abolition of a National Park or a Monument can be brought about only by a specific act of Congress.

There is yet another type of area in the National Park System: National Recreation Areas. Congress has expressed interest in the development of recreation areas to insure "adequate outdoor recreation resources" for all of our citizens. Beginning with appropriations for the Blue Ridge Parkway in 1933 and accelerated greatly in the decade of the sixties in response to the needs of our burgeoning urban population, Congress has included thirty recreation areas within the National Park System. These recreation areas have many different titles: National Seashore, National Lakeshore, National Waterway, National Riverway, National Scenic Riverway, National Parkway, National Recreation Demonstration Area, National Recreation Area, and similar names.

Thus the National Park System is not, and has not been, a static institution. Evolving since 1872, the National Park System includes now a wide variety of areas—national parks and national monuments of scientific significance to preserve some of the finest examples remaining of our natural inheritance; national parks, national monuments, national historic sites, memorials, and battlefields to preserve and commemorate the most cherished places and monuments of our Nation's history; and national recreation areas, which provide a variety of superior outdoor recreational opportunities.

2. Science and Scenery

The magic combination of science and scenery is the primary attraction for the millions of visitors who throng America's National Parks. Many people come to the Parks simply to get away from it all—to enjoy the solace and quiet of these superb wilderness areas and to escape for a short time the harried pace of everyday living. These visitors require little more than to sit quietly and contemplate the silent grandeur of Grand Canyon or the ice-draped flanks of Mount Rainier. Others want to get to know the scenery more intimately: to photograph the plants and animals; hike through primeval forests; fish in sparkling mountain streams; scale craggy mountain peaks. But regardless of one's reasons for visiting a National Park or how one spends his time there, the scenery will usually dominate his thoughts. And—directly or indirectly—scenery is the magnet that has drawn him to the area.

Fortunately, many visitors are not content to accept the Park's scenic splendors at face value. Instead, their curiosity is aroused; they want to know what is *behind* the scenery. Each year they ask: How were the mountains formed? Why are there so many different kinds of rocks? What causes the geysers to spout? Did the river *really* carve that canyon? Where did the glaciers go? How can flowers grow in the snow?

What, precisely, *is* scenery? How does it originate, and why may it differ so dramatically from one National Park to the next? The "what" of the above question is rather easily answered, for the dictionary states that scenery is "a picturesque view or landscape." But the "how" and "why" are not so simple, for landscapes and scenery may vary greatly from one place to another, and their differences arise from a variety of causes.

As is so often the case, we must rely on science to supply the missing answers. More specifically, we must turn to the sciences of *biology* and *geology,* for the combination of life forms and physical surroundings are the elements that form the scenery of an area. Thus, glistening glaciers and snowshoe rabbits, volcanic craters and silversword plants, snow-capped mountains and alpine fir, are among the biophysical elements that combine to compose scenery. In short, the scenery of an area is embodied in its *natural history:* its life forms and their natural physical surroundings.

Geology—the Science Behind Park Landscapes

What is geology? The word is self-defining, for it is derived from the Greek *geo,* "the earth," and *logos,* "study." Geology, then, is the study of the earth: its composition, its behavior, and its history. One need not, of course, be a geologist to appreciate and enjoy the scenery of a National Park. Yet—because most of our National Parks are areas of outstanding geologic significance—even a passing acquaintance with certain basic geologic principles can greatly enhance your visit. More important, your Park experience will be much more rewarding if you have some understanding of the meaning behind the scenery.

But, though geology and scenery are inseparable, the geologist must be more than a mere interpreter of landscapes, for geology is concerned with matters other than landforms. In fact, the geologist studies the earth within the framework of many other sciences. From astronomy he learns something about the origin of Earth, the planet; chemistry provides an understanding of the nature of rocks and minerals; the basic principles of physics help explain the mechanics of earthquakes and solve the mysteries of mountain building; and without the science of biology, the geologist could not hope to reconstruct the life forms of the geologic past. So geology is actually a synthesis of the sciences, an intriguing body of knowledge that has made possible a better understanding of the planet that is our home.

Because its scope is so broad, geology has been divided into two general fields: *physical geology* and *historical geology*. Physical geology deals with the structure and composition of the earth and with the physical processes which affect it.

Historical geology, on the other hand, is concerned primarily with the origin of the earth, the physical changes which it has undergone, and the history of life as recorded in its rocky crust. In his role as earth historian the historical geologist turns detective. For like the astronomer who scans the heavens searching for clues to solve riddles thousands of light years away, the geologist probes the earth—delving into its crust for evidence that will help unravel the mysteries of the geologic past. In so doing he traces the development of the earth from its cloudy beginnings more than five billion years ago through the frigid saga of the Great Ice Age. And he follows the slow march of life from the earliest known simple plants, more than two billion years old, through the "Age of Dinosaurs," to ancestral man—a relative newcomer on the geologic scene.

Rocks: the Raw Materials of the Landscape

Many of the more spectacular features of the natural landscape are composed of or have been carved into solid rock. This is especially true in the National Parks, for these scenic areas derive much of their natural beauty from the character of their exposed rock formations and the effect of geologic agents upon them.

Because *rocks* are the raw materials of geology and the stuff from which landforms are shaped, it will be helpful to know something of their general characteristics and their role in the development of the landscape. Rock is everywhere around us and is one of the commonest things in the world. Yet few people can actually define a rock. So, at the outset we should learn that rock is a naturally formed physical mass of one or more minerals. And because rocks are composed of *minerals,* we

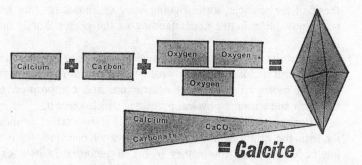

Most minerals are chemical compounds composed of a number of elements. From *The Story of the Earth*, by William H. Matthews III, Harvey House, Inc., 1968.

should also know that a mineral is a naturally occurring substance which has a fairly definite chemical composition (such as gold) and which occurs in a definite shape called a crystal. Perhaps we should also note that although some mineralogists cannot agree on the exact definition of a mineral, the above explanation will satisfy our purposes here.

It is generally believed that all the rocks of the earth's crust have originated in one of three general ways, and they have been classified accordingly. So in the broadest sense, there are three major classes of rocks: *igneous, sedimentary,* and *metamorphic* rocks.

Igneous Rocks. Rocks that have formed by the cooling and hardening of *magma* (molten rock) from within the earth's crust are called *igneous rocks* (from the Latin word *ignis* meaning fire). This type of rock becomes increasingly abundant at depth, for 95 per cent of the outermost ten miles of the earth's crust is composed of rocks of igneous origin. On the earth's surface, however, they are much less common. Some igneous rocks, for example, *granite* and *gabbro,* have cooled and solidified beneath the earth's surface. Due to the fact that they formed from magma injected or intruded into the surrounding rocks, these are called *intrusive,* or *plutonic,* igneous rocks. In general, those places where intrusive igneous rocks appear exposed on the earth's surface are areas that have undergone much erosion. The plutons, which were originally formed deep within the crust, have been gradually uncovered as the overlying and surrounding rocks have been worn away by wind, water, and weather. Rocks of this type are exposed in many of the National Parks, and you will see intrusive igneous rock in the bald, granite domes of Yosemite National Park, the sawtoothed peaks of the Tetons, and along the rocky coast of Acadia National Park in Maine.

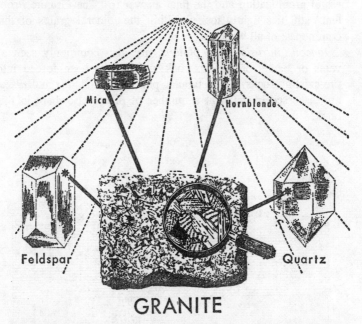

GRANITE

The most common igneous rock, granite, consists mostly of quartz and feldspar. However, other minerals, such as mica and hornblende, are commonly present. BUREAU OF ECONOMIC GEOLOGY OF THE UNIVERSITY OF TEXAS

Despite the fact that granite is generally classified as an igneous rock, there are some authorities who are not so sure that this is correct. They argue that many of the larger masses of granitic rock are much too large to have been intruded into the surrounding rock. For example, the granite rocks of the Coast

Ranges of British Columbia are found in a great linear belt at least 150 miles wide and about 1000 miles long. Furthermore, this granite is known to be some 7000 feet thick and there is no indication of any bottom to it. Because it is difficult to perceive igneous intrusions of such magnitude, there are those who believe that these igneous-looking granitic rocks were formed by *granitization* and that they are not of igneous origin. Unfortunately there is a great deal that is not known about the process of granitization and the final answer to "The Granite Problem" still lies tightly locked within the mineral grains of this commonest of all "igneous" rock.

Igneous intrusions, or *plutons,* are also commonly seen as veins or bands of igneous rock which have been forced into pre-existing rocks. These tabular plutons, called *sills* and *dikes,* are discussed in detail at a number of places elsewhere in this book.

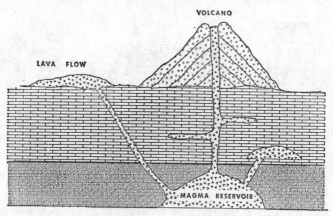

Cross section of a typical volcano.

When, as is often the case, magma spills out upon the surface, the molten rock is called *extrusive* igneous rock or *lava.* Extrusive, or *volcanic,* rock may reach the surface by forcing its way upward and erupting through volcanoes or by flowing out of great fissures in the earth's crust. Volcanic eruptions through a central vent typically build mountains or *volcanic cones,* of which there are excellent examples in Lassen Volcanic, Mount Rainier, Crater Lake, Hawaii, and Haleakala National Parks. But lava extruded as *fissure flows* spreads out rather evenly over the surface in thick, horizontal layers. Remnants of ancient fissure flows can be seen in Isle Royale and Shenandoah National Parks, and visitors to the Pacific Northwest will probably also notice the thick lava flows of the Columbia Plateau of Washington, Idaho, and Oregon. This extensive *lava plateau* covers more than 200,000 square miles and in places massive sheets of hardened lava called *basalt* attain a thickness of 4000 feet. Basalt, a black, dense extrusive rock is probably the most common rock type in the large lava flows throughout the Parks. However, *rhyolite,* a light-colored, fine-grained solidified lava, is also common. The yellowish rocks which form the walls of the Grand Canyon of the Yellowstone in Yellowstone National Park are composed primarily of rhyolite.

Lava commonly reaches the surface mixed with steam and gases which froth up leaving the rock filled with small bubbles. This hardened "rock foam" is called *pumice* and is filled with tiny *vesicles,* or air pockets. Pumice, which has a spongy appearance and is so light that it will float, is common in many of the National Parks which are associated with volcanic activity, especially in the Pinnacles Area of Crater Lake National Park, Oregon.

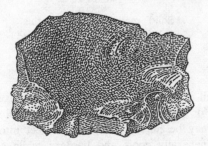

Obsidian is a glassy volcanic rock
formed from quickly cooling lava.

Jim Bridger's fabled black glass cliff of Yellowstone National Park is also composed of extrusive igneous rock. This rock, *obsidian,* is a natural volcanic glass which was formed from lava that cooled so quickly that neither gas bubbles nor mineral crystals had time to form.

In some areas, for example, Glacier, Hawaii Volcanoes, and Olympic National Parks, the solidified lava is occasionally found as lumpy, rounded masses called *pillow lava.* This type of rock structure is believed to have developed when hot lava was immersed in water or was extruded on the floor of a lake or sea. Another unique feature commonly associated with the cooling of lava is *columnar jointing.* Consisting of a set of fractures with a more or less hexagonal pattern, these vertical *joints* (or fractures) are generally considered to be shrinkage joints caused by contraction of the rock during cooling. This type of structure is classically developed at Devils Tower and Devils Postpile National Monuments and in a number of the National Parks including Shenandoah, Olympic, and Mount Rainier.

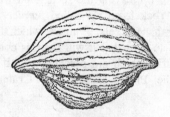

Solid material such as this volcanic bomb is produced by more explosive volcanic eruptions.

But not all of the extrusive igneous rocks consist of solidified lava flows, for when a volcano erupts explosively, molten rock and gases may be hurled thousands of feet into the air. The lava spray generally cools and hardens while still in the air and falls to the ground as solid particles of volcanic dust, ash, and cinders. This material is known collectively as *pyroclastic* —literally "fire-broken"—material. Pyroclastics are common in areas where vulcanism has taken place and the cinder cones of Lassen Volcanic National Parks and Wizard Island—a cinder cone that developed in Crater Lake—consist of great mounds of this violently ejected volcanic debris. The great destruction that can be wrought by heavy showers of pyroclastic material is dramatically displayed in the Devastation Trail area in Hawaii Volcanoes National Park.

Sedimentary Rocks. Even the most casual observer of the Grand Canyon of the Colorado in Arizona is likely to note that the walls of this awesome chasm consist of layer upon layer of rock. These layered, or *stratified,* rocks were originally loose deposits of sand, mud, and gravel. But in the course of time— vast periods of time—these unconsolidated *sediments* have been hardened into beds of *sedimentary rocks.*

Some of these sediments, those derived from the decay and disintegration of previously existing rocks, have been moved from their point of origin by some agent of erosion. Rivers, for example, carry sand and silt to the sea, wind may pick up finer sediments and carry them halfway around the globe, and the larger glaciers may transport boulders the size of a house. Because they consist largely of broken particles of pre-existing rocks, these rock fragments are called *clastic* or *detrital sediments.* Sandstone, clay, shale, and conglomerate (a coarse rock consisting of large to small, more or less rounded pebbles) are among the more common clastic sedimentary rocks.

The *chemical sediments,* on the other hand, originate quite differently. The *inorganic chemical sediments* were once dissolved in water, from which they were precipitated out of solution by water or evaporation. Gypsum, halite (rock salt), and certain types of inorganic limestone were formed by this process. *Biochemical,* or *organic chemical,* sediments are composed of the remains or products of ancient plants and animals. Coal, coquina (a mixture of loosely cemented shells), and fossiliferous limestone are examples of biochemical or organic sediments.

Although only about 5 per cent of the outer ten miles of the crust consists of sedimentary rocks, these rocks make up roughly 75 per cent of the rocks that are exposed on the earth's surface. Sedimentary rocks are also widely exposed in the National Parks and they have played an important role in reconstructing the geologic history of these areas. When ripple

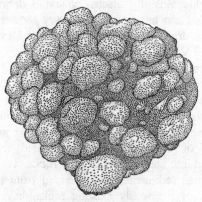

Conglomerate is a sedimentary rock made up of the rounded fragments of pre-existing rocks.

marks, cross-bedding, mud cracks, and fossils are present they provide valuable clues as to the conditions under which the sediments were deposited. These, and other features of sedimentary rocks, are treated in some detail later in this book.

Metamorphic Rocks. The third, and most complex, class of rocks are the *metamorphic rocks*. Derived from two Greek words which literally mean "change in form," these "made over" rocks can form from either sedimentary or igneous rocks. Thus the process of metamorphism may transform a fine-grained limestone (a sedimentary rock) into a harder, coarse textured, crystalline, metamorphic rock called marble.

What forces are great enough to bring about such drastic physical change? Actually, metamorphism comes about in a number of ways. During mountain building movements and other crustal disturbances, powerful forces squeeze, bend, break, and rub the rocks in the earth's crust. When this happens the heat and pressure produced while the rock is under stress may cause the minerals to become more closely crowded together forming a tightly interlocking mass. At other times rocks may be invaded by mineral-bearing gases and liquids boiling up from nearby magmas. When these materials seep into the surrounding rock they may dissolve some of the original minerals and deposit new ones in their place. As would be expected, metamorphic rocks are more common in areas that have been subjected to severe crustal intrusions. And, because they have had more time to undergo such change, metamorphic rocks are usually of very great age.

For convenience in study, geologists usually classify metamorphic rocks as *foliated* and *unfoliated* (the latter as either *dense* or *granular*). *Foliation* (from the Latin word meaning "leaved or leafy") is the tendency of the minerals to be arranged in virtually parallel layers of flat or elongated grains. Some foliated rocks exhibit a *banded* pattern like that seen in *gneiss* (pronounced *nice*), a coarse-grained metamorphic rock that may be derived from igneous rocks or clayey sedimentary rocks. *Schist* (pronounced *shist*), which commonly represents highly metamorphosed shale, has a more highly foliated texture and the "leaves" are relatively easy to cleave apart.

Gneiss is a banded metamorphic rock that may be formed from granite.

As indicated by the name, the minerals in nonfoliated rocks are not aligned in a leaflike arrangement. Instead, they may exhibit a *dense* texture wherein the individual mineral grains cannot be seen with the naked eye (such as hornfels), or *granular* texture in which case the interlocking mineral grains are clearly visible. *Quartzite*—recrystallized sandstone—and *marble,* metamorphosed limestone, are common examples of nonfoliated metamorphic rocks.

There are world-famous exposures of metamorphic rocks in the inner gorge of the Grand Canyon, and they also crop out in many other National Parks.

Geologic Processes: Shapers of the Landscape

If rocks are the materials from which landscapes are made, *geologic processes* are the tools which carve them. For example, Yosemite Valley, the major scenic attraction at Yosemite National Park, was first incised into the massive granitic rocks by streams which coursed down the west front of the Sierra Nevada about one million years ago. Much later in geologic time the valley was invaded by mighty glaciers which transformed the slant-sided, V-shaped, stream-cut canyon into the sheer-walled, flat-bottomed, U-shaped glacial valley for which this National Park is famed. So Yosemite Valley was produced primarily by two geologic processes: stream erosion and glaciation. But these are only two of the many geologic "tools" which sculpt the rocks. Because these processes are basic to the understanding of landscapes, let us briefly review them here. Later—in the chapters dealing with areas in which they have been most active—each of these processes is discussed in more detail.

Glaciation—the effect of glacial ice upon the land—has played a key role in creating the landscapes of many of our Parks. In Yosemite the changes wrought by ancient glaciers have assisted in the development of such prominent features as the lakes, waterfalls, Half Dome, El Capitan, and, finally, the valley itself. Mount Rainier is famous for the mighty rivers of ice that flow down its ice-scarred flanks, and evidence of glacial erosion and deposition abound in Glacier, Voyageurs, North Cascades, and Rocky Mountain National Parks. Many of the glaciers gouged out depressions that later became filled with water to form many of the Parks' lakes; indeed, most of the world's lakes owe their origin to the work of glacial ice.

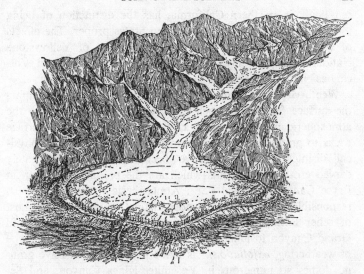

Tongues of ice called glaciers have played an important role in creating many of the National Park landscapes. PENNSYLVANIA GEOLOGICAL SURVEY

Stream erosion, the geologic work of running water, is also evident throughout the Parks. Working in concert with other geologic agents, stream erosion has given us the Grand Canyon of the Yellowstone, Zion Canyon, and the complexly dissected terrain of Canyonlands National Parks. But by far the most spectacular example of stream erosion can be seen in the Grand Canyon of the Colorado River at Grand Canyon National Park. Formed primarily by the work of the Colorado River, this great chasm measures approximately 1 mile deep, 9 miles wide, and 217 miles long. Where else can one find such eloquent testimony to the effect of running water upon solid rock? This awesome gorge is even more remarkable when we consider that it has developed during the last seven million years—a relative short time, geologically speaking.

Volcanism is another geologic activity that has produced some remarkably striking scenery. For example, the incredibly blue waters of Oregon's Crater Lake occupy the enlarged crater of an extinct volcano, a depression believed to have formed when the entire top of a mountain collapsed, and Las-

sen Peak, in northern California, has the distinction of being
the only active volcano in the United States proper. The effects
of volcanism can also be seen in the geysers of Yellowstone
National Park, the lava flows of Haleakala and Hawaii Volcanoes National Parks and on the slopes of Mount Rainier.

Weathering has also played an important part in shaping
the surface features of each Park. Continued exposure to the
atmospheric agents of wind, rain, and ice causes the surface
rocks to gradually undergo chemical decomposition and physical disintegration. Thus have been formed the intricately
carved erosional remnants which are the premier attraction in
Bryce Canyon National Park. Atmospheric weathering is also
responsible for some of the more distinctive features of a host
of other Parks including Zion, Big Bend, Canyonlands, and
Grand Canyon to name a few. In addition, a very special type
of weathering, *exfoliation,* has made possible the massive granite domes so numerous in Yosemite, Kings Canyon, and Sequoia National Parks.

Wave erosion has been confined to the relatively few National Parks that are located in or adjacent to large standing
bodies of water (lakes or seas). But in these areas the work of
the waves has produced some interesting scenery. Acadia National Park, on the rock-ribbed, wave-scarred Atlantic shoreline, is dotted with marine erosional features of many kinds,
while in Olympic, Redwood, and Hawaii Volcanoes National
Parks, similar landforms accent the Pacific Ocean shore. Nor is
wave erosion confined to the sea. It is interesting to note that
similar wave-cut features occur in Voyageurs National Park
Minnesota, and around the shores of Isle Royale National
Park in Lake Superior, the world's largest freshwater lake.

Mountain building, that great rearranger of rocks, has produced by far the most breath-taking scenery in the National
Parks. Some mountains, like those in the Grand Canyon, are
erosional remnants, but most mountains have been produced
by severe deformation of the earth's crust. During periods of
mountain building the rocks are commonly subjected to such
titanic forces that they will bend, buckle, or break. The *folds*
(flexures) and *faults* (fractures) thus formed are evident in

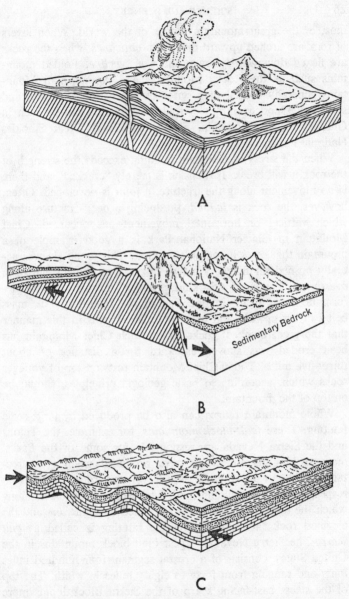

Mountain-building processes are responsible for much of the scenery in the National Parks. These cross sections through the earth's crust show (A) volcanic mountains, (B) fault-block mountains, and (C) folded mountains. U. S. GEOLOGICAL SURVEY most of the great mountain ranges of the world. When layers of rock are arched upward they form *anticlines;* when the rocks are flexed downward, they produce a *syncline.* Folded mountains such as the Rocky Mountains and Appalachian Mountains generally consist of a series of alternating anticlines and synclines. Good examples of folded mountains can be seen in Glacier, Rocky Mountain, Shenandoah, and Great Smokies National Parks.

When the stress of mountain building exceeds the strength of the rock it will break. If the rock is merely "cracked" and there is no movement along the fracture, a *joint* is developed. Often, however, the rock is *faulted,* producing a deep fracture along which vertical or horizontal movement has occurred. Chief Mountain in Glacier National Park is a good example of a mountain that owes its origin to a fault. This mountain was actually pushed for miles by a type of earth movement called an *overthrust.* Faults of this kind produce unusually severe deformation and result in one side of the fractured rock mass being thrust upward and over the other. It was in this manner that the great block of rocks from which Chief Mountain has been eroded was shoved eastward for a distance of about thirty-five miles. Today Chief Mountain rests on top of younger rocks which, according to basic geologic principles, should be on top of the mountain.

Whole mountain ranges can also be produced by large-scale faulting. These *fault-block mountains,* for example, the Tetons and the Sierra Nevada are responsible for some of the Parks' more striking terrain. They were produced when great rectangular blocks of the crust were thrust upward and tilted with respect to the surrounding land. The edge of the block along which the displacement occurred is called the *fault line* and the elevated rock face exposed by the faulting is called a *fault scarp.* The Sierra Nevada, largest fault-block mountains in the United States, consists of a crustal segment four hundred miles long, and ranging from forty to eighty miles in width. The top of the steep, east-facing scarp of the Sierra Block looms more than 11,000 feet above the floor of the Owens Valley which lies to the east.

Volcanic mountains are especially numerous in the National Parks and they vary considerably in size and shape. Some, like Mauna Loa, the rather broad, dome-shaped *shield volcano* of Hawaii Volcanoes National Park, are composed of innumerable lava flows superimposed one upon the other. Volcanic mountains of this type are built by volcanoes that erupt quietly. On the other hand, a steep-sided *cinder cone* like Wizard Island which rises from the surface of Crater Lake is typical of volcanic mountains formed by volcanoes that erupt explosively. Between these two extremes is the *strato-volcano,* or *composite cone.* This type of cone is composed of alternating layers of pyroclastics and lava that correspond with periods of quiet and explosive eruption. With the exception of the volcanoes of Hawaii and Iceland, most of the world's major volcanoes are of the composite type. In the National Parks, Mount Rainier and the remnants of Crater Lake's Mount Mazama are good examples of strato-volcanoes.

The Cascade Range, site of Crater Lake, Lassen, Mount Rainier, and North Cascades National Parks, is a classic example of a volcanic mountain range. The Cascades extend from California through Oregon and Washington into British Columbia. Many extinct volcanoes including Mounts Shasta, Hood, Baker, and Rainier attest to the fiery history of this majestic range.

The geologic phenomena discussed above are primarily the results of earth-shaping agencies which have operated on the face of our planet. Meanwhile, changes have taken place in the rocks beneath the surface of the earth. In some areas underground water has dissolved the rocks creating tunnels and subterranean chambers to form caverns. Later these same subsurface waters dissolved the limestone, picked up minerals in solution, redeposited them as countless layers of limy rock. These calcareous (limy) deposits have produced cave formations such as stalactites, stalagmites, and other cave deposits. Evidence of this type of geologic activity can be seen at Carlsbad Caverns, Mammoth Cave, and Wind Cave National Parks.

And so it is that over the millennia the agents of geology—ice, wind, weather, waves, running water, and mountain building movements—have brought forth the spectacular scenery that is the hallmark of our National Parks.

Unraveling Earth History

Like so many pages in a book, the rocky layers of the earth's crust tell a fascinating story of natural forces that have been operating for more than five billion years. From igneous rocks we learn of great periods of volcanic activity—of lava floods, violent explosions, and showers of glowing cinders and ash. The metamorphic rocks reveal bits of information that aid in understanding the great crustal disturbances that have wracked our planet since time began. They tell of a buckling crust, of minerals that were compressed and heated, and of new rocks made from old. Yet for all of the information that we glean from the study of igneous and metamorphic rocks, the geologist's most valuable clues to earth history are contained in the sedimentary rocks. These rocks, more than any others, reveal the remarkable series of events that have helped shape the modern landscape. From sedimentary rocks we learn of the comings and goings of vast inland seas, of flat lands and raging rivers, of restless winds and migrating sands, of powerful rivers of ice that have rasped and polished, gouged, and scraped.

In order to interpret earth history, the earth scientist must gather evidence of the great changes in climate, geography, and life that took place in the geologic past. He does this by studying the rock formations, the structural relationships and arrangement of the formations and the *topography* (general configuration of the land surface) of the area. The record of these ancient events is pieced together by studying the stony layers of the earth as one might study a giant history book. Indeed, the sedimentary rocks *are* the rocky "pages" of earth history, for in them we find the tracks, trails, bones, and stones which reveal the intriguing story of life long ago.

The first chapter of earth history begins with the most ancient rocks known. Because they were formed early in geologic time, these rocks are normally found deeply buried beneath younger rocks which have been deposited on top of them. It is for this reason that earth history is read from the bottom up, for the earliest formed rock layers correspond to the opening pages in our earthen history book. The later chapters are found in the upper, younger rocks which are located nearer the surface.

"Reading" earth history is not as simple as it might appear, however. In many areas the rock layers are not always found in the sequence in which they were originally deposited. Elsewhere, faulting and other structural disturbances have caused some of the rock "pages" to become shuffled and out of place and some are missing completely. Then, too, many of the rocks have been destroyed by erosion or greatly altered by metamorphism—their secrets are lost forever. These missing pages make the ancient story even more difficult to interpret so the geologist must uncover new clues that will permit him to "fill in the blanks."

Fossils—Silent Witnesses of the Past. Of all the features of sedimentary rocks, none is quite so informative as *fossils*—the remains or evidence of prehistoric plants and animals that have been preserved in the earth's crust. Through the science of *paleontology* (the study of fossils) life has been traced from its first clear, uninterrupted record—more than 600 million years ago—through its evolution into the more advanced forms of today. These fossils, which range in size from the remains of tiny one-celled plants, such as those in the "Garden Wall," overlooking Iceberg Lake in Glacier National Park, to the great fossil trees of Petrified Forest National Park, Arizona, are useful in deciphering the geologic history of the respective regions.

Curiously enough, one need not find the actual remains of an organism in order to have a fossil, for any trace or evidence of prehistoric life is also considered a fossil. In this type of fossilization there is no direct evidence of the original organism; instead, some trace or impression provides evidence of the plant or animal responsible for it. For example, interesting fossil footprints have been found in certain sedimentary rocks exposed in the walls of the Grand Canyon. By comparing these trails with recent tracks of similar nature, the paleontologist has deduced that the fossil footprints were made by an ancient lizard-like creature that roamed this area more than 200 million years ago. These and similar tracks have furnished the paleontologist with much otherwise unobtainable information about the characteristics and habits of the animal that made them.

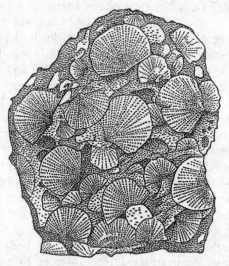

Fossils like those embedded in this ancient limestone formation tell us much about the life and physical conditions of the geologic past.

Unfortunately fossils are not equally distributed throughout the earth's rocky crust; rather, they are more likely to be found in areas where there are extensive exposures of sedimentary rocks. What is more, most fossils are found in *marine sedimentary rocks,* strata that were formed when saltwater sediments, such as limy mud, sand, or shell beds, were compressed and cemented together to form rocks. Conversely, only rarely does one find fossils in igneous or metamorphic rocks. This is not surprising, for as noted earlier that the igeneous rock was originally hot and molten and would have destroyed the organisms with which it came in contact. The metamorphic rocks have been so greatly recrystallized or distorted that any fossils that were present in the original rock have been de-

stroyed or so strongly altered as to be of little use to the paleontologist.

Fortunately the paleontologist can learn a great deal from
the study of fossils, for they usually furnish some clue as to
where they lived, when they lived, and how they lived. They
may even provide evidence as to how they died. In studying
fossils and reconstructing the past, the paleontologist relies on
the *Principle of Uniformitarianism*—the concept that ancient organisms lived under conditions similar to those of their nearest
living relatives or morphological counterparts. Or more simply
stated, it is believed that the present is the key to the past. For
this reason, paleontological studies require the knowledge and
techniques of both geology and biology.

Why is it helpful to know where an organism lived? We
know, of course, that most varieties of plants dwell on the
land, so it is reasonably safe to assume the rocks which contain fossilized land plants were probably formed on land. It
is also known that certain species of plants and animals live
only in sea water. Therefore, if we find the fossil remains
of sea organisms embedded in a rock, this suggests that the
rock was probably formed from sediments that were laid down
on the floor of an ancient sea. More important, certain species
of marine fossils can even provide such detailed information as
the approximate depth and temperature of the water the
original organism inhabited, whether the water was salty or
fresh, clear or muddy, or in which direction water currents
were flowing.

By the same token, a terrestrial environment can be postulated for those areas in which fossil tree stumps are found standing where they originally lived. The trees in the fossil forests of
Yellowstone National Park are good examples of this type of
fossil evidence. In many instances, however, fossil trees appear
to have floated into their present position. Fossil trees which
have accumulated in this manner often show effects of tum-
bling and abrasion, and lie essentially parallel to the enclosing
layers of rock. Log rafting of this type seems to account for
the presence of the great stone trees in Petrified Forest National
Park.

Geologists have learned that knowledge of ancient environments is especially useful in the preparation of *paleogeographic* (literally "ancient geographic") *maps*. By comparing the distribution of marine (saltwater) fossils to the distribution of terrestrial (land-dwelling) fossils, it is usually possible to draw

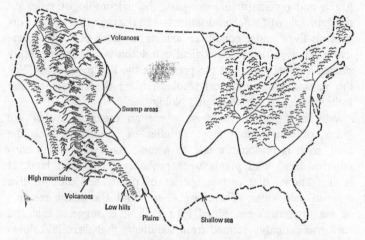

Geologists use information from rocks and fossils to construct paleogeographic maps. This one shows the United States as it might have appeared during Cretaceous time some 80 million years ago.
U. S. GEOLOGICAL SURVEY

maps that show the geography of these areas as it might have been at various times in the geologic past. So knowing where fossils lived makes possible the construction of accurate paleogeographic maps.

Because they can be used to trace the development of plants and animals from the dawn of life to the present time, fossils also offer one of the strongest lines of support for the theory of organic evolution. This is possible because the fossils in the older rocks are usually primitive and relatively simple. However, a study of similar specimens that lived in later time reveals that fossils became progressively complex and more advanced in the younger rocks. For example, the reeflike concentration of minute one-celled plant remains found in Glacier National Park are more than two billion years old, while the trunks of the more highly developed trees in the Petrified Forest are enclosed in rocks estimated to be only about 175 million years old.

Fossils have also been utilized in an attempt to re-create the climatic conditions of the past. The assemblage of fossil plants found in the Petrified Forest of Arizona, for example, is quite unlike the plants that now grow in this area. In fact, the fossil plants here are similar to plants that are now living in tropical or semitropical areas rather than in a semidesert environment such as exists in Arizona today. Likewise, the presence of fossilized remains of reindeer in Arkansas, ferns in Antarctica, and the musk ox in New York suggest quite different climates for these regions at previous times in earth history.

Although fossils are useful in a number of ways and have provided the earth scientist with invaluable information about the earth and its history, probably the most important use of fossils is for purposes of *correlation*—the comparison of ages of rock strata in widely separated areas. Correlating by means of fossils is a good way to locate gaps or breaks in the fossil record. These gaps—the "missing pages" in the fossil record— are called *unconformities,* and they represent sediments that may have been removed by erosion before later sediments were deposited. An unconformity may also indicate that there were no sediments laid down in the area during the particular portion of geologic time represented in the missing record.

The Geologic Time Scale. The earth historian, like the historian dealing with the development of civilization, must have some method of relating important events to one another. For this purpose the geologist has devised a special geologic time scale composed of named units of geologic time during which the various rocks were deposited. The largest of these *time units* are called *eras;* each era is divided into *periods,* which in turn may be subdivided into still smaller units called *epochs*. Arranged in chronological order, these units form a giant geologic "calendar" which provides a standard by which the

GEOLOGIC TIME SCALE

ERA	PERIOD	EPOCH	SUCCESSION OF LIFE
CENOZOIC "RECENT LIFE"	QUATERNARY 0-1 MILLION YEARS	Recent / Pleistocene	
CENOZOIC "RECENT LIFE"	TERTIARY 62 MILLION YEARS	Pliocene / Miocene / Oligocene / Eocene / Paleocene	
MESOZOIC "MIDDLE LIFE"	CRETACEOUS 72 MILLION YEARS		
MESOZOIC "MIDDLE LIFE"	JURASSIC 46 MILLION YEARS		
MESOZOIC "MIDDLE LIFE"	TRIASSIC 49 MILLION YEARS		
PALEOZOIC "ANCIENT LIFE"	PERMIAN 50 MILLION YEARS		
PALEOZOIC "ANCIENT LIFE"	CARBONIFEROUS PENNSYLVANIAN 30 MILLION YEARS		
PALEOZOIC "ANCIENT LIFE"	CARBONIFEROUS MISSISSIPPIAN 35 MILLION YEARS		
PALEOZOIC "ANCIENT LIFE"	DEVONIAN 60 MILLION YEARS		
PALEOZOIC "ANCIENT LIFE"	SILURIAN 20 MILLION YEARS		
PALEOZOIC "ANCIENT LIFE"	ORDOVICIAN 75 MILLION YEARS		
PALEOZOIC "ANCIENT LIFE"	CAMBRIAN 100 MILLION YEARS		
PRECAMBRIAN ERAS	PROTEROZOIC ERA		
PRECAMBRIAN ERAS	ARCHEOZOIC ERA		

APPROXIMATE AGE OF THE EARTH MORE THAN 4 BILLION 550 MILLION YEARS

The Geologic Time Scale. Reproduced from *Fossils: An Introduction to Prehistoric Life*, William H. Matthews III, Barnes & Noble, Inc., 1962.

age of the rocks can be discussed. Thus, rocks that formed during the Mesozoic Era are said to be Mesozoic in age.

Unlike days and years, the units of the time scale are arbitrary and of unequal duration, for we cannot be certain as to the exact amount of time involved in each interval. However, by referring to the geologic time scale it may be possible, for instance, to state that certain lavas were extruded during the Mesozoic Era in much the same manner that a historian might speak of a certain fort having been built during the Civil War. Each of these terms gives us a general idea as to the interval of time during which the lava was extruded or the fort constructed.

The geologic time scale has been developed upon the principle that unless a series of sedimentary rocks has been overturned, a given rock layer will be older than all the rock layers above it and younger than the strata which are beneath it. In short, this means that in a normal sequence of rocks the older rocks always underlie younger rocks. This basic geologic concept is called the *Law of Superposition*.

How can the geologist determine the age of the rocks? He does this in the field by comparing the relative positions of the rocks and by studying any fossils that may be present in them. This may provide some indication of the *relative* age of the rocks. But relative age does not imply age in years; rather, it fixes age in relation to other agents that are recorded in the rocks.

Within recent years, however, it has become possible to assign ages in years to certain rock units. This is accomplished by a system of very precise measurements of amounts of radioactive minerals (such as uranium or carbon 14) which change or decay at such a uniform rate that they are almost a natural radioactive "clock." This method of dating has made it possible to devise a measured time scale which gives some idea of the tremendous amount of time that has passed since the oldest known rocks were formed. It has also been used to verify the previously determined relative ages of the various rock units. This method will not work, of course, unless radioactive minerals are present in the rocks.

As you look at the geologic time scale you may wonder why some of the units have been given such odd and unusual names. Why have we not used simple English that would be easy to pronounce? In general, most of the names of the time units are derived from Greek and Latin words. This has been done for two reasons: first, ancient Greek and Latin are both "dead" languages and hence not subject to change as a modern language would be; second, they are international languages and mean much the same thing to geologists all over the world. For instance, if we look up the derivation of the word Paleozoic, we find that it is derived from the Greek words *palaios* meaning "ancient" and *zoikos,* pertaining to "life" (*zoe*). This is the geologist's way of saying that the Paleozoic Era was the time of ancient life. He expresses all of this in a single word that can be understood by scientists all over the world. Pronunciation? Simply break it into syllables: Pay'-lee-o-zo"-ik.*

Because geologic time terms are quite frequently used in describing certain features of the Parks, a knowledge of their origin and pronunciation should prove helpful to the reader of this book.

The five geologic eras and the pronunciation and meaning of their names are:

Cenozoic (see'-no-zo"-ik)=recent-life
Mesozoic (mess'o-zo"-ik)=middle-life
Paleozoic (pay'-lee-o-zo"-ik)=ancient-life
Proterozoic (prot'-er-o-zo"-ik)=fore-life
Archeozoic (ar'-kee-o-zo"-ik)=beginning-life

For convenience in reference, Archeozoic and Proterozoic rocks are commonly grouped together and referred to as *Precambrian* in age. Most Precambrian rocks, like those in the inner gorge of the Grand Canyon, have been greatly contorted and changed and the record of this portion of geologic time is difficult to interpret. Precambrian time represents that segment of geologic time from the beginning of earth history

* The principal accent is indicated by double accent marks ("), and the secondary accent by a single mark (').

until the deposition of the earliest fossiliferous Cambrian strata (see Geologic Time Scale, p. 34). If the earth is as old as it is thought to be, Precambrian time may represent as much as 85 per cent of all earth history.

In looking at the geologic time scale, you will note the *oldest* era is at the *bottom* of the list; successively younger eras are placed above it. Therefore, the geologic time scale is always read *from the bottom of the scale upward*. This is, of course, the order in which the various portions of geologic time occurred and during which the corresponding rocks were formed.

As noted above, each era has been divided into smaller intervals of geologic time called periods. Most of the periods derive their names from the places in which these rocks were first studied. Thus, the Devonian Period was named for the county Devonshire in England and the Pennsylvanian Period was named for rock exposures in the state of Pennsylvania.

The Paleozoic Era encompassed about 270 million years and has been divided into seven periods of geologic time, each with its corresponding system of rocks. With the oldest at the bottom of the list, these periods, the origin of their names, and a guide to their pronunciation are as follows:

Permian (pur″mee-un″)—for the province of Perm in the Ural Mountains of Russia

Pennsylvanian (penn′-sil-va″-ni-un)—for the State of Pennsylvania

Mississippian (miss′-i-sip″-i-un)—for the Upper Mississippi Valley

Devonian (dee′-vo″-nee-un)—for Devonshire, England

Silurian (si-lu″-ri-un)—for the Silures, an ancient tribe of Wales

Ordovician (or′-doe-vish″-un)—for an ancient Celtic tribe which lived near the type locality in Wales

Cambrian (kam″-bri-un′)—from the Latin word *Cambria,* meaning Wales

The Mesozoic Era lasted for approximately 170 million years and has been divided into the following geologic periods:

Cretaceous (kree-tay"-shus)—from the Latin word *creta,* meaning "chalk"; refers to chalky limestones such as those exposed in the White Cliffs of Dover on the English Channel

Jurassic (juu-rass"-ik)—for the Jura Mountains between France and Switzerland

Triassic (try-ass"-ik)—from the Latin word *trias* meaning "three"; refers to the natural threefold division of these rocks in Germany

The Cenozoic Era began about 65 million years ago and continues through the present day. The two periods of this era have derived their names from an old outdated system of classification which divided all of the earth's rocks into four groups. The two divisions listed below are the only names from this system that are still in use today:

> Quaternary (kwah-tur"-nuh-ri) implying "fourth derivation"
> Tertiary (tur"-shi"-ri)—implying "third derivation"

While the units discussed above are the major divisions of geologic time, the geologist usually works with smaller units of rocks called *formations.* A geologic formation is identified and established on the basis of definite physical and chemical characteristics of the rocks. Formations are usually given geographic names which are combined with the type of rock that makes up the bulk of the formation. For example, the Bright Angel Shale is a predominantly shale formation named for exposures along Bright Angel Creek in the Grand Canyon.

Geologic Formations in the National Parks

Like the wildlife and vegetation, the geologic formations of the National Parks are protected by federal law. Rocks, minerals, and fossils cannot be collected in these areas nor can the natural rock formations be disturbed in any way. Can you imagine what would soon happen to Carlsbad Caverns if each of its hundreds of thousands of visitors should take home even a small piece of one of the delicate cave formations? Or if each visitor to the Petrified Forest should remove a fragment of fossil wood?

It is especially important to remember that most geologic formations have been created over vast periods of time or by forces which are no longer at work in the area. Thus, thoughtless visitors and vandals can destroy in minutes those objects that it has taken nature many millions of years to create.

Not only do the geologic formations add to the beauty of the Parks, they also provide considerable information as to how and when the area's surface features were formed. As we have already learned, the landscapes of our National Parks have been produced by geologic changes which took place during prehistoric time, and the rocks commonly contain clues which help in the reconstruction of the geologic history of the area. Because of their outstanding geologic significance, the earth scientist considers most of the National Parks to be great outdoor classrooms—places where he can study the earth's surface in a near natural condition, relatively undisturbed by human activities.

Remember though, it is not necessary to be a geologist to appreciate the geologic phenomena in our National Parks. Each visitor should think of these areas as vast open-air geological laboratories where he can become acquainted with the forces and products of geologic change. Here with the help of expert "teachers"—the Park Naturalists and museums—one can learn to appreciate more fully the great age of the earth, its fascinating history, and the complexity of its structure and composition.

Biology—the Science Behind Park Life

Just as an acquaintance with geology will facilitate an understanding of the meaning behind Park landscapes, the science of *biology* provides a behind-the-scenes view of life in the National Parks. Most of the Parks have their own distinctive types of plants and animals and some, like the bears of Yellowstone and Yosemite's giant sequoia trees, are among these areas' better known attractions.

Ecology, the Science of Togetherness

The living community of each Park consists primarily of the *flora* (plant life) and *fauna* (animal life) native to that area. But the flora and fauna of each area are far more than a mere gathering together of a variety of plants and animals. Each of these organisms is closely interrelated with and frequently dependent upon the other plants and animals that are around it. These organisms are, moreover, dependent upon their physical surroundings. In fact, in most Parks it was the physical environment that originally attracted life to the area. In addition, by its activities each organism affects its environment and is itself affected by the ever-changing landscape. Thus, all the plants and animals of the living community are inextricably linked together in a great chain of being.

The study of the interrelations of plants and animals with one another and with the physical features of their surroundings is called *ecology*. Derived from the Greek word *oikos*, which literally means "home," this increasingly important branch of biology deals with the "home life" of organisms: their relation to one another and to their living and non-living environment. Because this environmental science addresses itself to all phases of our surroundings, it has sometimes been called "the science of everything." And, dealing as it does with the reasons behind plant and animal communities, ecology has also been called "the science of togetherness."

Most of our National Parks represent classic examples of an undisturbed natural environment. These range from alpine tundra to bone-dry deserts with a host of other environments in between. Some of the relationships between organisms and their environment are obvious. But certain of the more subtle—and important—relationships are not so noticeable. These relationships, subtle and obvious, bind all segments of each Park into a complex web called an **ecosystem.**

Scientists consider an ecosystem to be a natural unit of the landscape. Each ecosystem is a viable unit of nature in which plants, animals, and the physical environment are continually interacting. These interactions are based on an exchange of raw materials that produces a complex food web. Ranging in size from a mountain range to a mud puddle, there are land ecosystems and aquatic ecosystems. But although one is on land and the other is in the water, there may be important links between the two.

Ecology—a word that is appearing with increasing frequency in daily newspapers—does not limit itself to a study of the so-called "lower" plants and animals. Instead, it is becoming increasingly important to the future of mankind. Man, the dominant living species, has the ability to alter his own environment, but in so doing he has created serious ecological problems such as air and water pollution. Because of this and other man-generated imbalances in nature, it is becoming apparent that only through an understanding of basic ecological principles can we learn to use our natural resources intelligently for the welfare of all organisms, including man.

The National Parks are ideal places to observe natural plant and animal assemblages, for the National Park Service is dedicated to the preservation of normal ecological conditions within each area submitted to its charge. You can observe the fascinating story of ecology on many of the Park nature trails, where interpretive markers explain the relation of organisms to their biologic and physical environment. In addition, Ranger-Naturalists by means of conducted nature walks and other interpretive services, call attention to each area's special ecological significance and emphasize man's responsibility to help maintain proper natural conditions.

The natural areas of the National Park System are prized ecosystems, and each represents a different and unique form of land use. But, as always, the basic object of the National Park Service is to preserve these natural ecosystems for the education and enjoyment of our citizens. Although man is a part of the ecosystem—and is encouraged to enjoy each area —his presence is controlled in order that proper ecological balance may be maintained.

Wildlife in the Parks

In order to insure the proper balance of nature, the wildlife management policy of the National Park Service states "that the whole natural community of an area shall be subject to a minimum of human interference or manipulation, and that there shall be a free play of natural forces and processes." Thus, visitors have the opportunity to see the animals in their natural environment relatively undisturbed by man.

There are a number of reasons why the fauna and flora of the Parks should be preserved. First, the natural plant and animal life add greatly to the interest and beauty of each area; second, the plants and animals native to the area are able to live more safely in their natural habitat than elsewhere; and third, these natural game preserves are ideal field laboratories for biological research. Consequently, zoologists and other biological scientists visit the Parks each year in order to study these undisturbed animal communities under the most natural conditions possible. And, as a result of these scientific studies, biologists are commonly able to further the preservation of life forms native to the respective areas.

To further assure a natural environment for these wild creatures, the animal life in the Parks is protected by federal law, and Park Rangers are on hand to make certain that the law is enforced. Indeed, the preservation of certain species of wildlife has been one of the prime reasons for the establishment of certain Parks. For example, Mount McKinley National Park was established to preserve Dall sheep, Olympic National Park was established to protect the Olympic or Roosevelt elk, and Everglades National Park provides a sanctuary for a host of subtropical birds. In short, each National Park is a plant and wildlife refuge where hunting is not allowed and where special care is given to those species threatened with extinction.

Although hunting is not allowed in the Parks, fishing is permitted in many areas and each Park has its own fishing regulations which must be closely followed. State fishing licenses are usually required in those Parks in which the state reserved the right to require such licenses when the Park was established. However, fishing licenses are not required in National Parks that were established on public lands in territorial status.

Unfortunately, fishing has become less satisfactory as tourism has increased and at present the best fishing places are found in the more remote areas of the Parks. There is also a serious problem in conserving the natural fish population of each area, a factor which is making artificial stocking increasingly necessary.

Native Vegetation in the Parks

The plant life of the National Parks is also protected by National Park Service policy and picking wildflowers, removing plants and shrubs, and the indiscriminate cutting of trees and underbrush is not permitted. In those areas where it is necessary to thin vegetation in the construction of roads, trails, or campgrounds, this is done carefully and sparingly. Moreover, Park personnel take special precautions to maintain the proper ecological balance in order to allow the thinned area to look as natural as possible. In areas of historical significance the vegetation is usually maintained in, or restored to, a condition similar to that which prevailed at the time of the event being commemorated.

In addition to the prevention of damage by human activities and forest fires, Park Service personnel also take precautions to prevent losses from plant diseases and epidemic forest insects. Steps are also taken to eliminate non-native plants and to protect vegetation from damage by grazing and browsing animals.

Certain of the National Parks are famous for their trees. The huge sequoias of California's Yosemite, Kings Canyon, Redwood, and Sequoia National Parks are the largest trees in the world and some are estimated to be between three and four thousand years old. Other large trees, among them the western hemlock, Sitka spruce, and western red cedar, thrive in the wet rain forests of Olympic National Park on the Olympic Peninsula in Washington. California's Redwood National Park is an open-air botanical laboratory. Four main plant communities may be studied in relation to their animal life and physical environment. These outstanding ecosystems include a redwood forest, a north coastal scrub-plant community, a cutover forest, and a marine and shore habitat.

One of the outstanding features of the National Park forests is their natural and primitive character. In fact, these forests might properly be considered forest museums in which the normal processes of nature are given free rein. It should be noted that as time passes, the scientific, social, and economic importance of the National Park forests is increasing, for commercial lumber interests are steadily reducing the remaining undisturbed forests in unprotected areas elsewhere across the nation.

Because National Parks and National Forests are sometimes closely associated with each other, visitors frequently inquire as to the difference between these two natural preserves. The major distinction is that each of these areas is established under a different concept of land use. The Forest Service regards their forests primarily as a renewable resource which can be harvested periodically if proper conservation principles are followed. National Parks, on the other hand, are units of federal land that are to be used only for recreation and education. In addition they are preserved, as nearly as possible, in a natural condition. Although National Forests are maintained primarily to insure a continuous supply of forest products and to protect watersheds, in the course of such management it is often possible for recreational, educational, and other benefits to be obtained. Therefore, many fine public campgrounds, lakes, and scenic areas are to be found in many of our National Forests.

Forest fires are an ever-present source of danger and one which Park personnel must continually guard against. For this reason the forested areas of the Parks are under intensive and constant fire control and Rangers are on 24-hour fire-call duty. During the "fire season," from about late June to October, the watch is even more rigidly maintained. Yet despite the vigilance and precautions of the Rangers, the civic education of the public, and general cooperation of the visitors, forest fires do occur. There is no doubt, however, that such fires would be much more prevalent if the forests were not under National Park Service supervision.

But trees and forests are not the only forms of plant life for which the parks are famous. Many are well known for a specific type of vegetation. Big Bend National Park is famous for its desert plants, and Everglades National Park for its lush tropical and subtropical vegetation. Still other Parks are noted for their colorful displays of wildflowers. Mount Rainier and Olympic National Parks are especially famed for their colorful flower "shows." Because of marked differences in elevations and soil and the short, intense growing seasons, these two Parks are unique for the abundance and variety of its species of beautiful wildflowers. Each year these lovely flowers blanket the mountain meadows with an extraordinary profusion of color from May through September. Some of these alpine wildflowers do not even wait for the spring thaw; instead, they push right up through the snow. (Visitors are, of course, requested not to pick the wildflowers. If each visitor picked but a single bud, these natural flower gardens would soon be barren.)

Thus, the physical features and life forms of the National Parks are the basic ingredients of the Parks' spectacular scenery. Luckily for you and me, the biology and geology of these scenic areas are blended together in just the proper proportions.

3. The Parks Are Yours—
Enjoy and Protect Them

Thanks to the foresight of a small but dedicated group of early conservationists and lawmakers, a host of our nation's most beautiful and unusual natural wonders have been permanently set aside for the enjoyment, recreation, and education of the American people. These many-splendored areas are commonly a paradise for the angler, the birdwatcher, the hiker, or the tired city dweller who simply wants to relax and enjoy nature in solitude. There are towering mountains, sparkling lakes, verdant forests, unusual wildlife, geologic phenomena, and archeological ruins—in short, there is something for everyone in the wonderlands that are America's National Parks.

Remember though that the beauty and primitive atmosphere of our National Parks exist today only because of the diligence of National Park Service personnel and the cooperation of the visiting public. Remember, too, that the Parks belong to your children and their children. Would you deny them the same pleasures that these remarkable areas have afforded you? More than anything else the National Park Service wants you to enjoy your visit, but they also want those who come after you to enjoy *their* visits. It is for this reason that great care has been taken to see that the scenery, the rocks, the plants, and the animals remain carefully preserved under the most natural conditions possible. Yet at the same time these natural features have been made accessible for your enjoyment and edification.

You will find that the regulations posted in each Park are few and reasonable. Moreover, they are intended not only for protection and preservation of the Park, but also for your own personal safety. Please obey them.

Preparing for Your Visit

More often than not, the most enjoyable trip is the well-planned trip. This is especially true of journeys to the National Parks because these are usually relatively long trips into somewhat unfamiliar areas.

Books and Maps. Your trip will be more enjoyable and you will understand more of what you see if you will take time to read some of the material that has been written about the Parks before you actually visit them. For most of the units in the system the National Park Service issues illustrated informational publications that are presented to visitors at the Park's information centers or entrance stations. However, these brochures may also be purchased by mail from the Superintendent of Documents, U. S. Government Printing Office, Washington, D.C. 20402. By obtaining these publications prior to your visit, you can have a preview of the area's attractions and will be able to plan how your time might be most profitably spent. The National Park Service also issues a number of specialized publications that deal with the natural and human history of certain areas or provide camping and recreational information. (Price lists of National Park Service publications sold by the Superintendent of Documents may be obtained from the Government Printing Office.)

In addition, there are other informative publications on sale at Park Headquarters, visitor centers, and other points within the Park. These range from detailed geologic and biologic studies of each area, to books dealing with the history and folklore of the region. Many of these materials have been published by Natural History Associations which are affiliated with the various National Parks. These private, nonprofit organizations encourage scientific investigations and are pledged to aid in the scenic and scientific features of the respective Parks. The titles of many of these publications are listed in the various informational brochures.

It will also prove helpful to consult a map of the Park before your visit. You will then be able to plan a route through the area that will allow you to see as many features as possible. Maps showing roads, trails, and places of interest within the Park are included in the descriptive brochures and a variety of special maps are normally sold in each Park. Topographic maps are available for many of the Parks. These maps, which show valleys, rivers, hills, and mountains, as well as roads, towns, houses, and other artificial features of the region, are especially useful in planning hiking and mountain-climbing trips. Topographic maps can be purchased at a nominal cost from the United States Geological Survey, Washington, D.C. 20242, or Denver, Colorado 80225. To assist in locating the maps, the Survey supplies, without charge, a key sheet, or index map, showing all of the maps available for each state. For an index map or special information about maps, write the Office of Map Information, Washington, D.C. 20242.

Of course, this book should also prove helpful in planning your stay in the Parks. "The Parks at a Glance" section that accompanies the chapters on the respective Parks will provide you with a condensed "thumbnail sketch" of what each Park has to offer.

What to Bring. As a rule, most visitors are inclined to take along more clothing and luggage than they actually need. Elaborate garb is not necessary, but comfortable clothing and shoes are. Moreover, the type of clothing required varies with the unit that is being visited, for the visitor to Glacier National Park near the Canadian border would not wear the same type of clothing as the person visiting the Everglades in Florida.

In the western parks, light jackets and sweaters will be comfortable for cool nights around the campfire and do not forget to tuck in a raincoat for that unexpected shower. Rubber-soled shoes are almost a necessity for walking on rocks and slippery trails. Those planning extensive hikes should wear sturdy walking shoes or hiking boots and a brimmed cap or hat and sunglasses are desirable in many areas.

Binoculars, or field glasses, can greatly add to your appreciation of the Park, especially of its birds and other animals. And by all means do not forget your camera; concessioners can supply you with film and can often help with any special photographic problem that might arise.

What to Do in the Parks

What can I see and do here? How can I get the most out of my visit? What should I do first? These are some of the questions that usually cross the visitor's mind as he passes through the Park entrance station. Naturally the answers to such questions will differ considerably from one visitor to another, for people visit the Parks for many different reasons. But whatever the purpose of your visit, *first* read the descriptive brochure that has been given you by the Park Ranger at the entrance station. The brochure is specifically designed to help answer these initial questions, plan your visit, and to make your Park experience more meaningful. In it you will also find background material on the Park, and general information about where to stay, where to eat, the location of campgrounds, and the various services that are available.

By all means take a few moments to study the map of the area. On it you can locate the visitor center, museum, ranger stations, lodges, motels, campgrounds, roads, trails, and places of particular interest. Then, as you read the text and study the map, you will find it much easier to decide where you want to go and what you want to see and do. When you have finished reading the brochure be sure to keep it with you for ready reference while in the area.

If the Park has a visitor center this is the place to start your visit. Here you will find a Park Naturalist, Ranger, or other uniformed employees who will welcome your questions and help you plan your stay. Moreover, they will provide you with specific information about the various conducted tours, nature hikes, evening campfire programs, and other scheduled interpretive activities. They will also advise you of some of the things that you can see and do for yourself. While in each Park make a practice of consulting the bulletin boards at ranger stations, campgrounds, and lodges. The latest "Naturalist Program" and other announcements of activities, services, time schedules, and other events are posted here and information of this type will keep you advised as to what is going on and help you to make the most of your time.

The Naturalist Program

Park Naturalists and their assistants, aided by Park Rangers, provide certain interpretive services to help you understand and appreciate the Parks and their special features. These include exhibits that explain the natural history of the area, campfire programs, illustrated lectures, and guided trips to places of interest. Regardless of your age or interests, there is an activity for you—and, best of all, these services are free.

Museums. Most of the Parks have exhibits and educational displays depicting the more interesting features of the area. These may be located in the visitor center or in a separate museum; more often, however, the two are in the same building. In the exhibit areas you may enjoy displays of rocks, wildlife, native vegetation, and natural features. And there are commonly maps and models portraying the geologic story and illustrations and dioramas dealing with the history of the area.

In certain of the larger Parks (for example, Yellowstone, Yosemite, and Grand Teton), there may be several museums or branch visitor centers, each of which is situated in a different part of the area. Some of these have special themes that deal with certain features of the Park (Indian history, geological formations, native plants, etc.).

As mentioned earlier, the visitor center museum is the logical place to start your visit. Here you can get some indication of what you can expect to see in the area and also become better prepared to look for those features that will heighten your appreciation of the Parks. In addition, some museums present regularly scheduled orientation films or lectures by the interpretive staff. You may find that after your tour of the Park you will want to return to the exhibit room to obtain answers to questions which are almost sure to arise.

Campfire Programs. You will recall that the idea which led to the founding of the first National Park was conceived during a campfire discussion in what is now Yellowstone National Park. Campfire talks are still held in Yellowstone and most other Parks. At these interesting programs, Park Ranger-Naturalists interpret the wonders of each area in terms of its plants and animals, geology, human history, mountains, etc. These program topics change frequently and the lectures are usually accompanied by colored slides and movies. In some of the Parks evening programs are sometimes held in the visitor center or at one of the lodges. But whenever possible they are conducted outdoors under the stars in a natural setting where rocks and fallen trees may serve as seats and the sky is the ceiling. The location and schedule of subjects of the campfire talks are normally posted on bulletin boards throughout the Park or they may be obtained from the visitor center or museum. Many people consider these informal and informative twilight talks to be the most memorable part of their visit and these programs should be a "must" on anyone's list of things to do.

Conducted Tours. If you really want to become familiar with the Parks and their natural history, join the Naturalists or Rangers on a conducted tour. Check the bulletin boards and visitor center for schedules that will inform you of the destination and the times and places to assemble for these trips. Scheduled several times daily in most areas, these tours will lead you to some of the Park's most outstanding areas and features. They may range from leisurely one-hour nature walks to more vigorous all-day hikes into the back country; but in each instance you will benefit from the comments and lectures of the guide and he will be more than happy to answer your questions. There is no better way to get "close to nature" or to more fully enjoy the scenery, plants, and wildlife of a Park and you will also gain a better understanding of the natural history or human history of the area.

Self-guiding Trails. Made to order for the "do-it-yourself naturalist," these short nature trails are found in most of the National Parks. At the start of many trails there are leaflets which contain explanatory nature notes with numbered paragraphs in the guide corresponding to numbered markers along the trail. Thus, the hiker can identify the natural objects that are indicated by the markers and the guide leaflet will explain their significance. Those trails for which guide leaflets are not available are usually provided with trailside interpretive signs which explain the meaning of the various natural features. The self-guiding trails have the advantage of permitting the hiker to set his own pace and to spend more time at the attractions in which he is most interested. Then, too, you can begin the trip at your convenience rather than waiting for a regularly scheduled departure.

The self-guiding trails are usually designated on the Park maps and discussed in the brochures. They are also normally listed and described on the schedule of events or the *Naturalist Program* posted throughout the Park. *A suggestion:* On those trails provided with descriptive folders, please take only the number that you need. If, upon your return to the starting point, you decide that you do not want to keep the folder, please place it back in the rack or give it to another hiker. On busy days the leaflets are likely to be in short supply and some visitors may have to make the trail without them.

Fun in the Parks

There is ample opportunity for recreation in the Parks and these activities are usually in keeping with the natural surroundings of the area. Golf courses, tennis courts, and bowling alleys are generally lacking, but a host of relaxing, healthful, outdoor activities await your pleasure.

Along the Roads. It has been said that most of the roads in our National Parks do not follow the most direct routes. In general, this is true but it is also true that the Park roads provide us with some of the world's most magnificent scenery. There are good roads leading to most places of interest in each Park and interpretive markers have been erected at many of the turnouts and lookouts along the way. In addition, there are attractive and informative roadside exhibits which will explain some of the natural features that are located along the road. In some of the Parks you may purchase descriptive road logs that will permit you to take self-guided auto tours of certain parts of the area.

Along the Trails. It is one thing to see one of the Parks from an automobile window on a paved road; but seeing its beauty from a park trail is a never-to-be forgotten experience. Even though you have only a day or two in the Park you should plan at least one hike into the wilderness. You will soon realize that hiking brings you into close contact with nature and you will be amazed to learn how many things you notice while hiking that you probably would not otherwise see.

National Park trails are well marked, regularly maintained, and carefully planned to take you into the more remote parts of the area. For most hikes you will need only walking shoes, comfortable clothing, and a reasonable amount of energy, but for back-country or wilderness travel you will find it helpful to take along a topographic map that shows trails, streams, lakes, glaciers, and other natural features. Topographic maps can be purchased at most visitor centers or can be ordered by mail, and special trail maps are available for many areas. Inquire about these and other maps at Park Headquarters or a visitor center.

Along some of the trails open shelters have been provided for overnight hikers. These are available on a first-come first-served basis and the hiker must provide his own sleeping and cooking gear. But before leaving on an overnight trip be sure to visit a ranger station to check the latest trail conditions and obtain a fire permit and camping information. Finally, be sure to notify friends and a Park Ranger of your route, your expected times of departure and return, and also when you have reached your final destination.

Picnicking. Picnic areas are located at convenient and picturesque spots throughout most of the Parks (refer to Park map for specific locations). These are provided for short stops and camping is not permitted in them. Tables, benches, and running water are supplied in most of these places and some have other facilities including rest rooms, fireplaces or open grills, and occasionally a limited supply of firewood. Picnickers are requested to build fires only in fireplaces and to use waste receptacles to dispose of all paper and trash.

Camping. Within recent years there has been a reawakened interest in camping, for many families have learned that their vacation dollars will stretch further and that their trips are more enjoyable if they camp out along the way.

The family that likes to live out-of-doors will find camping at its best in the National Parks—outdoor living in the undisturbed natural surroundings of one of these great scenic areas is a most relaxing and memorable experience. Recognizing the need to provide camping facilities for the burgeoning numbers of family campers, the National Park Service has provided excellent camping facilities in most of the National Parks.

Camping is permitted only in designated areas and campsites are assigned on a first-come first-served basis. However, exceptions are made in the case of large groups which can make advance reservations for one of the group campsites. In certain of the more heavily used areas, use of the campgrounds may be subject to a time limit in order to provide campsites for as many people as possible.

Camping in the National Parks is basically of two types: *wilderness* camping and *campsite* camping. Wilderness camping is for the more experienced outdoorsman and permits the camper to live in back-country areas far from human habitation. Because of the hardship and possible dangers involved, wilderness campers must inform a Park Ranger of their plans and, where necessary, obtain a fire permit.

For obvious reasons, most campers use the camping facilities that have been provided by the Park Service. These include (1) *campgrounds,* (2) *camping areas,* (3) and *group camps.*

Campgrounds are rather well-developed areas that have clearly marked roads, parking spaces, and designated campsites. Drinking water is available as are toilets and waste disposal cans. Facilities for each campsite typically consists of tent and parking spaces, bench and table, and fireplace or grill. In some areas campsites are assigned by the Rangers and campers are asked to register at campground entrances. This procedure makes it possible for Park Rangers to deliver emergency messages.

Camping areas are relatively unimproved areas and may be accessible by either road or trail. Facilities in this type of camping accommodations are minimal and consist essentially of access roads, basic sanitary facilities, and a limited number of fireplaces and tables.

Organized groups, such as school groups, Boy Scout troops, or other large parties are assigned to *group camp* areas. Such places are provided with several tables, large fireplaces, and ample parking space for buses, trucks, or several cars.

You may write to the Superintendent of the areas you plan to visit, requesting full information about the number and type of camping accommodations that are available. Or you may consult with one of the Park Rangers when you arrive in the Park. Additional helpful information may be obtained from *Camping in the National Park System,* a pamphlet which can be purchased from the Superintendent of Documents, U. S. Government Printing Office, Washington, D.C. 20402.

Horseback Riding. Of the many ways to see the Parks one of the more effortless and effective is by horseback. In some areas experienced riders can rent horses without a guide for a leisurely ride along bridle paths or other trails; or you may take a regularly scheduled horseback trip escorted by an experienced guide. Trips of this sort may range from several hours to several days in duration. Check the visitor center or one of the lodges to see what trips are available and when they are scheduled.

Fishing. You are welcome to try your luck at fishing in most of the National Parks and fishing information, maps, and guides are available at concessioner facilities in many of the areas. Because fishing regulations vary from one area to another and are subject to change, inquire at, or write to, the respective Park Headquarters before you start to fish. (As previously noted, state fishing licenses are normally required in those Parks in which the state reserved the right to require such licenses when the Park was established.)

Boating. Boating is a popular pastime in the waters of many of the National Parks. You may bring your own boat, subject to individual Park regulations, or rent one from concessioners. In some areas special boat trips are provided (for example the launch trip around Crater Lake at Crater Lake National Park and the cruise around Jackson Lake in Grand Teton National Park).

Swimming. Swimming is allowed in the lakes and streams of many of the Parks and is a popular form of recreation. Although swimming pools are not usually provided (Yosemite National Park is an exception) they can generally be found in conjunction with public accommodations outside the boundaries of the Park.

Mountain Climbing. The mountainous terrain of certain of the Parks offers unlimited mountain-climbing opportunities. These normally include peaks that can be scaled safely by the beginner and others that will challenge the most experienced alpinist using specialized equipment. For the visitor who would like to try his hand at mountaineering, climbing instructions, rental equipment, and guide service are available in several of the Parks.

Each area has its own rules and regulations for climbing, which may be obtained from the office of the Park Superintendent. In general, they state that mountaineers must (1) register before ascending any peak; (2) show evidence of good physical condition, sufficient experience, and proper equipment before undertaking hazardous climbs; and (3) never climb alone. Because of the element of personal danger involved, the mountaineer—novice or veteran—should carefully follow these regulations.

Winter Sports. Certain of the National Parks afford excellent opportunities for winter recreational activities. These include skiing, tobogganing, and ice skating. Most of the Parks that have well-developed winter sports areas are provided with ski lifts and rope tows to help you uphill. Ski trails are also available in some of the Parks.

Tours. During the tourist season scenic bus tours are conducted through certain of the Parks. Some are all-expense tours of several days duration; others last for only a few hours. In most instances, guides point out the more interesting features of each area and stops are made at various scenic spots along the way. Such tours are especially recommended to visitors who come to the Parks without their own automobiles, and tour information can be obtained from your travel agent or the Park Superintendent.

Photography. The only shooting permitted in the Parks and Monuments is with a camera and no one should visit without one. Opportunities for picture taking are unlimited and offer a challenge to the amateur and an opportunity for the professional. Although many visitors prefer to use color film, dramatic black and white photographs can also be taken. Some of the Park brochures contain pertinent information about picture taking in the respective areas and Park concessioners can usually help with photographic problems. As an added interpretive service a few of the visitor centers have special exhibits which provide the photographer with specific instructions for photographing the various features of the area. Special equipment is not really necessary to capture the beauty of the Parks, but the more serious "shutterbug" will find that a light meter, accessory lenses, and filters will help produce the best pictures. A haze filter will prove to be an especially good investment. Not only will it sharpen up long-distance scenic shots, it will also protect the camera lens.

To Make Your Stay More Pleasant

In order to make your visit as enjoyable as possible, Park concessioners and/or the National Park Service have provided a variety of public services in most areas. Although the services in the various Parks may differ considerably, those listed below are typical. For more specific information pertaining to available services consult the descriptive brochures for the respective Parks.

Accommodations. Living accommodations in the various Parks range from simple housekeeping tents to luxurious lodges. All such facilities are operated by licensed concessioners and rates are subject to government approval. Because of the large number of visitors to most areas, it is generally advisable to make reservations in advance; a list of concessioners and their addresses is provided under the "Parks at a Glance" section for the appropriate area.

Meals and Supplies. Meals can be obtained in cafeterias and restaurants in most of the Parks and box lunches and sandwiches can also be purchased. Most units also have stores where groceries, film, camping supplies, curios, and post cards are available.

Service Stations. Gasoline stations are strategically located throughout most of the Parks and many of them have facilities for minor auto repairs.

Rental Services. A few areas have provisions for renting camping supplies, tents, cots, blankets, and cooking utensils. In winter, skates, sleds, and ski equipment are also available on a rental basis in some Parks. In addition, pack and saddle animals, boats, automobiles, bicycles, and mountain-climbing equipment may be rented in many of the Parks.

Communications. Post offices and/or mail service are available for all areas. Long-distance calls can be made from public telephones where available. Check the various Park brochures for more specific information.

Transportation. Concessioners offer bus service in certain of the larger Parks, and automobiles may be rented in some areas.

Medical Care. During the summer season, some of the more heavily visited Parks have resident nurses and/or physicians in attendance. A few (such as Yellowstone, Grand Canyon, and Yosemite) maintain modern, well-equipped hospitals. Some of the larger hotels have trained nurses in residence and in case of emergencies first aid may be obtained from ranger stations and visitor centers.

Religious Services. Church services, both Roman Catholic and Protestant, are conducted in most of the Parks. Times and locations of the services are posted on bulletin boards or may be obtained from Park personnel.

Miscellaneous Services. The larger Parks offer a variety of miscellaneous services. These include laundry and dry-cleaning services, self-service laundries, baby sitting, photographic processing, bath houses, and special instruction in mountain climbing, skiing, and photography, to list but a few.

Tips for Tourists

When you visit the National Parks you will live under conditions that are likely to differ considerably from your normal routine. First, you are on United States Government property that is to be conserved and protected. It is therefore necessary to follow certain rules and regulations that have been established for the protection of the Park and the personal safety of the visitor. Second, you will most probably be in a physical environment to which you are unaccustomed and one should consider this when planning recreation and other physical activities. The following brief suggestions are intended to help you have a more enjoyable and carefree visit.

For Yourself. Although your vacation should be as relaxing and carefree as possible, you could spoil it if you become careless. Observe Park regulations—they are posted in conspicuous places—stay on designated trails, and avoid trips alone. If you are in mountainous country accustom yourself gradually to your surroundings and in all cases avoid overexertion.

For Your Car. As noted earlier, Park highways are not high-speed thoroughfares but are designed primarily to permit maximum enjoyment of the scenery. All Park roads are safe if you drive carefully and do not exceed the maximum speed limits which are posted and enforced by Park Rangers.

Practice the usual courtesies of the road: avoid parking on curves, keep to the right of the center stripe, signal when leaving the road to park on overlooks and turnouts, and pass only when road signs or center striping indicate that it is safe to do so. In addition, restrict vehicular travel to constructed roadways and park your car in those areas set aside for this purpose.

When driving on mountain roads always drive on the right side. Many of those roads are crooked and steep, so drive slowly and shift into a lower gear to avoid overheating your engine and excessive brake wear. More importantly, you will see more and have a more enjoyable tour if you drive slowly.

For Your Camp. If you decide to camp, you will be asked to study the camping regulations and to keep a clean camp. Campfires should be kept small and under control at all times; in picnic areas and campgrounds fires must be confined to fireplaces. Use only fallen dead wood for fuel and be sure your campfire is out before you leave it or retire for the night. Firewood and fuel for gasoline-operated stoves are commonly sold by concessioners but it is generally advisable to carry fuel with you at all times.

Do not feed the wild animals in the campground and keep food supplies locked up or hung out of their reach. Rubbish and garbage should be burned and noncombustible trash should be placed in cans provided for this purpose. In addition, you are asked to avoid making unnecessary noise—especially late at night and early in the morning.

Fires and Smoking. Fire is the greatest enemy of the National Parks—*report all brush, forest, or grass fires to the nearest Park Ranger, ranger station, or visitor center*. As a safety precaution build fires only in specified places and never near or on roots of trees, dry leaves, or dead wood. One should never leave a campfire unattended and before leaving or retiring always extinguish it with *water* and make *sure* it is out. Be equally careful with any other burning material and do not throw matches, cigarettes, cigars, or pipe ashes from your automobile or along the trails. Remember, a single act of carelessness can start a fire that could produce a trail of desolation and destruction that might take hundreds of years for nature to repair.

Litter. Within recent years the Parks have been infested with human pests called the "litterbug." Unfortunately a few inconsiderate visitors consistently manage to mar the natural beauty of the Parks by scattering about paper, picnic refuse, and other waste material. Waste receptacles are numerous and placed at convenient locations throughout the area—please use them.

On the Trails. The hiker should confine himself to established trails—shortcuts between zigzags or switchbacks can cause destructive erosion to the trail and may be dangerous to you and to any persons below you. Vehicles are not permitted on Park trails or bridle paths and horses and pack trains have the right-of-way on all roads, trails, and bridges.

Pets. Although you may bring your pets to the Parks, dogs and cats are permitted in those areas only if they are on leash, crated, or otherwise under physical restraint. Under no circumstances are they allowed in public buildings, in boats, or on the trails.

Wildlife. In some of the Parks you may see bears, deer, elk, and other animals; all are wild although they may appear to be tame. It is unlawful to feed, tease, touch, or molest these creatures: observe and enjoy them, but for your sake—and theirs —do not approach them.

Preservation of Natural Features. All of the Parks' natural features—wildlife, native plants, rocks, and minerals—belong to you. They also belong to your neighbor and federal law prohibits their destruction, defacement, or removal. The following motto is a good one to remember: "Take from the Parks only pictures and pleasant memories; leave only traces of your footsteps."

Fishing and Boating. Fishing and boating are permitted in most areas subject to local rules and regulations that differ from one Park to another. Sport fishing is provided in more than sixty areas of the National Park System. In all areas except ten, anglers are required to buy state fishing licenses and, generally, to conform to regulations established by the states for fishing in waters adjacent to the Park. The only exceptions for a state fishing license are in nine areas in which the appropriate state legislatures have waived the requirement for a state fishing license by ceding exclusive jurisdiction to the federal government and in Yellowstone National Park, established prior to the admission into the Union of Idaho, Montana, and Wyoming.

Hunting and Trapping. The National Parks are wild game preserves in which hunting and trapping are prohibited.

Firearms. The use or display of firearms within Park boundaries is prohibited by law. Possession of firearms must be declared at entrance stations and they must be sealed, cased, or otherwise packed to prevent their use while in the Park.

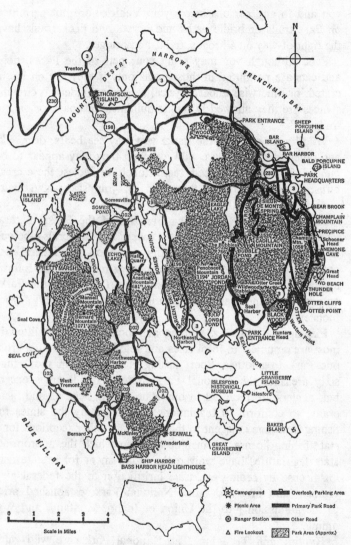

Acadia National Park, Maine.

4. Acadia National Park, Maine

Where Mountains Meet the Sea

In Acadia National Park one is immediately exposed to the charm and enchantment of the rugged New England coast. Precipitous wave-cut cliffs, tidal pools teeming with marine life, large sea caves, tree-fringed lakes, and deep forests are but a few of the Park's many attractions.

Situated primarily on **Mount Desert Island** off the coast of Maine, Acadia National Park also includes parts of **Isle au Haut, Schoodic Point** on the mainland, and several smaller islands. Mount Desert Island (originally *l'Isle des Mont-deserts* —"Isle of the Solitary Mountains") was discovered and named in 1604 by Samuel de Champlain, a French explorer. **Cadillac Mountain** located on the eastern side of the island, rises 1530 feet above sea level and is the highest point on the Atlantic Coast. From here one is greeted by a panorama of unusual beauty; Isle au Haut looms to the southwest, to the south the restless Atlantic stretches as far as the eye can see, and, although not visible, Nova Scotia lies to the east.

Ice, Water, and Rocks

As in all our National Parks, the rocks of Acadia provide clues to the geologic history of the area. Here the rocks tell a story of igneous activity, of the deposition and subsequent erosion of sedimentary rocks, of glaciation, and, finally, of spectacular marine erosion.

Geologists have learned that the pink granite core of the Mount Desert Range is composed of rocks that cooled slowly beneath the surface several hundred million years ago. Today these originally deep-seated rocks are exposed because the overlying sedimentary rocks have been stripped away by erosion. In parts of the Park, *dikes* (tabular bodies of intrusive igneous rock that cut across the structure of adjacent rocks) have been intruded into the older pink granite. These, and the lavas that are found in some places, are evidence that the area has undergone several periods of igneous activity.

More recently—geologically speaking—the original Acadian landscape was reshaped by a massive sheet of glacial ice. During Pleistocene time, this great glacier flowed slowly out of Canada, spread southward across New England, and eventually completely covered this part of North America. As the glacier thickened and continued to advance, the Mount Desert Range was buried beneath a thick blanket of ice.

Glaciers Shape the Landscape

Because glaciation has played such an important part in developing the scenery in Acadia, and many other National Parks, an understanding of their nature and occurrence will be helpful to the reader. *Glaciers* are great masses of moving ice and granular snow that form in areas where the rate of snowfall and the refreezing of melted snow exceeds the annual rate of melting. Today glaciers occur primarily in regions of high latitudes such as Antarctica, Greenland, and Iceland. In addition, there are smaller glaciers at high elevations at middle and relatively low latitudes in such areas as the Rocky Mountains, Cascade Range, the Alps, and the Andes.

Although now restricted in their occurrence, glaciers were once much more widespread. For example, about one million years ago during the Pleistocene Epoch glaciers formed at much lower latitudes and elevations than they do today. This was the so-called "Ice Age"—a period of extensive glaciation when much of North America was covered with ice. It was during this time that the mighty glaciers did much to shape the scenery of Acadia and many other National Parks.

There are three types of glaciers: (1) valley, or mountain, glaciers, (2) piedmont glaciers, and (3) ice sheets, or continental glaciers. *Valley glaciers* are rivers of ice that flow out of the mountains. Such ice masses normally follow the courses of pre-existing river valleys and are confined between the valley walls. Because they occur only in mountainous regions, these glaciers are also called *alpine* or *mountain* glaciers. Much of the spectacular scenery in Grand Teton, Mount Rainier, Glacier, Rocky Mountain, North Cascades, and Olympic National Parks has been created by the geologic work of valley glaciers.

Piedmont (literally "at the foot of the mountain") *glaciers* develop when the terminal part of a mountain glacier expands at the point where it emerges from its valley. In some areas two valley glaciers may unite to form large piedmont glaciers. During Pleistocene time glaciers of this type were probably present in several parts of North America, including Yosemite, Mount Rainier, and Olympic National Parks.

But the largest of all glaciers—and the type that once covered Acadia National Park—are the *ice sheets* or *continental glaciers*. These huge sheetlike ice masses have gently sloping surfaces from which the ice flows outward in all directions. Some of these, for example, the Antarctic ice sheet, are capable of covering all or much of an entire continent. And because of their great size (hundreds or thousands of square miles in area and from 1000 to 10,000 feet thick), continental glaciers generally flow over any obstacle that may get in their paths.

Glaciers are among nature's most efficient and powerful earth-shaping tools. As forces of destruction and erosion they plow, file, quarry, polish, and gouge; as depositional or constructional agents, they carry rock debris of all shapes and sizes, and deposit it when they melt. In Acadia National Park one can see eloquent evidence of the geologic efficiency of a continental ice sheet. At a number of places you will see great *glacial erratics*, large boulders that have been transported considerable distances from their original source and, hence, commonly of different composition than the rocks upon which they have been deposited. At the north end of **Jordan Pond** on the east slope of the **South Bubble**, there is a large balanced boulder which is a classic example of an erratic. Elsewhere there is evidence of ice *abrasion.* This is evident from the presence of *glacial polish*—lustrous rock surfaces that were ground and polished by the ice and large and small scratches called glacial *grooves* or *striations.* These scars were carved into the bedrock by sharp fragments of rock that were embedded in the under surface of the ice sheet. As the ice moved over the land, this combination of ice and rock formed a natural abrasive tool which scoured and gouged the rocks beneath it.

Acadia's valleys and mountains provide even more conclusive evidence of the ability of glaciers to alter the landscape. As the great mass of ice spread slowly outward and southward across New England, the Acadia area was eventually covered by an ice sheet thousands of feet thick. It was then that the flowing ice carved the typical V-shaped stream-cut valleys into the flat-bottomed, glaciated, U-shaped valleys that characterize the area today. Moreover, the profiles of the mountains are a direct result of this remodeling by ice. For example, the northwestern mountain faces are marked by smooth slopes; the southeastern slopes are typically steep and are characterized by sharp cliffs and great steplike ledges. Evidence of this type is of considerable geologic importance for it enables the geologist to determine the direction of ice flow. In Acadia, the ice followed a southeasterly course and ground smooth the northwestern slopes as it ascended the mountain. But the ice cliffed the descending (southeastern) side as it moved downslope. The cliffs were formed by *plucking* or *quarrying,* a process whereby huge pieces of granite were pulled from the southeastern slopes as the moving ice pulled away from them. The effect of cliffing is clearly shown on the two peaks called **The Bubbles.**

As the ice plowed steadily through Mount Desert Island's valleys, they were deepened and widened to form great basins. Later, when the ice disappeared from the land, these became filled with water. One of these, **Somes Sound,** has become a *fiord*—a narrow elongate body of water formed when the sea invaded the abandoned glacial valley. Some geologists believe that Somes Sound is the only true fiord on the Atlantic Coast of the United States. Elsewhere in the Park area other ice-scoured rock basins became filled with fresh water to become *glacial lakes;* **Jordan Pond** is typical of these.

Because of the great weight of this massive ice sheet, Mount Desert Island is believed to have sunk as much as 220 feet. Later, as the ice slowly melted, the island gradually emerged from the sea. Even so, it still remains partly submerged, and forms one of the islands and headlands which mark the summits of the ancient mountains which once dotted the coast of Maine. Moreover, its many coves and inlets suggest the presence of valleys and low places, that were flooded as the ancient coastal plain sank beneath the sea.

Sea Versus Land—a Never-ending Struggle

Although glaciation and volcanic activity are no longer active in the Park, the shaping of Acadia's face continues. The most important and most obvious of present-day geologic forces is the work of the sea.

The dramatic conflict of sea versus land is almost as old as the earth itself yet this battle still continues along Maine's rocky coast. Acadia National Park proudly bears the "battle-scars" of this timeless struggle—scars which have greatly enhanced the rugged beauty of the rocky New England coastline.

Anyone who has observed the ceaseless churning of the sea can readily understand its power as an agent of erosion. The sea's most effective work is done by means of waves and wave-produced currents, for each wave that breaks on shore erodes by a combination of processes. During storms great waves explode violently against the shore pounding the rocks with force measured in thousands of pounds. Sand and rock are rolled to and fro gradually eroding the shore by *abrasion.* Those waves that batter the rocky cliffs compress the air within the rock crack. At times they produce such great pressure that the cracks are widened and blocks of rock may be lifted bodily out of the cliff. By this means the sea actually quarries rock from the cliffs through the force of *hydraulic action.* The blocks thus removed fall to the base of the cliff where they are further broken along with the other wave-pounded rock fragments that have accumulated there. Thus, through the combined action of wave abrasion and hydraulic action the coastal rocks are slowly but steadily being worn away.

Some of Acadia's wave-produced land forms are good examples of *wave-cut cliffs*—for example, the near-vertical westward-facing cliffs of **Schoodic Peninsula, Otter Cliffs,** or the east face of **Great Head.** These typical features of marine erosion were developed as waves eroded the underlying rocks thereby causing the overlying rocks to fall into the sea. At the base of the 107-foot-high **Otter Cliffs** on the southeastern shore of Mount Desert Island, one can see a *wave-cut bench*—another example of marine erosion. This relatively flat surface slopes gently seaward and has been produced as Otter Cliffs slowly retreats under continued marine erosion.

At nearby **Thunder Hole** the sea continues to gnaw away at the island cliffs. Here, the surf dramatically "explodes" in a narrow chasm developed along fractures in the weathering granite. During heavy seas the resounding "boom" that is produced as compressed air is suddenly displaced from the crevice is not unlike a nearby clap of thunder. At extreme low tides the sound from Thunder Hole is more reminiscent of the roll of distant thunder. This effect is produced as large cobbles and boulders knock against one another as waves move them back and forth at the end of the chasm.

At a number of places along the shore of Mount Desert Island there are finger-like projections of resistant rock extending out into the water. These features known as *promontories* or *headlands* are often called *"points"* or *"heads."* **Great Head, Schooner Head,** and **Otter Point** are typical features of this type. Many of the sea cliffs which form the sides of the headlands contain *sea caves* which were formed as waves removed the rock from weak places in the cliff face and when the tide is out one can explore these caves. **Anemone Cave,** carved eighty-two feet into the steep cliffs of the eastern side of the island, is easy to reach and contains a profusion of marine life.

The water-filled indentations between the headlands are called *coves* or *inlets.* **Otter Cove, Seal Cove,** and **Ship Harbor** are but a few of the many coves which indent the coast of the island. The last derives its name from an incident during the Revolutionary War when a local ship, attempting to escape a British war vessel, ran aground and was stranded in the shallows of this inlet.

Although marine erosion is taking place on most of the Acadian coast, the sea is actually building up the land in other areas. For example, **Sand Beach** on the eastern side of the island is a sandbar which has been deposited across a once larger cove.

Thus, over a period of thousands of years, two dynamic geologic agents—ice and sea—have joined forces to produce the rugged but beautiful Acadian landscape.

Life in Acadia National Park

An unusually fine system of roads, trails, and bridle paths enables the visitor to explore each facet of this multifeatured Park. As you drive—or better still, hike—through the Acadian forest, you will see fine stands of spruce, fir, pine, beech, and maple. Noticeable also are the effects of the great forest fire of 1947—a fire that ravaged almost seventeen thousand acres of Mount Desert Island's finest woodlands. Such areas serve as grim reminders of the ever-present danger of forest fires and their devastating effect and emphasize the need to observe every precaution to prevent their occurrence.

From early spring until autumn, Acadia is bedecked with a profusion of wildflowers including, to name but a few, the pitcher plant, rhodora, sea lavender, and the bunchberry. Forest animals likely to be seen include deer, chipmunks, squirrels, beavers, and a large variety of songbirds, especially warblers.

But forest life makes up only a part of the living community in Acadia National Park. Perhaps the most interesting organisms of this area are those that live in the sea. The marine environment of the Gulf of Maine is one of cold, shallow water —an ideal situation for a thriving saltwater population.

Sea anemones (close relatives of the stony corals), starfish, sea urchins, barnacles, and marine snails inhabit the tidal zone. These and a host of other saltwater plants and animals are best seen in the rocky tidal pools along the shore. For a good look at the tide-pool fauna be sure to visit **Anemone Cave,** one of the more interesting parts of the area.

Lobsters, crabs, and an infinite variety of fishes live both near shore and in deeper waters. Overhead one is apt to see such typical ocean-loving birds as the osprey, the gull, and the black guillemot. The latter, a member of the auk family, is expert at diving for fish. In addition, the bald eagle is occasionally seen in the Park.

Tips For Tourists

The interpretive program at Acadia National Park is designed to help you better to see, enjoy, and understand the natural and human history of the area. You may choose from a number of interesting activities which are listed on the *Naturalist Program* which is posted on bulletin boards throughout the Park or that is available at Park Headquarters.

Plan to visit the **Abbe Museum of Stone Age Antiquities** at **Sieur de Monts Spring.** This museum, which is open from May 30 to about September 20, contains stone relics of Indians who lived on the island hundreds of years ago. In the nearby **Nature Center** (open from about May 1 to October 16) there are interesting and educational displays that explain how glaciation and the work of the sea have affected the scenery of Acadia.

Regularly scheduled "seagoing" sightseeing trips are available at all coast towns or you can hire boats for private cruises around the islands. In addition, Park Naturalists accompany certain of these cruises to more interesting coastal areas in the Park. Although there is no charge for the service of the Naturalists, a fee is charged for boat fare on these privately operated cruise vessels. These sea trips include the **Naturalists Sea Cruise,** which leaves the Frenchman Bay Boating Company pier at Bar Harbor for an interesting cruise around Frenchman Bay. Along the way you will see such Acadia highlights as **Anemone Cave, Great Head,** the precipitous sea cliffs of **Ironbound Island,** and the **Porcupine Islands.** A variety of sea birds and an occasional seal or porpoise are sometimes seen.

Another pleasant and scenic boat trip is the **Islesford Historical Cruise** and **Museum Tour.** This features a cruise on the **"Great Harbor of Mount Desert"** and there is also a stop on **Little Cranberry Island.** While on the island you will receive a guided tour of the Museum, which is open every day except Sunday. Additional information about this trip may be obtained at Park Headquarters.

The **Baker Island Cruise** combines an ocean trip with a landing on **Baker Island,** providing, of course, that weather and sea make this possible. Once on land you will walk across the island to see the lighthouse and the south shore sea wall, which is composed of massive blocks of granite.

Although seascapes are the order of the day for most photographers, an infinite number of photographic opportunities abound in Acadia. Close-up views of marine life and flowers, typical New England fishing and boating villages, and sweeping land-sea panoramas are but a few of the more attractive possibilities. Because the light is brighter than you might suspect, be careful not to overexpose and by all means use a haze filter on those long-distance sea shots. Even more dramatic and exciting subjects await you along the rocky coast, but when working near the water take special precautions to protect your camera from the corrosive effects of salt water. A good way to protect the camera lens is to keep a haze filter in place at all times.

Acadia National Park at a Glance

Address: Superintendent, Box 338, Hulls Cove, Maine 04644.

Area: 41,642 acres.

Major Attractions: Rugged coastal area showing effects of marine erosion and glaciation; beautiful display of seacoast, lake, mountain, and eastern forest; marine life.

Season: May through November.

Accommodations: Campgrounds; trailers are welcome but no utility connections are available. For list of nearby accommodations contact: Chamber of Commerce at Bar Harbor, Northeast Harbor, Southwest Harbor, Maine.

Activities: Camping, fishing, hiking, horseback riding, nature walks, picnicking, swimming, motor trips, carriage rides, sea cruises, skiing, and snowmobiling.

Services: Food service, gift shop, saddle horses, and carriages.

Interpretive Program: Campfire programs, guided hikes, museums, nature walks, naturalist sea cruises, self-guiding trail.

Natural Features: Erosional features, forests, geologic formations, glaciation, lakes, mountains, seashore, unusual birds, volcanic features, wilderness area, wildlife, sea cave, sea cliffs, and a fiord.

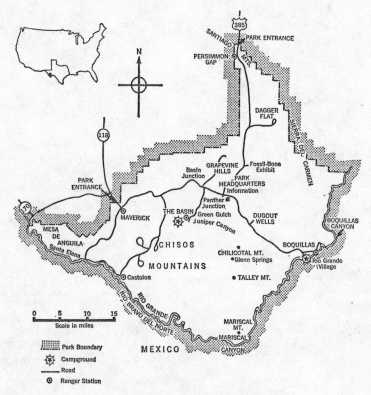

Big Bend National Park, Texas.

Old Faithful Geyser—most famous scenic attraction in Yellowstone,
America's oldest National Park. NORTHERN PACIFIC RAILWAY

Mount Rainier's extensive system of alpine glaciers can be seen well in this photograph taken from Yakima Park. WASHINGTON STATE DEPARTMENT OF COMMERCE AND ECONOMIC DEVELOPMENT

At Glacier Point, in Yosemite National Park—high above the center of visitor activity in the valley below—a Ranger leads a conducted nature walk and explains the geology and plant life of this part of the Park. Beyond Half Dome (in the right middleground) are the bald, billowing granite domes of the High Sierra. JENNIE. A. MATTHEWS

Fortunately, most visitors are not as thoughtless as the vandals who threw the above debris into Yellowstone's lovely Morning Glory Pool. It behooves each visitor to protect the natural wonders of *our* National Parks. NATIONAL PARK SERVICE

Fossilized leaf imprints such as these from the Petrified Forest have provided geologists with much information about the nature of ancient plant life. NATIONAL PARK SERVICE PHOTO BY GEORGE A. GRANT

Volcanic activity—both recent and ancient—has played a significant part in developing the scenery of many National Parks. Here, in Hawaii Volcanoes National Park, we see a great lava fountain associated with the 1959 eruption of Kilauea Iki. NATIONAL PARK SERVICE PHOTO BY BOB HAUGEN

The Olympic Rain Forest is one of the most remarkable and heavily populated plant communities in the National Park System. NATIONAL PARK SERVICE

The effect of the constant interaction of weather, water, and the earth's rocky crust is dramatically displayed at Utah's Bryce Canyon National Park. UNION PACIFIC RAILROAD

These little creatures in Crater Lake National Park are typical of the many animals that live in the Parks. JENNIE A. MATTHEWS

The effect of stream erosion on bedded layers of sedimentary rocks is clearly evident in this view of Arizona's famed Grand Canyon. FRED HARVEY

The scene above—familiar to all who have visited Yellowstone National Park—marks the beginning of a "bear jam." Park bears, spoiled by feeding from thoughtless visitors, become a problem when they hold up traffic. Park regulations expressly forbid the feeding of bears—because it is injurious to the animals, and visitors often suffer bites and scratches. Left alone, bears seldom molest humans.
NATIONAL PARK SERVICE PHOTO BY JACK E. BOUCHER

Virgin Islands National Park's Trunk Bay beach is considered to be one of the ten most beautiful beaches in the world. This is the site of the Trunk Bay Marine Trail—a self-guiding underwater nature trail.
NATIONAL PARK SERVICE

These great terraces in the Mammoth Hot Springs area of Yellowstone National Park are composed of travertine that was deposited by mineral-laden waters issuing from hot springs. JENNIE A. MATTHEWS

View through window of Hawaiian Volcano Observatory at Hawaii Volcanoes National Park. Seismographs in the foreground are recording tremors that might warn of an impending eruption. U. S. GEOLOGICAL SURVEY

5. Big Bend National Park, Texas

A Land of Contrasts

Big Bend National Park, located at the southwestern tip of Texas, is one of the most remote and unspoiled areas of the United States. Here, where the lazy Rio Grande makes the big elbow bend that gives the Park its name, is a country that is part mountain, part desert—a land more like Mexico, a stone's throw across the river, than like the United States. There is wild beauty and dramatic contrast in the vast complex of deserts, canyons, and mountains that comprise this rugged Park.

The heart of the Big Bend country is the **Chisos Mountains,** a jagged mountain range which looms some 4000 feet above the barren plains. Most of the visitor activity is located at the **Basin,** a natural bowl located high in the Chisos about 5400 feet above sea level. To get there you travel seven miles on a good road that winds steadily upward through **Green Gulch.** As you gain altitude and leave the desert flora behind, you will encounter oak, piñon, juniper, and other typical mountain vegetation. But even more noticeable is the freshness of the mountain air—a welcome relief from the warmer temperatures of the arid plains below. Finally, at **Panther Pass,** an elevation of 5800 feet, the road begins its sharp winding descent into the **Basin.** Here, high above the desert, there are food and lodging, campgrounds, service station, and other facilities.

High Mountains and Deep Canyons—Why?

The Chisos—an Indian name which is believed to mean phantom or ghost—look like a rocky island emerging from the flat Texas plains which surround them. The geologic history of this ghostlike range and the surrounding plains can be read from the rocks which are found there. These rocks tell of prehistoric seas in which lived many unusual forms of life, of erupting volcanoes, violent disturbances in the earth's crust, and finally of a period of profound erosion.

The geologic story of Big Bend began many millions of years ago when the area was covered by a warm shallow sea. Slowly, but steadily, great thicknesses of limy mud and sand accumulated on the bottom of this ancient ocean. Mixed with these sediments were the bones and shells of the animals which inhabited the sea. As time passed, the deposits grew thicker and heavier and were gradually converted into layers of sedimentary rocks. The animal remains in the sediments were also turned to stone—today they can be found as fossils in many parts of the Park. For example, the observant visitor will notice these ancient shells along the trails which enter **Santa Elena** and **Boquillas Canyons.**

Eventually, powerful forces within the earth caused the land to rise and the seas gradually receded. But Big Bend was not yet ready to "dry up." Instead, the area was characterized by extensive marshes and lush tropical forests. This was the home of the giant reptiles of Cretaceous time. Here has been found the skull of *Phobosuchus,* a true "Texas-sized" crocodile that attained a length of fifty feet. The area has also yielded the bones and teeth of dinosaurs and other unusual prehistoric reptiles. It was during this same time (about 75 million years ago) that some of the trees became petrified and other plant material was converted into coal.

The Mountains

Once again there was renewed disturbance within the earth's crust, this time accompanied by "fireworks" in the form of volcanic activity. Much of this igneous activity took place deep within the earth as great masses of molten rock (called *magma*) pushed upward toward the surface by means of openings in the crust. This material—lava, cinders, and ashes—was distributed over a wide area. Unable to reach the surface, other concentrations of magma cooled at depth. These slowly cooling magmatic bodies formed the *igneous plugs* that can be seen in many of the Big Bend mountain peaks. They are visible today because the overlying less resistant rocks have been removed by erosion.

The volcanic activity of this part of earth history was accompanied by great structural deformation of the rocks. In many places the rocks were riddled with *faults,* fractures in the rocks along which there has been displacement. One of these faults was of such magnitude that it produced a great *rift valley* some hundred miles long, thirty-five miles wide, and almost a mile deep. The great fault scarp that marks the western side of this valley can be seen in **Mesa de Anguila.** The **Sierra del Carmen** ranges of Mexico mark the eastern side of the valley.

Finally, within relatively recent time—geologically speaking —wind, water, and weather have added the finishing touches to the Big Bend landscape. The softer rocks have been removed by erosion, thus producing wide valleys in the lowlands and leaving behind the sharp rugged profile that typifies the Chisos. The larger mountains, for example, the imposing **Casa Grande** and saddle-like **Pummel Peak,** are the remnants of more resistant masses of granite-like rocks which have not yet succumbed to the elements. Others, like **Tule Mountain,** are flat-topped erosional remnants. Erosion has also produced numerous canyons and pinnacles such as those in the **Grapevine Hills.**

Even the Basin, which was once interpreted as an extinct volcanic crater, and more recently—and erroneously—as the result of glaciation, is now believed to be the product of erosion. Running water and atmospheric weathering have worked together to carve this crater-like depression out of the resistant igneous rocks. When it rains, water leaves the Basin by means of the **Window,** a sharp V-shaped notch which is one of the better-known Big Bend landmarks.

Some of the more prominent mountains which can be seen from within the Basin are **Emory Peak** (at 7835 feet, the highest point in the Park); **Ward Mountain,** which forms the south side of the Window; and **Vernon Bailey Peak,** which marks the north side of the Window. **Pulliam Peak,** which forms the Basin's northern rim, **Toll Peak, Casa Grande,** and **Lost Mine Peak** are also visible. **Lone Peak,** a rather small mass of igneous rock which has stubbornly resisted the forces of erosion, dominates the central part of the Basin.

The southern extent of the Chisos Mountains is marked by the **South Rim.** Believed by many to be the most interesting part of the Park, the South Rim is accessible only by foot trail or by horseback. From this 7200-foot cliff which rises approximately 5000 feet above the plains, the visitor is treated to a breath-taking panorama of much of Texas and even more of Mexico.

The Canyons

Although the mountains are the dominant feature of the Big Bend landscape, some of the most striking scenery is to be found in the great canyons that have been carved by the Rio Grande River. The canyon that is best known and most easily reached is **Santa Elena Canyon,** which is located thirty-five miles southwest of the Basin. Here a yawning chasm has been cut through the great fault scarp of the Mesa de Anguila. This canyon is some seven miles long. It is most spectacular at its mouth, where Terlingua Creek flows into the Rio Grande, for at this point the sheer steep canyon walls tower more than fifteen hundred feet above the river. As you look into the mouth of the canyon, the wall on your left is in Mexico, while the right-hand wall is in the United States. If the creek waters are not too high, take time to hike the trail that leads back into the canyon. Along this relatively easy path you will notice great blocks of Cretaceous limestone which have fallen from the perpendicular canyon walls. Many of these boulders contain well-preserved fossil oysters and clams which lived in the warm shallow seas which occupied this area during the Cretaceous Period.

The Rio Grande has carved a similar gorge thirty-five miles southeast of the Basin in the face of the Sierra del Carmen. This is **Boquillas Canyon,** which is longer (twenty-five miles) and wider than Santa Elena Canyon. As you follow the trail into the canyon, note the great sand slide that is piled up along the canyon wall. This huge pile of sand is a good example of a deposit of *eolian,* or *wind-blown,* sediments. Further evidence of the work of the wind as a geologic agent can be seen in the small cave at the top of the slide. Carved out of the canyon wall by sand-laden winds, this cave shows well the abrasive action of nature's highly efficient "sandblasting machine."

Mariscal Canyon (presently accessible only by rough dirt roads) is located in the southernmost part of the Park. This canyon, with walls that are steeper and higher (about 1650 feet) than Santa Elena Canyon, is believed by many to be the most beautiful canyon in the Park. This great gorge, like Santa Elena and Boquillas Canyons, offers eloquent testimony to the geologic work of running water, for the Rio Grande has scoured each of these out of hard, massive beds of limestone.

Plants and Animals of Big Bend National Park

Persons who are not familiar with the desert generally think that such an area is virtually devoid of life. This is certainly not so in the Big Bend region, for more than a thousand different plants have been noted there. Most of these plants are typical desert species, but some are found only within the confines of this particular Park.

Because of varying climatic conditions and differences in altitude, the Big Bend plants occur in five distinct plant communities—the river flood plain, the desert scrub, the desert grassland, the piñon-juniper-oak woodland, and the ponderosa pine-Douglas fir-Arizona cypress forest communities.

The *desert scrub* community includes the predominant vegetation in the Park and occupies the lower parts of the area. Here the most common plant is the creosote bush; however, ocotillo, lechuguilla, the yucca, or Spanish dagger, and a wide variety of cacti can also be seen. In addition, during the rainy season the desert is covered with flowers of many different colors.

Along the lower slopes and foothills of the mountains, there is a noticeable change in the flora. In this, the *desert grassland* community, chinograss and sotol are typical, but a variety of other plants also thrive here. The *piñon-juniper-oak woodland* is found on the middle and upper slopes of the Chisos. Piñon pine, the drooping and alligator junipers, the madrone, or "naked Indian," several species of oaks, and a variety of smaller plants typify this group.

Growing in the ravines and canyons of the higher elevations of the Park, the *Douglas fir-ponderosa pine-Arizona cypress forest* community is a plant assemblage of great beauty. Unfortunately this segment of Big Bend plant life is of limited extent and appears to be steadily diminishing in size.

In contrast to the vegetation mentioned above, totally different types of plants are found growing along the banks of the Rio Grande River. These lowland, river valley forms include mesquite, willows, cottonwoods, and reeds.

The wildlife of Big Bend is equally interesting and varied and includes several species that are restricted to this area. Mammals that live in the Park include desert mule deer, Del Carmen's white-tailed deer, coyotes, foxes, skunks, ringtails, jack rabbits, javelinas (the collared peccary), and mountain lions. Present also are more than two hundred species of birds. As you hike along the trails be on the lookout for Scott's oriole, the Colima warbler, white-winged dove, scaled quail, Couch's

Mexican jay, cactus wrens, and hummingbirds. While driving keep your eyes peeled for a view of the roadrunner. This unusual bird is commonly seen dashing madly along the roadside, its long tail streaming along behind.

And, as would be expected, there are reptiles—both lizards and snakes. Stay clear of the last; leave them alone and they will get away from you just as quickly as you try to escape them.

Tips for Tourists

A variety of recreational activities await the visitor to Big Bend National Park and a number of interpretive services are also available.

Visitor Center Museum. An orientation exhibit is on display at the **Panther Junction Administration Building.** Featured here are general collections pertaining to the natural and human history of the Big Bend Country.

Good roads make most of Big Bend easily accessible to the motorist. You will want to visit **Santa Elena Canyon** (in the morning if possible) and **Boquillas Canyon** later in the day. An interesting side trip can be made to the abandoned mercury mines and ghost towns of **Terlingua** and **Study Butte.** If you are interested in desert plants you will certainly want to take the self-guiding auto road to **Dagger Flat.** Roadside markers identifying the various plants add to your enjoyment on this trip. Also of interest is the **Fossil Bone Exhibit** east of the Park road and north of the **Tornillo Creek** bridge. The more adventurous visitor may want to drive over the unimproved road to the abandoned frontier store at **Hot Springs.**

Excellent photographic subjects can be found among the flowering desert plants, colorful mountain panoramas, and ghost towns of the Big Bend area. For distant scenes take pictures in the morning and late afternoon—a haze filter is advisable if there is dust in the air. The best pictures of Santa Elena Canyon are taken in the early morning, but save Boquillas Canyon for afternoon when the setting sun strikes the face of the Sierra del Carmen.

Big Bend National Park at a Glance

Address: Superintendent, Big Bend National Park, Texas 79834.

Area: 708,211 acres.

Major Attractions: Spectacular mountain and canyon scenery, cactus-covered plains, and desert wasteland. A unique blending of Mexican and United States flowers, trees, and wildlife.

Season: Year-round.

Accommodations: Cabins, campgrounds, group campsites, motels, and trailer sites, and lodging are also available in Alpine (110 miles northwest) and Marathon (80 miles north). *For reservations contact:* National Park Concessions, Inc., Big Bend National Park, Texas 79834. NOTE: Because of the relatively small number of accommodations that are available it is advisable to make advance reservations as early as possible.

Activities: Boating, camping, fishing, hiking, horseback riding, nature walks, picnicking, scenic drives, and pack trips.

Services: Food service, gift shop, guide service, post office, Protestant religious services in summer, service station, telegraph, telephone, general store, saddle horses, rest rooms, and picnic tables.

Interpretive Program: Campfire programs, visitor center exhibits, nature trails, roadside exhibits, and self-guiding trails.

Natural Features: Canyons, deserts, erosional features, forests, fossils, geologic formations, hot springs, mountains, rivers, rocks and minerals, unusual birds, unusual plants, volcanic feature, waterfalls, wilderness area, wildlife and Indian sites.

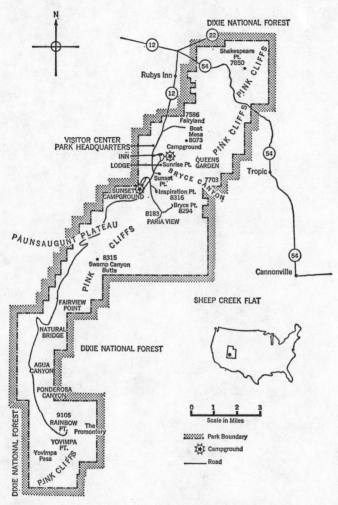

Bryce Canyon National Park, Utah.

6. Bryce Canyon National Park, Utah

The Pink Cliffs of Utah

In south-central Utah at the eastern edge of the Paunsaugunt Plateau, there is one of the world's most colorful and unusual canyons. This is Bryce Canyon—a veritable fairyland of fantastic rock sculptures of every imaginable shape, color, and size. Bryce Canyon is not a canyon in the strict sense; rather, it is a horseshoe-shaped basin or amphitheater which has been intricately carved out of white and pink limestone, shale, and sand. This great depression—it is roughly two miles wide, three miles long, and several hundred feet deep—contains a fascinating assortment of multicolored rock formations in the shapes of countless spires, pinnacles, walls, natural bridges, arches, and windows. Present also are formations that resemble castles, domed cathedrals, temples, and miniature cities; some even resemble animals and men.

How the Canyon Was Formed

Although the many-hued stone formations of Bryce Canyon have been carved during relatively recent geologic time, we must reach far back into the dim reaches of the geologic past to trace the development of this unusual natural phenomenon.

Sand, Mud, and Gravel. About fifty-eight million years ago during the Eocene Epoch of the Tertiary Period, Utah was quite unlike it is today. Large, shallow inland seas and lakes covered what is now the south-central part of the state. These bodies of water received large amounts of sediments which were washed in from the surrounding lands and in some places as much as two thousand feet of gravel, sand, and limy mud accumulated in these low-lying basins. As time passed and sediments were piled upon sediments, the tiny rock particles became firmly cemented together by minerals and were tightly compressed by the great weight of overlying deposits. Finally, these sediments were transformed into solid rock. The sedimentary rocks thus formed comprise the *Wasatch Formation*—the colorful rock strata which are seen in the famous Pink Cliffs of Utah. Many of these rocks are very fossiliferous and the remains of a variety of organisms including trees, clams, snails, turtles, and primitive mammals have been collected from them.

The coloration of the rocks is due to the presence of certain minerals which were present in the sediments. Oxidation of these minerals—a chemical change that is produced as oxygen, assisted by moist air, combines with minerals to form oxides—is responsible for the various shades of color that have stained the Wasatch rocks. For example, various iron compounds such as *hematite,* which produces different shades of red and brown, and *limonite,* which provides a wide range of yellows, are especially common. Those rocks which are purple or lavender probably contain *manganese oxides* while rocks which are white have probably had much of the mineral content bleached out of them.

The Land Rises. After deposition and consolidation of the Wasatch beds, the area that is now southern Utah was slowly elevated until the ancient sea and lake bottoms were almost two miles above sea level. This gradual uplift, which began about thirteen million years ago, was caused by strong *tectonic* or *diastrophic* movements inside the earth. Such movements, resulting from great forces within the crust, are responsible for many of the major elevations and depressions of the earth's surface.

As the lands slowly rose, the rocks were subjected to great stress and strain and this caused large fractures in the surface. In places the crust was broken into tremendous blocks, some of which are several miles in dimension. The Paunsaugunt Plateau is one of nine great fault blocks which were produced by this tremendous upheaval.

The Carving of the Amphitheaters. Because the fracturing of the rocks on the Paunsaugunt Plateau left them riddled with fractures and particularly vulnerable to the forces of erosion, it was not long before the landscape began to undergo considerable alteration. The main geologic agent in bringing about this change was the erosive force of *weathering:* the process whereby rock materials are altered during exposure to air, moisture, and organic matter.

Weathering may be brought about by physical change, chemical change, or more commonly a combination of the two. Both physical weathering (*disintegration*) and chemical weathering (*decomposition*) have played active parts in creating the striking erosional remnants in Bryce Canyon. Included among the agents of weathering that are responsible for the Park's bizarre formations are *frost action* (which takes place as the rocks are weakened by the alternate freezing and thawing of water in cracks and pores in the rocks), and *organic activities* such as burrowing animals, pressure exerted by growing plant roots, and more recently by thoughtless hikers, who take short cuts on the canyon trails. The material thus loosened is carried away by streams formed from melting snow and rain.

Not only did weather sculpt the rocks, weathering processes helped color Bryce Canyon's remarkable landscape.

But why, you many ask, are the canyon's formations so diversely shaped? Here, again, the answer is to be found in the rocks themselves. Because the various rock strata are of varying degrees of hardness, they tend to erode at different rates of speed. Hence, the harder, more resistant limestones form the shelves, ledges, and "caps" of the rock sculptures; the softer shales and sands are more readily removed and thus form grooves, recesses, and small caves. In addition, many of the beds contain large cracks and fissures which may eventually develop into *arches* and *windows*. Elsewhere, the walls may become more deeply dissected thereby developing rows of sharp *pinnacles* such as those seen in the **Silent City.**

Interestingly enough, the same forces that have created these unusual formations are also destroying them. As time passes and erosion progresses, pinnacles are eroded away, and their remnants are washed into the valleys below. But even as the old landforms are being destroyed, wind, water, and ice are attacking the canyon rim to produce still more of these amazing stone sculptures.

Plants and Animals of Bryce Canyon National Park

The plant life of Bryce Canyon, like that of the Big Bend area, is controlled by factors of elevation and precipitation. At the higher altitudes (between 7000 and 8500 feet) along the rim of the canyon there are forests of Douglas fir, Rocky Mountain juniper, and ponderosa, bristlecone, and limber pine. Piñons and junipers occur on the slopes below the rim, and the open valley floors support growths of grasses and sagebrush. At elevations above 8500 feet, there are fine stands of ponderosa pine, blue spruce, white fir, and quaking aspen. Among the many wildflowers that grow in the area you will find the yellow evening primrose, violets, blue columbine, "Indian paintbrush," goldenweed, and western yarrow.

Animals living in the Park include deer, porcupines, skunks, ground squirrels, prairie dogs, yellow-bellied marmots, and chipmunks. The latter are probably the most common animals in the Park, and they are certainly the most entertaining. You are likely to see them scampering among the rocks at almost any place along the **Rim Drive.** Park regulations prohibit feeding the animals, but these little "beggars" apparently are unaware of this.

Tips for Tourists

The best way to see the many wonders of Bryce Canyon is to take a hike—preferably with a Park Ranger—to the floor of the canyon. As you walk along the bottom of the canyon you will see some of the more unusual rock formations and your guide will tell you how they have been formed. Be sure not to forget your camera, for there are countless opportunities for some interesting photographs. However, a few suggestions are in order: First, watch your exposure—the tendency here is to overexpose for the light is considerably brighter than you think: use your light meter. Although dramatic photographs may be taken at most any time of day, the light will probably be "flat" from 10 A.M. to about 3 P.M. so plan your schedule accordingly. While you will certainly want views looking down into the amphitheaters, don't overlook shooting *up* from along the canyon trails. Finally, be sure that you have an ample supply of film before you "go below," for you will probably take more pictures than you plan, and it is a long way back to the rim. For more specific photographic tips consult *Crawford's Trail Guide* which is available at Visitor Center and other points throughout the Park.

Bryce Canyon National Park at a Glance

Address: Superintendent, Bryce Canyon National Park, Utah 84717.

Area: 36,010 acres.

Major Attractions: A brightly colored horseshoe-shaped amphitheater containing innumerable fantastically carved erosional forms including arches, pinnacles, walls, spires, and windows.

Season: Year-round. National Park Service Visitor Center open all year. Four major viewpoints open all winter. No public accommodations in winter.

Accommodations: Cabins, campgrounds, group campsites, and trailer sites. *For cabin reservations contact:* Utah Parks Company, P.O. Box 400, Cedar City, Utah 84720.

Activities: Hiking, horseback riding, nature walks, scenic drives, and bus tours.

Services: Food service, gift shop, health service, post office, public showers, religious services (summer only), service station, telegraph, telephone, transportation, picnic tables, general store, saddle horses, and registered nurses (summer only).

Interpretive Program: Campfire programs, museum, nature trails, nature walks, roadside exhibits, and self-guiding trails.

Natural Features: Canyons, erosional features, forests, geologic formations, and wilderness area.

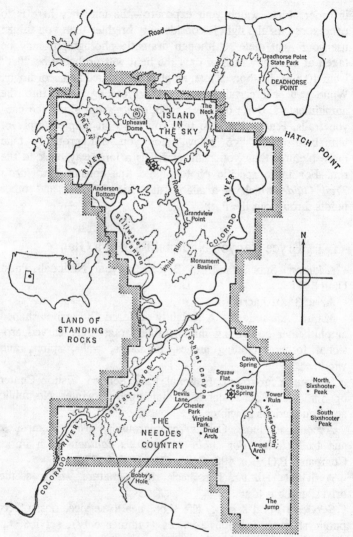

Canyonlands National Park, Utah.

7. Canyonlands National Park, Utah

The Sculptured Earth

"As far as the eye can see, the canyonlands country of south-eastern Utah presents an array of visual wonders. Over millions of years, the mighty lashes of wind and water gouged canyons, stripped rock layers until they bled in a fury of color, eroded the land into startling pinnacles and arches, and sliced away at the plateaued surface."* These words, written by Stewart L. Udall, Secretary of the Interior when Canyonlands became our thirty-second National Park in September 1964, convey in part the mystery and fantasy of this spectacular expanse of red rock canyon country.

Masterpieces of Erosion

Located within this Park's 257,640 desert acres is a land-scape so incredibly diverse that one's first reaction is likely to be disbelief mingled with awe. Most visitors are not prepared for the panorama that lies within Canyonlands, for the land-scape, as seen from the highways that pass near the Park, is essentially that of a rolling, rather monotonous open desert: there is no hint of the scenic grandeur that can be viewed from the rim of the canyons. But once the perimeter of the canyons has been reached, the observer will not soon forget the bewildering maze of steep-sided chasms, the majestic arches,

* *Western Gateways* Magazine, "Canyonlands Highway Issue," Vol. 4, No. 1, Autumn 1964.

crenelated fins, tapering needles, and natural stone monuments that distinguish this untouched wilderness—a landscape which some geologists consider to contain a greater diversity of erosional features than any comparable area in the nation.

Running Water—the Master Sculptor

The face of Canyonlands, like that of Zion and Grand Canyon National Parks, has been profoundly affected by running water. Here two mighty streams—the Colorado and Green Rivers—have slashed across the land incising a fantastic succession of labyrinthine, down-plunging gorges. These powerful rivers have done their work well, for as they have carved the surface of the vast Colorado Plateau, they have also stripped away great thicknesses of the relatively soft, flat-lying sedimentary rocks, thereby exposing an unusually fine cross-section of the earth's crust.

The Canyons. This National Park is most appropriately named, for Canyonlands is truly a land of canyons. Although there are canyons of all shapes and sizes, none are as large and spectacular as those that have been carved by the Green and Colorado Rivers.

The Colorado River Canyon was—and is—being cut by the Colorado River, the master stream in the drainage of the Colorado Plateau. The Colorado enters the area from the northeast and flows southwest through its sheer-walled, rather narrow canyon. Equally striking is the canyon that has been cut by the Green River, the largest tributary of the Colorado River. Entering the northwest corner of the Park, the Green follows a meandering course to the heart of Canyonland where it joins the Colorado River. At the confluence of these two great streams, the waters of the Green River are "captured" by the Colorado, which then enters Cataract Canyon and continues southwestward, toward the Grand Canyon. Cataract Canyon is appropriately named, for on its trip through this rocky chasm, the Colorado begins one of its wildest and most turbulent passages.

Thus, these three spectacular canyons—Green River, Colorado River, and Cataract Canyons—divide the Park into three distinct natural units, which are visually and geologically continuous but which are accessible only from three different directions (see map).

The Island in the Sky. The dominant feature in the northern sector of the Park is **The Island in the Sky,** a cliff-bordered upland formed where the plateau rim extends into and dominates the land between the rivers. The Island terminates at **Grandview Point,** below which stretch the winding canyons of the Green River on the west and the Colorado River on the east; to the south lies massive **Junction Butte.** In the distance, and to the east of Cataract Canyon, are **The Needles,** while the **Land of Standing Rocks** is situated to the west. Approximately 1500 feet below Grandview Point lies **The White Rim,** a resistant ledge of white sandstone of Permian age (see Geologic Time Scale). The White Rim is an excellent vantage point from which to view the rivers, their inner gorges, and the various basins filled with isolated erosional remnants. As we shall learn later, it is this segment of the Park that is destined to become the primary visitor-use area of Canyonlands.

The Island in the Sky section is also the site of **Upheaval Dome,** one of the Park's more mysterious and peculiar landforms. This immense crater—it measures about 4500 feet from rim to rim and is some 1600 feet deep—is a "must-see" for the Canyonlands visitor. Although structurally this feature is a *dome* (the rock strata have been arched upward into a conical, roughly circular, upfold), it more closely resembles an immense volcanic crater. However, it is obvious that the crater is not volcanic in nature, because it is composed exclusively of sedimentary rocks.

But although it is not difficult to rule out a volcanic derivation for this great, jagged depression, an alternate explanation for its origin is not so easily conceived. Yet because it is so unique—there are no similar surface structures in the area— Upheaval Dome has long been of interest to geologists, several of whom have attemped to account for its occurrence. One of the earlier theories of origin held that Upheaval Dome was a *cryptovolcanic structure*—a small, nearly circular area of highly disturbed strata in which there is no evidence of volcanic rocks to substantiate a volcanic origin; hence, the term "crypto" which means hidden. This theory postulates the explosion of steam or hot gas arising from a subterranean source such as a mass of hot igneous rock resting at a shallow depth beneath the center of the crater. Unfortunately this theory is based largely on the resemblance of Upheaval Dome to crypto-volcanic structures in other parts of the world, rather than on more concrete geologic evidence in Utah.

It was later suggested that Upheaval Dome is a *salt dome* formed by the upward plastic flow of a roughly cylindrical mass of salt. The salt is believed to have been derived from the Paradox Formation of Pennsylvanian age that underlies this area and which contains a large amount of salt. This theory is further supported by experiments which prove that when rock salt is subjected to the pressure of thousands of feet of overlying sediments, it may flow plastically. It is also known that in some areas, for example the Gulf Coast region of Texas and Louisiana, the salt may be thrust upward into the overlying sediments in great plugs known as salt domes. Moreover, the structural attitude of the rock strata in Upheaval Dome, lends further support to the salt dome theory of origin.

More recently, however, a new theory has been proposed as a result of maps constructed from data compiled by an airborne magnetometer, an instrument used to measure variations in magnetic intensity in the rocks. Information derived from the magnetometer, which was flown over Upheaval Dome and its immediate vicinity, indicates that a large dome-shaped mass of igneous rocks is buried about 4800 feet beneath Upheaval Dome. Further detailed geophysical and geologic studies based on these maps indicate that this feature is a large salt dome which has been affected both by localized uplift and the intrusion of masses of igneous material from below. The igneous rocks are not, however, believed to be of volcanic origin.

Although it has not yet done so, the salt and/or igneous core of Upheaval Dome may someday be exposed. Meanwhile, an undetermined thickness of sedimentary rocks have already been stripped from the dome, thereby producing the volcano-like crater that characterizes this peculiar structure today.

Land of Standing Rocks. The western section adjacent to the Park is much more isolated and is characterized by an extensive, broad bench that lies west of the rivers (see map). Two features dominate this area: the **Maze,** a wild, as yet unexplored wilderness area consisting of an intricate, interlocking series of rugged canyons; and the **Land of Standing Rocks**—an aggregation of oddly shaped, upright erosional remnants.

The Needles Country. The southeast segment of Canyonlands is generally referred to as the **Needles** area, because of its bristling array of close-set pinnacles, spires, and delicately balanced rocks. These striking formations have developed as a result of erosion along sets of closely spaced joints in the Cedar Mesa Sandstone member of the Cutler Formation of Permian age. These rocky needle-like features are particularly well developed in **Virginia** and **Chesler Parks,** local expanses of flat grassland which can presently be reached only by jeep, horseback, or foot. A short distance west of the Needles, the Cedar Mesa Sandstone has been faulted in places, thus producing a series of parallel sunken valleys called *grabens*. These well-defined, rather linear features consist of down-faulted blocks of rock between upthrown blocks. This is the origin of **Devils Lane** northwest of Chesler Park and **Bobby's Hole,** which is located near the southern boundary of the Park.

The southeastern section of the Park is also the site of **Druid** and **Angel Arches,** towering, sandstone arches which are hallmarks of Canyonlands National Park. Angel Arch, graced by a naturally carved, sandstone "statue" of an angel, is located in Salt Creek Canyon. Majectic **Druid Arch** stands in rock-ribbed Elephant Canyon. Both of these arches are situated in isolated areas of very rugged terrain and are accessible only by trail or jeep. Another interesting attraction in this part of the Park is **Tower Ruin,** a crumbling cliff-dwelling built by prehistoric Indians who once inhabited this area.

Wind and Weather Add the Final Touch

Nature's mighty sculptors, the Green and Colorado Rivers, have made the boldest cuts on Canyonland's face, but the finishing touches have been added by more subtle—yet equally effective—geologic agents. One of the more important of these is weathering, an erosional process that has been greatly accelerated by the diverse climate and topography of southeastern Utah. This is in many ways a land of extremes. Surface elevations range from a low of 3600 feet in the depths of Cataract Canyon to almost 8000 feet on Cathedral Butte, and there are also extremes in temperature: on a hot summer day the mercury may soar to 110 degrees; in winter the temperature may fall to 20 degrees below zero. And, as in most desert areas, the daily, as well as seasonal, temperature variations can be most severe. Thus on hot days the air may cool quickly after sundown, drop to a low shortly before sunup, and then be warmed rapidly as the sun rises. This cycle may be repeated every twenty-four hours and rocks subjected to changes of this type repeatedly expand and contract in response to these drastic fluctuations in temperature. As time passes and this process continues, the fabric of the rock is eventually weakened and their destruction is accelerated. Canyonlands is also dry country. The average annual precipitation is from five to nine inches, yet water derived from late-summer thunderstorms may produce flash-flooding that further erodes and reshapes the arid red rock-canyon country. Extremes such as these provide an ideal environment for continued erosion by wind, rain, frost, and other agents of weathering. That these agents have done their work well is evident at a glance, for the fins, pinnacles, and arches of the Park remain as monuments to the geologic effectiveness of these weathering processes.

Other geologic factors have played supporting roles in shaping Canyonlands' landscape. For instance, the texture and composition of the rocks has determined to a great extent the type of landforms that have been developed. Some formations are relatively resistant to erosion and have formed prominent benches and ledges. Examples of these formations are the Wingate Sandstone that forms the upper surface of the plateau and the canyon rims, and the White Rim Sandstone that supports the White Rim intermediate plateau. Others, such as the Cedar Mesa Sandstone, are riddled with countless close-set, parallel, vertical joints believed to have been formed when the sandstones of the area were broken during crustal uplifts that occurred during fairly recent geologic time. These joints, some of which are intersecting, are of considerable geologic significance for they have controlled to a large degree the development of many of the Park's most spectacular formations.

Fins, Needles, and Pillars. Among the landforms that heighten the visual excitement of Canyonlands are the myriad bizarre erosional remnants that occur throughout the Park. Rare, indeed, is the visitor who fails to wonder at these unique natural monuments and to speculate as to how they might have been formed. As noted earlier, several geologic factors have been involved in developing the terrain of Canyonlands; but two of these—weathering and jointing—have been the controlling elements in the formation of the fins, pinnacles, and pillars that are commonly seen throughout the Park. The influence of joints and weathering on landscape development is especially obvious at the **Needles, Monument Basin,** the **Fins,** and in the **Land of Standing Rocks.** In these areas erosion has been hastened by the chemical action of water and moist air that penetrates deep into the rock along the joint planes. Continued erosion by running water and the destructive activity of alternate freezing and thawing further enlarges the narrow, parallel, vertical cracks, allowing the decomposed minerals and rock fragments to be washed or blown away. As erosion proceeds, the cracks in the rock may develop into deep fissures with narrow vertical slabs of bedrock between them. These tall, narrow rock walls are called *fins.* Most fins contain vertical fractures and as these cracks become progressively wider, the fin will slowly be reduced to a linear series of freestanding *columns, pinnacles,* and *spires.* Eventually, of course, the fins and pillars will be completely destroyed as the unrelenting force of erosion steadily gnaws at the canyon walls. The development of fins and standing rocks is particularly evident in **Monument Basin,** where one can observe these scenic features in various stages of development, ranging from their creation to their ultimate destruction.

In some areas weathering has proceeded along horizontal

bedding planes between rock layers of varying thickness and resistance to erosion. This type of differential erosion has given rise to wondrous displays of remarkably diversified standing rocks such as those of Standing Rock Basin and the Needles Country. Certain of these are slender, sharp-pointed needles, whereas some are flat- or dome-topped columns. Still others appear as massive pillars with rounded remnants of harder rock precariously perched atop their crowns.

The Arches. Like famed Arches National Monument to the northwest, Cannyonlands is the site of several interesting *natural arches*. Few erosional features are more awesome than these gigantic, weather-hewn archways of stone. The largest and best-known arches in Canyonlands are **Druid Arch** at the head of Elephant Canyon and **Angel Arch** in Salt Creek Canyon.

Geologists have long been interested in these unusual structures and their origin has been studied in considerable detail. It has been learned, for instance, that the first stage in the creation of an arch is the development of a fin, a process which has already been discussed. But not all fins develop arches. This is obvious when we compare the number of arches that are present in Canyonlands with the arches of Arches National Monument, for although fins are extremely numerous in Canyonlands, arches are not so common. In the Arches area, on the other hand, at least eighty-eight stone archways have formed within the fifty-three square miles that constitute the Monuments. Why, then, if there are such large numbers of fins in Canyonlands National Park, do we not find more natural arches? The answer lies in the nature of the geologic formations exposed in these two areas. At Arches National Monument the arches have been carved out of the Entrada Sandstone of Late Jurassic age. After fins have developed in the Entrada, their sides may be undercut and perforated by more rapid weathering in softer, less resistant rock layers within the sandstone. Once the fin has been perforated, the erosive effects of ice, snow, wind, rain, and the pull of gravity gradually expand the opening until *windows* are developed. As weathering continues, the window is enlarged and the rough edges of the rock are worn smooth by the abrasive action of wind-blown sand. The final product of this cycle of erosion is a graceful, smoothly contoured natural arch. The developmental sequence of stone arches is well displayed at Arches National Monument where arches in all stages of development and collapse can be seen.

But what of the arches in Canyonlands? The major arches have formed in the Cutler Formation, a rock unit that tends more toward the developments of standing rocks and one that does not seem to have the arch-forming tendencies of the Entrada. There are, nevertheless, twenty-five known arches in Canyonlands National Park.

To appreciate fully the beauty of natural stone arches, the visitor to Canyonlands National Park should, if at all possible, plan a side trip to Arches National Monument, which is located about six miles northwest of Moab, Utah. There a good system of roads permits you to drive within easy walking distance of some of the world's finest examples of natural arches, windows, and balanced rocks.

Plants and Animals of Canyonlands National Park

The native plants and animals of Canyonlands are characteristic of organisms that have adapted themselves to a rather harsh, arid environment. Cottonwoods and other water-loving plants grow around water seeps or springs and there are a few Douglas fir trees in scattered pockets along the upper plateau rim; however, the typical trees are in the piñon and Utah juniper. On the lower benches and in grassy flats such as Chesler and Virginia Parks, there are various desert shrubs and sparse grass.

The animal population includes a few larger mammals such as mule deer, coyote, and foxes; however, these are rarely seen. You are more likely to see rabbits, ground squirrels, and kangaroo rats, plus a variety of birds and reptiles. The Park also supports one of the United State's few remaining natural populations of desert bighorn sheep.

Tips for Tourists

Because Canyonlands is one of the newer areas to be acquired by the National Park Service, its visitor facilities and interpretive program are still in the developmental stage. However, improvements are being made as quickly as time and available funds permit, and within a few years there will be access roads to most major points of interest. In addition, steps are being taken to develop a more extensive interpretive program to effectively explain the natural and human history of this interesting area.

According to present plans, the **Island in the Sky** area will be the principal visitor-use area of Canyonlands. As the area is developed, the National Park Service will provide roads, short trails, overlooks, and interpretive devices including observation buildings, exhibit shelters, and a visitor center. Campgrounds and picnic areas are already established, and concessioner-operated meal facilities will also be made available. Attractions to be featured in the Island in the Sky section are **Upheaval Dome** and **Grandview Point.** Also proposed is a road along the **White Rim** that will circle the Island and lead into the lower part of **Monument Basin** and the **Green River Bottoms.** Eventually there would be a campground, ranger station, overnight visitor accommodations, and boat launching facilities along the river.

The **Needles Country,** or southeastern sector of the Park, is entered near Cave Spring, from which point roads will be constructed to permit vehicular travel to attractions such as **Confluence Overlook, Chesler Park,** and **Devils Lane.** Improved campgrounds, picnic areas, foot and bridle trails, and jeep roads will also be maintained. Activity here will center about **Squaw Flat** where there is now an improved campground and where a visitor center, trailer park, lodge, and store are projected for the future.

The western side of the river is vast and wild; it will be left very much "as is" in order to retain its true wilderness appeal.

After entering this isolated area by foot, jeep, or horseback, the more adventurous visitor may explore the **Land of Standing Rocks** or perhaps venture into the fantastic erosion-scarred world of the **Maze.**

But in the meantime, there is still much to do and see in Canyonlands.

Canyonlands National Park at a Glance

Address: Superintendent, First Western Bldg., Moab, Utah 84532.

Area: 258,640 acres.

Major Attractions: Spectacular eroded area featuring natural arches, standing rocks, canyons, and other objects of geologic interest.

Season: Year-round.

Accommodations: Campgrounds. Lodging and meals available in nearby towns.

Activities: Camping, fishing, hiking, horseback riding, guided jeep trips, picnicking, and scenic drives.

Services: Picnic tables, rest rooms.

Natural Features: Canyons, erosional features, fossils, Indian ruins, and wildlife.

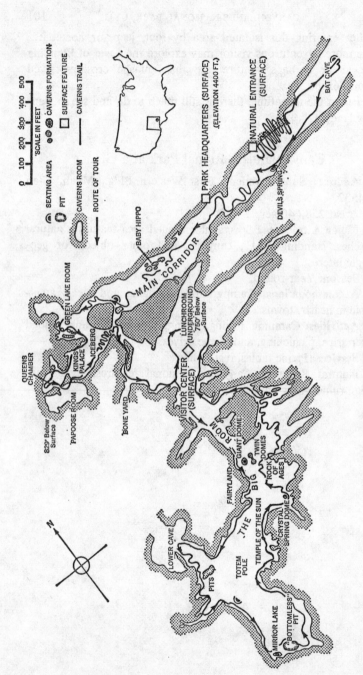

Carlsbad Caverns National Park, New Mexico.

8. Carlsbad Caverns National Park,

New Mexico

A Subterranean Wonderland

Man has long been interested in caves and has used them for shelter and protection since the beginning of his life on earth. In more recent times the eerie darkness and subterranean beauty of caverns have captured the fancy of scientists and "spelunkers" (cave explorers), as well as tourists and sight-seers. Each year hundreds of thousands of people visit commercial caverns to see the amazing changes that have been brought about by the work of underground water.

Caverns are found in many parts of the country, but none can compare with the magnificent Carlsbad Caverns which lie adjacent to the arid New Mexico plains. Here, in the largest and most beautiful caverns yet discovered, you enter a veritable underground wonderland. Imagine, if you will, a single underground chamber with a ceiling as high as a 22-story building and floor space equal to fourteen football fields, and all more than seven hundred feet underground.

Formation of the Caverns

Carlsbad Caverns have been dissolved out of the Tansill Formation and the Capitan Limestone, rocks which originated from limy muds and a large organic reef somewhat similar to the Great Barrier Reef off the coast of Australia. This reef grew in the warm, rather shallow seas which covered southeastern

New Mexico during Permian time, approximately 250 million years ago (see Geologic Time Scale, p. 34). With the passage of time the marine plants and animals that built the Capitan reef began to die and their remains were buried beneath a great thickness of sediments. As these deposits accumulated, their great weight caused the reef-rock to become fractured by the pressure of the overlying sediments. Groundwater invaded crevices in the fractured rocks, causing them to be dissolved. This process, known as *solution,* is the most important factor in the formation of caverns.

Solution in the Carlsbad Caverns area was particularly effective because the rocks in which the caverns have been formed consist primarily of soluble limestone. Calcium carbonate, the principal constituent in limestone, is only slightly soluble in chemically pure water. However, water containing dissolved oxygen and carbon dioxide forms *carbonic acid,* a substance which readily attacks carbonate rocks such as limestone. Thus, the rocks containing the caverns were dissolved when groundwater, charged with carbon dioxide derived from the atmosphere and the soil, seeped slowly downward along cracks or between beds of limestone. As the water descended, the openings were slowly enlarged to form an irregular network of underground tunnels such as those seen in the **Boneyard.** At first only small fissures were dissolved out of the limestone; then, as solution progressed, more extensive cavities developed. As the rocks surrounding these large cavities, or *solution pockets,* continued to dissolve, the floors and ceilings collapsed, leaving openings that connected them to other solution pockets. Finally, this process, continuing for millions of years, produced the great underground chambers and the connecting passageways which we see today.

The last stage in the development of the caverns is believed to have taken place between one and five million years ago when another series of earth movements raised the area above the level of groundwater, thus allowing the water that filled the caverns to drain out and the chambers to become filled with air. Although this stage brought an end to the solution of the limestone, weakened rocks along the walls and ceilings of some of the rooms collapsed under their own weight. You will note these great jumbles of rocks (called *collapse blocks*) in many parts of the cavern. The **Iceberg,** a massive collapse block weighing about 100,000 tons, is a spectacular reminder of this phase of the caverns' development. Today, however, one need not worry about falling rocks—the last such collapse took place many thousands of years ago.

Decoration of the Caverns

After the caverns had been drained of water and became filled with air, the slow process of deposition began. During this phase of the caverns' development, innumerable secondary lime deposits called *flowstone* or *dripstone* were formed.

The material composing most of these secondary limestone or calcium carbonate deposits has been called *onyx* or *cave onyx*. This is a misnomer, for true onyx is composed of chalcedony, a translucent form of the mineral *quartz*. When sectioned and polished, however, many pieces of dripstone and flowstone show the characteristic translucency and banding of true onyx. Much of the so-called "Mexican onyx," actually translucent banded calcium carbonate, has been obtained from cavern deposits.

The term *travertine* is more properly used to describe the material comprising dripstone and flowstone deposits. In addition to being deposited in the subsurface, travertine may also be deposited on the surface by hot and cold springs.

Dripstone Deposits. *Speleothems* (cave formations) formed of materials deposited by dripping groundwater are called dripstone. Deposits of this type include stalactites, stalagmites, columns, or pillars, and draperies; all of these features are present in Carlsbad Caverns.

Dripstone was deposited as surface water seeped downward through cracks in the limestone and eventually emerged on the ceiling of the cave. As the water passed downward through the rocks, it dissolved calcium carbonate from the limestone through which it passed. Drops of mineral-laden water were often suspended for some time on the ceiling and when they fell to the floor they left behind a thin layer of calcium carbonate. Thus, over a long period of time, successive accumulations of these limy layers formed hollow tubes which are suspended from the cave ceiling. These pendant-like deposits, called *stalactites* (from the Greek word meaning "oozing out in drops") literally cover the ceilings in many parts of the cavern. The simplest form of stalactite is called a *soda straw;* these long, straight, hollow pendants are quite fragile and represent an early developmental stage in stalactite formation.

A stalactite normally begins as a thin-walled hollow tube; the water dripping from it flows through the interior canal and drops from its end. Eventually, however, it may become choked with calcium carbonate. The water then trickles down the outside, and deposition of travertine causes an increase in the length and width of the stalactite.

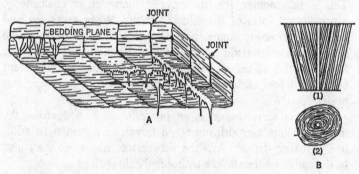

(A) Bedding planes and joints provide access routes for surface waters to penetrate limestone formations, thus bringing about solution; subsequent deposition of dissolved material takes place under certain conditions. (Reproduced from *The Geology and Physiography of Mammoth Cave National Park* by permission of the Kentucky Geological Survey.) (B) Diagrammatic representation of the development of stalactites and stalagmites. With growth the central tube often becomes closed and water reaches the tip by way of the outside. Longitudinal (1) and transverse (2) cross sections of stalactite showing banded arrangement of travertine. (Reproduced from *Behind the Scenery in Kentucky* by permission of the Kentucky Geological Survey.)

Sometimes a row of stalactites join together as they grow downward from a narrow crack in the limestone ceiling of a cave. These may produce blade-shaped *curtains* or *draperies* which hang from the ceiling in folds such as the **Papoose Draperies** in the **Papoose Room** or the "teeth" in the **Whale's Mouth.** If the seepage water contained impurities such as iron it will produce a yellow, red, or brown stain in the rock; the color will be determined by the percentage of mineral impurity in the water. Such discolorations often produce interesting effects in the draperies and some of these deposits resemble slabs of bacon.

When seen in cross section, dripstone deposits show distinct concentric layers which resemble growth rings of trees. These layers, normally of varying widths and colors, mark interruptions in the growth of the speleothem due to changes in mineral composition or periods of deposition. Should the water stop dripping for some reason, the surface of the formation may dry out, and become discolored by weathering. When deposition is resumed, a layer of clear calcium carbonate is then deposited over the stained weathered layer.

At times water dropped from the ceiling at a rate so rapid that most of the carbonate solution fell to the floor. Calcium carbonate deposited by water built up a deposit called a *stalagmite* (from the Greek word meaning a "dripping"). In general, most of these rather broad-based, blunt inverted cones will have a stalactite "feeding" it from above.

When a downward-growing stalactite meets an upward-growing stalagmite a *column* or *pillar* is formed. You will see many of these columns in each of the scenic rooms such as the **King's Palace,** the **Queen's Chamber,** and the **Hall of Giants.**

Flowstone Deposits. Deposits of travertine formed when mineral-laden water flowed over the walls or floor of a cavern called *flowstone*. This type of deposit is responsible for many of the caverns' more unusual formations. The **Rock of Ages,** the **Frozen Waterfall,** and other attractions are due in part to the deposition of flowstone. Each of the above deposits was produced over a long period of time as successive layers of travertine were deposited one upon the other. As a result of continued deposition some of these deposits have grown to be quite thick. Flowstone, like dripstone, will commonly display a banded effect as the results of intermittent periods of growth.

Among the more interesting cave formations present in the caverns are *cave pearls,* small rounded accumulations of calcium carbonate which formed around a nucleus of sand or some other tiny object. Cave pearls originated in small pools of water which were constantly agitated by dripping water. Present also are *helictites* (from the Greek word *helix,* meaning a spiral), peculiar curved or branching formations which appear to develop in defiance of the laws of gravity. Originating as very thin stalactites, by some process not agreed upon by geologists, they begin to grow outward and/or upward, taking on a variety of weird shapes. You can see fine examples of helictites on the ceiling of the **Queen's Chamber.** Other interesting speleothems present in Carlsbad Caverns include *"cave coral," "cave popcorn,"* and *"lily-pads."* The latter are rather thin, irregular, scalloped deposits of limestone formed around the edges of subterranean pools. You will see some "lily-pads" along the trail near the **Frozen Waterfall** and in **Fairyland.**

A number of the cave formations appear to be quite dry and discolored—some are almost dusty. These are said to be "dead" formations and are found in the drier parts of the cave where deposition by groundwater is no longer active. Certain other speleothems literally shine as they reflect light from their wet glistening surfaces; these are usually "live" formations. Such speleothems are actively growing and are found in the wetter sections of the caverns. **Crystal Spring Dome,** probably the largest single live formation in Carlsbad Caverns, has the glistening appearance described above. However, it is interesting to note that not all wet cave formations are alive and "growing"; some are actually being eroded by the water that passes over them.

Although visitors to the cavern often ask about the growth rate of speleothems, there are so many variable factors involved that an intelligent estimate is not possible. But one thing *is* certain—cave deposits do form *very slowly* and some are many thousands of years old. Because of their slow rate of deposition and great fragility, visitors are not permitted to mar the natural beauty of the cave by handling or breaking them.

The Underground Tours

Jim White, the now-famous cowboy who first reported the caverns in 1901, did not explore the caverns in the same manner as the visitor of today. He used a kerosene torch for light and rope-ladders for access; later, visitors were lowered in a huge iron bucket. Today, you can walk the three miles of well-lighted trails with a minimum of effort, stop briefly to enjoy lunch in the huge **Underground Lunchroom,** and then enter a high-speed elevator to be whisked quickly to the surface some 750 feet overhead. Or, if you do not care to take the walk-in tour, you can descend by elevator to join the less strenuous—and less rewarding—tour of the **Big Room.**

The Walk-In Trip. Within recent years a new, "continuous" tour system has been put into operation. Rather than having to wait for regularly scheduled tours, visitors may enter the Cavern on their own and see the cave at their own pace. During the winter, visitors may walk in through the natural entrance anytime between 8:00 A.M. and 2:30 P.M. or make the Big Room trip anytime between 8:00 A.M. and 3:15 P.M. In the summer, the walk-in trip is available continuously from 7:00 A.M. to 4:30 P.M. and the Big Room trip from 7:00 A.M. to 6:00 P.M.

A full staff of Park Rangers are on duty daily throughout the Cavern to answer visitors' questions, as well as to protect the Cavern features and to assure visitor safety. However, rather than leading visitors into the cave in groups at scheduled times, they now greet visitors along the trail. New interpretive signs have also been installed to aid visitors in their enjoyment and understanding of the Cavern features. This system provides much greater flexibility and has eliminated large, concentrated groups, which often numbered more than a thousand people under the old system.

Walking time for each tour is about three hours, but this will vary considerably from one visitor to the next. The distance covered is three miles. The paths are solid, but may be some-what slippery in spots; there are no ladders to climb, but, be-cause of the steepness of certain parts of the trail, visitors should wear low-heeled shoes. As the temperature in the cav-erns remains near 56 degrees year-round, a light coat or sweater will probably make your visit more comfortable.

Admission tickets to the caverns may be purchased at the ticket booth in the lobby of the Visitor Center (children un-der sixteen are admitted free). Pets are not allowed in the cav-erns but may be left in the kennel which adjoins the Visitor Center. Nursery facilities for babies and small children also are available.

Features of the Tour. Although space does not permit a de-tailed description of the many unusual features of the complete tour, a few words about the more outstanding attractions are in order. You will start your tour at the **Natural Entrance,** the site of the original discovery of Jim White. Soon you are in the im-mense **Main Corridor** of the caverns. Here you get your first impression of the true immensity of the caverns as you descend the mile-long trail into the depths below. Although the more scenic chambers lie ahead, you will pass such interesting forma-tions as the **Whale's Mouth** with its "teeth" of limy draperies, the **Baby Hippo,** a "live" and growing stalagmite. Incidentally, at this point, the ceiling is some 214 feet above you.

Near **Iceberg Rock,** the massive collapse block mentioned earlier, you will probably stop for a "breather" while the Ranger provides some details about the geology of the caverns. Here, too, he will probably turn off the lights momentarily in order that you can experience total and absolute darkness.

The first scenic chamber to be visited is the **Green Lake Room,** so-named because of a small eight-foot-deep pond which appears deep green under the artificial light that illuminates the caverns. Other highlights here are the **Veiled Statue,** a tall stately column, and a limy cascade-like deposit appropriately named the **Frozen Waterfall.** Next you enter the **King's Palace,** an elaborately decorated gallery containing a fantastic variety and number of graceful formations. You are at the "bottom" of the tour now for you have descended 829 feet into the depths of the earth. Before leaving, look for the **King's Draperies,** a complex grouping of stalactites and draperies which display a variety of colors. The remarkable 7½-foot soda-straw stalactite called the **King's Bell Cord** is above the trail opposite the draperies.

Conveniently nearby is the **Queen's Chamber** which is festooned with an amazing array of lancelike stalactites, graceful draperies, and the unusual curiously shaped helictites.

Now comes the **Papoose Room**—a fitting name for the smallest scenic chamber in the caverns. In keeping with the size of the room the formations here are small but you will quickly notice their delicate and fragile beauty. The **Papoose Draperies** drop gracefully from the ceiling and, because they are "alive," they glisten and shine.

Leaving the **Papoose Room** by means of a man-made tunnel you re-enter the **King's Palace** to start the steep hike up **Appetite Hill** which leads to the **Underground Lunchroom** and a well-deserved rest stop. But before you get there you will pass the **Boneyard** and the partially dissolved limestones which are exposed there.

The Big Room Tour is especially designed for those visitors who, because of time limitations, physical inability, or any other reason, are unable to take the walk-in tour. This trip, which is over an easy trail one and a quarter miles long, lasts for about one hour and ten minutes. Take the elevator to the lunchroom and begin your tour there.

Features of the Tour. As you leave the lunchroom and begin the trail around the perimeter of the **Big Room,** you are entering the largest subterranean chamber known to man. Perhaps you are wondering exactly *how* big the so-called **Big Room** actually is. The great T-shaped gallery has a floor space of about 14 acres and at its highest point the ornately festooned ceiling towers as much as 255 feet above you. The stem of the "T" is about 1800 feet long; the cross bar at the top of the stem runs for a distance of 1100 feet.

Much is to be seen in this tremendous and beautifully decorated chamber for it contains some of the better-known formations of the caverns. The tour first enters the **Hall of Giants,** where Carlsbad's larger dripstone formations are located. This chamber is dominated by the **Giant Dome,** the largest depositional feature in the caverns. Rising majestically sixty-two feet above the floor, it has a maximum diameter of about sixteen feet. As you can see in the photograph, the **Giant Dome** is actually a column for it is attached to the ceiling of the chamber. Close by can be seen the **Twin Domes** formations; although they have formed in close proximity to each other, they are not as much alike as their name might suggest.

A short distance downtrail is **Fairyland** with its unusual stalagmites, many of which are coated with popcornlike clusters of cave coral. Close at hand is multihued **Temple of the Sun.** As the trail continues, it skirts many crystal-clear ponds of water fringed with "lily-pads," and the floor is covered with stalagmites of every shape, size, and variety. Overhead the ceiling is hung with graceful draperies and myriad stalactites ranging from tiny needles to giant inverted spires.

Following the trail you pass the **Totem Pole,** a gracefully sculptured stalagmite thirty-six feet tall, and a stone wall which marks the **View of Lower Cave**—a precipitous drop leading into a lower level which is not open to the public.

Next comes Mirror Lake, a beautiful reflectional pond of unusual clarity which is shaped like a hand mirror. Shortly thereafter you can look down into the well-known **Bottomless Pit** —an eerie but not-so-bottomless hole about 140 feet deep.

Soon you see **Crystal Spring Dome,** the most actively building formation in the caverns; this is followed by the **Chinese Temple,** an unusual pagodalike stalagmite. After a relatively short distance, the winding trail approaches colorful **Painted Grotto** with its mineral-stained draperies and stalactites. Here also is the world-famous **Rock of Ages**—a massive delicately colored stalagmite which rises majestically from the cavern floor. Finally the tour enters the **Doll Theater**—a charming little chamber containing numerous fragile stalactites, stalagmites, and graceful columns.

All too soon, this underground adventure is over and you find yourself in the elevator lobby adjacent to the lunchroom, and in a matter of minutes you will quickly travel the 754 feet to the surface in complete safety and comfort.

Plants and Animals of the Caverns Area

A variety of plant and animal life inhabit the area around the caverns. The vegetation present is of the desert variety and includes yucca, bear grass, creosote bush, century plant, mesquite, prickly pear, sotol, and ocotillo. To learn more about these unusual plants, take the self-guiding desert nature trail located near the Visitor Center.

On the surface you are likely to see such animals as mule deer, antelope, raccoons, jack rabbits, ground squirrels, skunks, foxes, and ringtails. The latter are rather unlikely looking creatures with a squirrel-like body, a face like a fox, and a tail similar to that of a raccoon. These are nocturnal (night foraging) creatures so look for them along the caverns road as you leave the Park after the Bat Flight.

Present also are a variety of birds including quail, dove, woodpeckers, owls, and the cave swallow which builds its nest on the walls of certain caves within the Park. The state bird of New Mexico, the fascinating roadrunner is also commonly seen. Reptiles present include snakes and a variety of lizards, all of which are typical desert species.

But life forms in this Park are not limited to surface-dwellers. Deep within the dark recesses of the caverns live peculiar wingless insects, gnats, and small crustaceans called isopods. Other animals such as certain crickets, beetles, and mice spend their time engaged in both surface and subsurface activities. However, the most interesting cave-dwellers are the bats which spend the daylight hours of the warmer months in the undeveloped parts of the caverns. These unusual flying mammals leave the cavern at sunset each night in search of insects in the low-lying river valleys some distance away; this **Bat Flight** and the accompanying lecture by a Park Ranger is one of the highlights of a visit to Carlsbad Caverns. It is interesting to note that this natural phenomenon was instrumental in the discovery of the caverns. Jim White was first attracted to the caverns' natural opening by a great black column of "smoke" which he soon learned consisted of millions of bats making their evening exit from the caverns.

Among the fourteen species of bats which inhabit Carlsbad Caverns, there are pallid, lump-nosed, western pipistrel, and fringed myotis bats. The most abundant, however, is the Mexican free-tail bat, a migratory form that spends the winter months in semitropical regions.

Tips for Tourists

There is an excellent museum in the Visitor Center which deals largely with the geology of the caverns. Be sure to see these exhibits (including a scale model of the caverns) before the trip and, if at all possible, again *after* you have made the tour. In this way you can better appreciate and understand the wonders of the caverns. There is no campfire program as such, but a naturalist lectures on the Bat Flight each evening during the summer. The Desert Nature Walk is an interesting trail which will acquaint you with the desert flora of the area.

Carlsbad Caverns National Park at a Glance

Address: Superintendent, Box 1598, Carlsbad, New Mexico 88220.

Area: 46,753 acres.

Major Attractions: The largest and most spectacular caverns that have been discovered, Carlsbad Caverns contain an infinite number of stalactites, stalagmites, and other cave formations.

Season: Year-round; continuous tours daily.

Accommodations: No accommodations in Park, but available in nearby towns—White City (just outside Park) and Carlsbad (twenty-seven miles northeast). Contact Chamber of Commerce, Carlsbad, New Mexico, for list of available accommodations.

Activities: Cavern trips, nature walks, and Bat Flight.

Services: Food service, gift shop, guide service, kennel, nursery, telephone, rest rooms, elevator service, and underground lunchroom.

Interpretive Program: Museum, nature trails, nature walks, self-guiding trails, cavern trips, Bat Flight with lecture by Park Ranger.

Natural Features: Canyons, caverns, desert, erosional features, geologic formations, and desert flora.

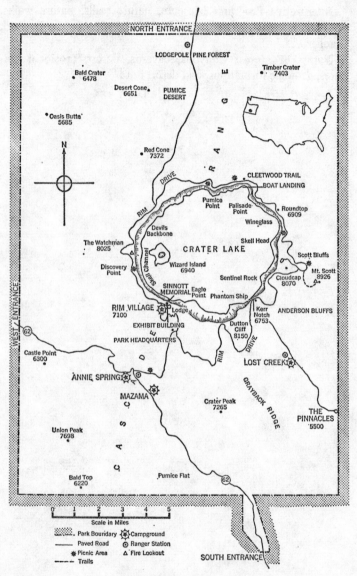

Crater Lake National Park, Oregon.

9. Crater Lake National Park, Oregon

The Lake in the Top of a Mountain

Crater Lake National Park is the site of one of the most beautiful lakes in the world. Here, in the collapsed summit of an ancient volcano, is a lake that is 1932 feet deep, six miles wide, and possesses twenty miles of shoreline. This unusual body of water is surrounded by steep walls of igneous rocks of many colors; in places the cliffs tower as much as 1980 feet above the lake. But the most impressive feature of Crater Lake is its incredibly blue water—water so intensely blue that it defies reproduction by painter or photographer, yet so crystal clear that an object seventy-five feet underwater appears almost at your fingertips.

The Geologic Story of Crater Lake

Once you have seen Crater Lake you will never forget it; and perhaps you, like countless visitors before you, will decide that surely this is one of the most peaceful and beautiful scenes in the world. But the present serenity of Crater Lake is not in keeping with its geologic history. The deep basin that is occupied by the lake is the product of violent volcanic activity and the thunderous collapse of a mountaintop.

Mount Mazama

Thousands of years ago the location now occupied by Crater Lake was the summit of a huge volcano. This tremendous mountain, which is now called Mount Mazama, is part of the

Cascade Range and came into being at about the same time as Mount Hood, Mount Shasta, Mount Rainier (p. 285), and Lassen Peak (p. 233).

Geologic studies of the remnants of Mount Mazama suggest that it was one of the most impressive peaks of its time; it is estimated to have had a maximum height of about 12,000 feet—comparing favorably with lofty Mount Rainier. Like Mount Rainier, Mount Mazama had great glaciers flowing down the valleys on its flanks; typical U-shaped glacial valleys such as Munson Valley, Kerr Notch, and Sun Notch, glacial polish and striations, and deposits of glacial sediments, provide conclusive evidence that glaciation played an important part in shaping the face of Mount Mazama.

But for all its majestic beauty, Mount Mazama, like all volcanoes, had a very humble beginning—it started as a crack in the earth's crust through which was extruded the igneous rock which formed the mountain. This opening, more properly called a *vent,* was connected with an underground reservoir of molten rock material called *magma.* The reservoir, or *magma chamber,* contained the hot molten rock which was either intruded into the earth's crust or extruded upon the surface (see p. 18).

During periods of volcanic eruption, steam, dust, ashes, stone, and molten rock (called *lava*) emanated from the vent. Volcanic products erupted in this manner eventually built the towering cone-shaped peak that was Mount Mazama.

Because its cone was composed of alternating layers of lava flows and solid volcanic debris such as ash, cinders, and pumice, Mount Mazama is thought to have been a *composite volcano* or *strato-volcano.* Volcanic cones of this type are constructed during intermittent periods of quiet eruption—at which time lava pours from the volcano—and explosive eruptions which throw out great quantities of solid volcanic materials. This type of volcanic mountain differs considerably from those seen in Hawaii Volcanoes (p. 205) and Haleakala (p. 199) National Parks. Because the Hawaiian-type volcanoes are composed almost exclusively of lava and are not so steep-sided, they are commonly referred to as *lava domes* or *shield volca-*

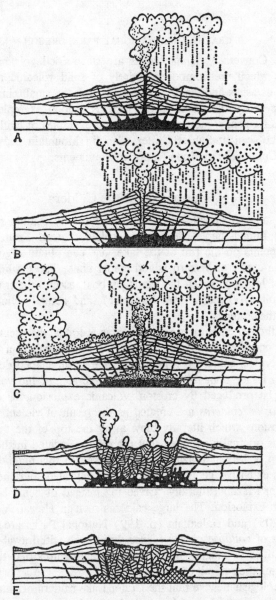

Reproduced from *Crater Lake: The Story of Its Origin* by permission of Howel Williams and the University of California Press. The evolution of Crater Lake. (A, B) Beginning stages of the great eruptions. (C) Glowing avalanches sweep down the sides of the volcano. Magma chamber is being rapidly drained. (D) Summit collapses into magma chamber. Gas vents appear in the caldera floor. (E) Crater Lake today. Wizard Island and lava are shown on the lake floor. Magma in underlying chamber is in large part, or entirely, solidified.

noes. Conversely, *cinder cones* are steep-sided volcanic structures which are composed entirely of solid volcanic materials such as ashes and cinders. **Wizard Island,** the small cinder cone rising above the water of the lake, is a good example of this type of volcanic mountain. Additional examples of cinder cones can be seen at Sunset Crater, Capulin Mountain, Lava Beds, and Craters of the Moon National Monuments.

Formation of the Caldera

The summits of volcanic mountains typically contain relatively large funnel-shaped depressions called *craters.* From various viewpoints on the rim of the lake, one can obtain a good view of the crater of Wizard Island cinder cone. Those who want a closer look can reach the island by boat, ascend the trail to the summit (760 feet above the water), and actually climb down into the crater.

Although commonly referred to as a crater, the great depression occupied by Crater Lake is properly called a *caldera.* The typical caldera is a nearly circular basin-shaped depression much larger than the original vent or crater of the volcano; it may be produced by erosion, volcanic explosion, or collapse. *Explosion calderas* are created as the result of violent volcanic explosions which literally blow away the top of the mountain. It was originally assumed that the caldera located in the hollow stump of Mount Mazama was produced in this manner. However, extensive geologic studies have revealed that the top of Mount Mazama literally "caved in" instead of being blown off by an explosion. The large calderas seen in Hawaii Volcanoes (p. 205) and Haleakala (p. 199) National Parks are also the result of *collapse* or *subsidence* due to the withdrawal of large quantities of magma which had originally supported the rocks forming the summit of the mountains.

Geologists believe that the Crater Lake caldera originally contained some twelve cubic miles of volcanic material and that the overlying cone of Mount Mazama was composed of about five cubic miles of lava, ashes, and cinder. Thus, approximately, seventeen cubic miles of material disappeared during the eruption and subsequent development of the caldera. Much of this material poured out of the volcano in great glowing avalanches of gas-charged, frothy, liquid lava some of which was deposited as much as thirty-five miles from Mount Mazama. Part of this *pumice,* as this lightweight, glassy, cellular lava is called, was picked up by the wind and carried as far north as British Columbia in Canada.

These great eruptions, which probably occurred not more than seven thousand years ago, are believed to have ejected approximately five cubic miles of lava and one and a half cubic miles of older rock from around the vent. At the same time, yawning fissures apparently developed along the sides of the volcano permitting an estimated ten cubic miles of molten rock to flow out as *fissure eruptions.* Without the underlying support of the material that had been ejected, the summit of Mount Mazama collapsed under its own weight. Thus, one of ancient Oregon's loftiest mountains was converted into one of the world's deepest and most beautiful lake basins.

Birth of an Island

But igneous activity did not stop with the death of Mount Mazama; as recently as one thousand years ago, three small cinder cones were formed by volcanic eruptions from vents which opened in the caldera floor. Although rising more than 1000 feet from the lake bottom, two of these cones are submerged and cannot be seen; the third, **Wizard Island,** rises more than 2600 feet above the floor of the caldera. This cone was formed so recently that it has not been greatly affected by erosion; the 90-foot-deep, 300-foot-wide crater at its summit is still clearly discernible. After formation of the cinder cone, lava poured from the west side of the island and built a platform extending from the cone almost to the western wall of the caldera. This one final outpouring of lava apparently marked the final episode in the volcanic history of Crater Lake. However, the geologic story continues to unfold as the rocks forming the rim of the caldera undergo continued weathering and erosion.

Origin of the Lake

Crater Lake has no inlet or outlet; thus all of its waters have accumulated from rain and snow. It must have taken a long time to fill this tremendous pit with the 1932 feet of water which occupy the lake today, and there is evidence to indicate that the surface of the lake may have been even forty to fifty feet higher than it is at present. It is interesting to note that the water level of the lake remains relatively constant from one season to the next. The loss of water by seepage (from the porous sides of the caldera) and evaporation are offset by the average annual precipitation of sixty-nine inches, most of which is snow.

As mentioned earlier, Crater Lake is 1932 feet deep in its deepest part; this makes it the deepest lake in the United States, the second deepest in the Western Hemisphere, and the seventh deepest in the world.

Of course, the crowning glory of the lake is its sapphire blue water, an optical effect believed to be produced as sunlight is scattered in the deep clear waters of the lake. The water apparently absorbs all of the sunlight except the blue rays—these are reflected upward. (You will notice that the water appears green in some of the shallower areas near the shore of the lake where the bottom may be seen through the water.)

Because of its unusual beauty, it is not surprising that early visitors to Crater Lake referred to it variously as Blue Lake, Deep Blue Lake, and Lake Majesty. It acquired its present name in 1869.

Battleground of the Gods

There is much evidence to suggest that there were human witnesses to the fireworks that accompanied the destruction of Mount Mazama and the birth of Wizard Island. Legends of the Klamath Indians refer to this area as a "battleground of the gods" and tell of earth-shattering explosions and mountains that spewed fire. One can only imagine the terrifying thoughts that raced through the minds of these early people as great clouds of smoke and volcanic ash blotted the sun from the sky and their homes were covered with falling pumice. These Indian legends are also supported by archeological discoveries; Indian artifacts have been discovered beneath Mount Mazama pumice in certain caves in central Oregon.

Life in and Around Crater Lake

A variety of plants and animals lives both in and along the shores of Crater Lake. Fish living in the lake include rainbow trout and kokanee, a landlocked form of sockeye salmon. Present also are salamanders, newts, crayfish, snails, and freshwater shrimp. Among the plants living in the lake are algae, moss, and a variety of microscopic forms.

More than 570 species of ferns and flowering plants have been reported from within the Park. The plants are distributed according to elevation and soil type and are thus restricted to three rather distinct life zones. Plants living on the lower flanks of Mount Mazama fall within the *Transition* life zone (3977 feet to 5500 feet); Douglas fir, ponderosa pine, white fir, and sugar pine typify the flora of this zone. The *Canadian* life zone (5500 to 6250 feet) is characterized by dense stands of lodgepole pine, as well as western white pine, subalpine and Shasta fir, and mountain hemlock. Currants, honeysuckle, whortleberry, and manzanita comprise most of the shrub community.

The *Hudsonian* life zone is restricted to the upper reaches of the Park (6250 to 8926 feet) but around the rim of the caldera there is a mixture of plants common to both the Hudsonian and Canadian zones. Plants likely to be encountered in either zone are Shasta red fir and mountain hemlock. At higher elevations the Hudsonian zone is marked by thick stands of whitebark pine.

During the months of July and August, the meadows of Crater Lake are decorated with a colorful array of wildflowers. Look for western pasqueflower, phlox, painted cup, fireweed, aster, monkey flower, columbine, and violets. To really become acquainted with Crater Lake's plant communities, take the self-guiding **Castle Crest Nature Trail.** In addition, be on the lookout for identifying signs near plants on the **Discovery Point** and **Garfield Peak Trails.**

Equally fascinating and varied is the wildlife around Crater Lake, which is composed primarily of an amazing variety of mammals and birds. Although flying, silver-gray, and golden-mantled ground squirrels are all present, the latter form is the most commonly seen. They will be found at almost any place where people congregate. The porcupine, yellow-bellied marmot, chipmunk, fox, deer, and black bear are also likely to be seen. As in all of our Parks, you are reminded not to feed the animals and especially to stay clear of the bears—they are wild animals and should be treated as such.

The most common birds are Clark's nutcracker, Steller's jay, and the gray jay; they (along with the golden-mantled ground squirrel) are common at many points along the rim. Other birds known from here include owl, gull, hawk, raven, golden and bald eagles, osprey, robin, and thrush, to name but a few.

Tips for Tourists

Visitor activities at Crater Lake National Park are centered on the lake; the hikes, drives, trails, and boat rides are designed so as to permit maximum enjoyment of the area in the shortest possible time.

Interesting exhibits dealing with various phases of the Park's natural history may be seen in the **Exhibit Building** and **Sinnott Memorial Overlook;** these two structures are at **Rim Village** midway between Crater Lake and the cafeteria. Representative examples of rocks, plants, and wildlife can be seen in the Exhibit Building and the Park Naturalists on duty there will be glad to answer any questions that you might have about the Park. At nearby Sinnott Memorial Overlook you will be treated to the finest possible view of the lake and see additional exhibits and illustrations pertaining to the origin of the caldera. On the terrace of Sinnott Memorial Overlook are a series of mounted monoculars trained on some of the lake's more noteworthy features; these, and their accompanying explanatory labels, provide interesting background information for the visitor. In addition, Park Naturalists give talks on the origin of Crater Lake several times each day. At Crater Lake, as in the other Parks, the museum is the logical place to begin your visit.

To truly appreciate the quiet beauty of Crater Lake and to realize fully the vast size of the caldera, take the launch trip around the lake. These trips are made in concessioner-operated motor launch and are accompanied by a Ranger-Naturalist who points out and explains features of interest in and around the lake. (Although there is a concessioner's charge for transportation on this trip, no charge is made for the guide service of Ranger-Naturalist.) Although it is neither feasible nor legal to launch private boats on Crater Lake, row-boats can be rented at the boat landing.

Crater Lake is truly a photographer's dream—the peaceful beauty of fleecy white clouds drifting above the caldera and reflecting in the sapphire waters of the lake will cause even the most hardened shutterbug to shoot that spare roll of film. If you have a little patience and a quick shutter finger you can also get some good shots of the golden-mantled ground squirrels.

Wizard Island, Phantom Ship, the **Pinnacles,** and various points along the caldera rim also provide worthwhile photographic subjects. And don't overlook the opportunity to photograph wildflowers if you are there while they are in bloom.

Winter photographs of the Park are especially striking; the high snowbanks along the Park roads, trees burdened with their mantle of snow, and the deep blue lake water framed in white provide unmatched opportunities for beautiful winter landscapes.

Finally, a special word of caution: In order that your visit may be as safe and enjoyable as possible, please restrict hikes and walks to Park trails. Under no circumstances should you venture over the caldera wall—the soil and rock here are often loose and crumbling and are not likely to support your weight.

Crater Lake National Park at a Glance

Address: Superintendent, Crater Lake National Park, Box 7, Crater Lake, Oregon 97604.

Area: 160,290 acres.

Major Attractions: A deep lake of incredibly blue water located in the collapsed summit of an ancient volcano; the walls encircling the lake range from five hundred to almost two thousand feet in height and display a variety of color.

Season: West and South Entrance roads are open year-round; the North Entrance road is closed during the winter.

Accommodations: Cabins, campgrounds, hotel, and trailer park. *For reservations contact:* Crater Lake Lodge, Inc. Summer: Crater Lake National Park, Crater Lake, Oregon 97604.

Activities: Boating, boat rides, camping, fishing, guided tours, hiking, mountain climbing, nature walks, picnicking, scenic drives, swimming, and winter sports.

Services: Boating facilities, boat rentals, food service, gift shop, post office, religious services, service station, ski trails, telegraph, telephone, transportation, limited groceries, picnic tables, and rest rooms.

Interpretive Program: Campfire programs, museum, nature trails, nature walks, roadside exhibits, self-guiding trails, daily lectures on geologic story of the lake, scenic boat and bus tours.

Natural Features: Canyons, "deserts," erosional features, forests, geologic formations, lake, mountains, rivers, rocks and minerals, unusual birds, unusual plants, volcanic features, waterfalls, wilderness area, and wildlife.

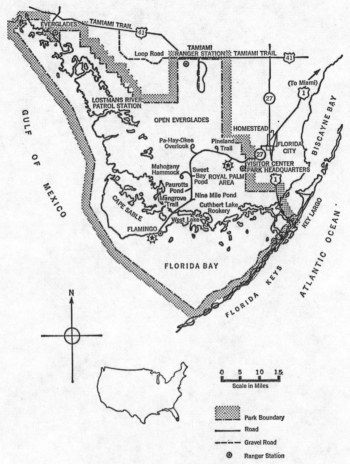

Everglades National Park, Florida.

10. Everglades National Park, Florida

Land of the Grassy Waters

Near the southern tip of Florida is one of our nation's most unusual National Parks. Here are the Everglades—far-reaching expanses of grassland, open areas of water, scattered islands of trees and bushes, and a host of subtropical plants and animals.

To the Indians the Everglades were "Pa-hay-okee," a word which literally means "grassy water." This describes well the sprawling open vastness of this low-lying, unusual, subtropical wilderness which is characterized by open meadows of sawgrass.

The Sea Floor Rises

Geographically the Everglades are situated in the Coastal Lowlands province of the Floridian Plateau—that great southern projection of North America which separates the deep water of the Atlantic Ocean from that of the Gulf of Mexico (see map). A broad, relatively level region, the Floridian Plateau at its highest is little more than 325 feet above sea level. In addition, as much or more of it lies submerged beneath the waters of the Atlantic Ocean and the Gulf of Mexico.

Geologically the Everglades occupy some of the newest land in North America for this region has been elevated above sea level within relatively recent geologic time. But the geologic story of the Floridian Plateau began long before the seas were drained from the Florida Peninsula. Evidence for this is found in ancient metamorphic rocks obtained from deep well borings in southern Florida. These rocks, not unlike those of the Piedmont region of Georgia, suggest that this huge continental projection has been in existence since Paleozoic time. Moreover, throughout its existence it has alternately been covered by shallow seas or exposed as low-lying dry land. Thus today the deeply buried Paleozoic "basement" rocks are covered by more than four thousand feet of sedimentary strata. These beds represent the seaward extension of the rocks which comprise the coastal plain of Georgia and Alabama. In the vicinity of the Everglades, the uppermost rocks of this sedimentary sequence consist of a Pleistocene limestone called the *Miami oölite*. This *oölitic limestone* (limestone characterized by tiny calcareous spheres which resemble fish roe) underlies much of the Everglades. In places, especially near the edges of the Everglades, the Miami oölite occurs near the surface. Here the limestone exhibits a rough pitted surface produced by the solvent action of rain and subsurface waters. In the lower, generally flooded parts of the Everglades, the limestone is covered by *peat* (dark-colored partially decomposed plant remains) or muck.

One can get a good look at the Miami oölite on the Park road about three miles from the Pineland Trail parking area. At this point the road passes through an outcrop of the oölite which can be seen on either side of the road. This place, which has been named **Rock Reef Pass** has an elevation of three and one-tenth feet—a rather significant elevation when viewed in relation to the surrounding lowlands.

To what does the Everglades owe its existence? The very slight elevation of southern Florida, the gradual southward sloping of the Floridian Plateau, and the presence and character of the oölitic limestone are the major geologic factors involved. These, coupled with the subtropical climate and the rapid growth of sawgrass and other water-loving plants, have all contributed to the formation of this interesting area.

Doorway to the Tropics

Despite the fact that it has an interesting geologic background, Everglades National Park was not established as an area of geologic interest. Rather, its geographic position (the latitude of the area almost places it in the tropics) and climate have attracted a host of tropical and subtropical plants and animals. Here, in a fascinating blend of climatic and geographic ingredients, there is a mixture of land and sea, salt water and fresh water, and tropical and more temperate species of plants and animals.

One's first view of the Everglades typically results in impressions of distance and a remote, untamed wildness probably never before experienced. Scanning the landscape, two features seem dominant: the expansive open grasslands with their "islands" of trees called *hammocks,* and open areas of salt and fresh water with their small, heavily vegetated islets.

The hammocks, relatively small elevated mounds rising out of the swamps, support a variety of tropical plants and animals. Hardwood trees, mosses, ferns, orchids, royal palms (found only in this part of North America), gumbo-limbo, and a variety of other tropical and subtropical species thrive on many of the moist hammocks. Animals attracted to the humid, jungle-like hammocks include tree snails, birds, reptiles, and a host of insects.

Other natural communities within the Everglades include bald cypress swamp-forests, the pine forests, or pinelands, and the mangrove swamps. The latter support heavy growths of mangrove trees with strongly arched prop roots and spreading intertwined branches.

Mammals commonly seen in the Everglades are raccoon, opossum, fox, otter, muskrat, and several varieties of squirrel. Larger mammals include black bear, cougar, or panther, deer, and bobcat. Two aquatic mammals—the manatee and the porpoise—also inhabit Everglade waters. Manatees, or seacows as they are commonly called, somewhat resemble a seal. They are timid, elusive animals which have two front flippers and a broad horizontal tail like that of a whale; some attain a length of as much as twelve feet and weigh almost a ton. The more familiar porpoise is sometimes seen jumping out of the water; they may be as much as ten feet long and weigh in excess of two hundred pounds.

As might be expected, the warm humid Everglades are a haven for reptiles of many kinds. Snakes, such as the water moccasin, Everglades swamp snake, banded water snake, and tree snakes inhabit the swampy areas. Coral snakes, rattlesnakes (both pigmy and diamondback), indigo snakes, green snakes, and the Everglades racer, or black snake, are found in the drier parts of the Everglades. Turtles can be found in fresh and salt water as well as on the land. These include soft-shelled turtles, box turtles, and the seagoing loggerhead and green turtles. The loggerhead turtle may weigh as much as eight hundred pounds.

Of special interest are the alligators and crocodiles. Although their numbers have been greatly reduced by hide-hunters, alligators are occasionally seen in freshwater sloughs and swamps. The much rarer American crocodile prefers the saltier water of mangrove swamps and salt marshes. Visitors commonly ask how to distinguish the alligator from the crocodile. In general, alligators are darker in color, have a relatively short blunt snout, and rather broad heads. Conversely, the crocodiles have a longer, more slender snout, a thinner body, and some of their teeth are exposed when their mouths are closed. Like all of the Park's plants and animals, the crocodiles and alligators are protected by federal law; these interesting reptiles should not be fed, teased, or otherwise disturbed.

Fishes, both saltwater and freshwater species, abound in the Park's waters; among the saltwater forms are redfish, sea trout, snook, snapper, and grouper. Freshwater fish include bass, sunfish, and gar. Florida tarpon, some weighing more than a hundred pounds, also inhabit Park waters.

Of all the Park's animals, the birds are the feature attraction. In fact, many people visit Everglades for the express purpose of bird watching; about three hundred species, many of them quite rare, have been recorded within the Park. Here is to be found what is probably the largest and most unusual bird fauna in North America and there are species here that are found nowhere else in the United States. The Everglades are especially famous for their numerous rookeries, or nesting grounds. Among the many birds to be observed swimming or wading in Park waters or flying above the trees are herons, ibises, storks, coots, grebes, gulls, hawks, and bald eagles. Present also are the anhinga, or water turkey, purple gallinules, ducks, pelicans, and cormorants. Of special interest are the egrets and roseate spoonbills, both of which were slaughtered to near-extinction in order to secure plumage which was used in decorating twentieth-century millinery. Consult Park personnel to determine the best time and place to see the various types of birds, or participate in one of the special conducted tours organized for this purpose.

Tips for Tourists

The visitor to the Everglades has the opportunity to see one of the world's great wildlife sanctuaries. The fauna and flora of the Park can be observed along one of the several self-guiding trails, while on a guided boat tour, or at many points while driving through the Park. In the Visitor Center at the Park Entrance there are exhibits on the factors creating the Everglades and the history of the Park itself. An introductory slide program is shown in the auditorium and Park guides are on duty to give information. At the Royal Palm Visitor Center there are exhibits of the birds, tree snails, and the Indians of the Everglades. As in all of the Parks, a stop at one of the Visitor Centers is the proper way to start your visit. There are also interpretive exhibits in the Visitor Center at Flamingo.

Educational guided boat trips from Flamingo and West Lake will take you into the backwaters and on visits to the rookeries where possible. The **Cuthbert Rookery** boat trip, conducted from February until May (if the rookery forms), offers excellent opportunities to see anhingas, cormorants, herons, graceful egrets, and the majestic wood stork or flinthead.

Everglades National Park at a Glance

Address: Superintendent, Box 279, Homestead, Florida 33030.

Area: 1,400,533 acres.

Major Attractions: Largest remaining subtropical wilderness in United States; extensive freshwater and saltwater areas, open Everglades prairies, mangrove forest, abundant wildlife including rare and colorful birds.

Season: Year-round (best in winter).

Accommodations: Cabins, campgrounds, motels, and trailer sites. *For reservations contact:* Everglades National Park Company, Inc., 28494 South . Federal Highway, Miami, Florida 33157.

Activities: Boating, boat rides, camping, canoeing, fishing, guided tours, hiking, nature walks, picnicking, and scenic drives.

Services: Boating facilities, boat rentals, food service, laundry, gift shop, guide service, post office, religious services, service station, telephone, and picnic tables and rest rooms.

Interpretive Program: Campfire programs, museum, canoe trails, nature trails, roadside exhibits, and trailside exhibits.

Natural Features: Erosional features, lakes, marine life, coastal prairies, rivers, sand dunes, seashore, swamps, unusual birds, unusual plants, wilderness area, wildlife, Indian ruins, and unusual reptiles.

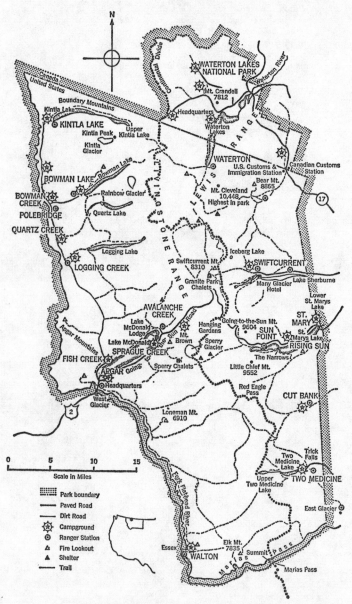

Glacier National Park, Montana.

11. Glacier National Park, Montana

The Crown of the Continent

Located in northwestern Montana and straddling the Continental Divide, Glacier National Park is comprised of more than one million acres of magnificent scenery. The surface of the Park presents an unusually rugged and varied panorama of lofty, angular mountain peaks, slowly moving "rivers" of glacial ice, and more than two hundred sparkling, gemlike lakes. It is not surprising that an early explorer of this beautiful region called it "The Crown of the Continent."

Few of our National Parks offer the visitor more spectacular beauty or a better chance to "get away from it all," for Glacier has more than a thousand miles of well-marked trails which lead into the heart of unspoiled wilderness. You may take your choice from short strolls along one of the self-guiding nature trails or day-long back-country treks to picturesque chalets. But whether you hike for an hour or a day you will find that each of the trails has been planned to bring you closer to nature and to further an understanding and appreciation of this peaceful mountain refuge.

The Story in the Rocks

The earliest chapter in the Park's geologic history began about one billion years ago during Precambrian time (see Geologic Time Scale, p. 34). It was then that a long narrow arm of the sea extended from the Arctic Ocean as far south as Colorado, or possibly Arizona.

This was a time of great erosion; plant life had not yet invaded the land and great quantities of sand, silt, and gravel were dumped into this elongated, rather shallow body of water. As time passed and more sediments were deposited, the floor of the sea gradually sagged downward to form a long troughlike depression called a *geosyncline*. Geosynclines play an important part in the geologic history of most mountainous areas, for the development of almost all mountains requires an accumulation of thousands of feet of sediments in one of these shallow, subsiding sea-filled basins. These sediments will later form the rock of which the mountains are composed. In the Glacier National Park region about twenty thousand feet of sediments were deposited in this ancient geosyncline.

Much later in geologic time these sediments became *lithified* or converted into solid sedimentary rock. Thus, sand grains were cemented and compacted together to form sandstones, limy muds were converted to limestone, and the more silty, clay-bearing muds became siltstones and shales. In some areas certain of the shales and siltstones contain large quantities of crystallized quartz and have been made harder by deep burial and pressure. Such rocks, called *argillites,* can be seen in many parts of the Park. For example, the greenish gray argillite of the formation called the Appekunny Argillite is conspicuous on the face of Appekunny Mountain, on the side of Singleshot Mountain, and on the lower portions of Allen Mountain and Grinnell Point. Good exposure can also be seen at certain places along the Going-to-the-Sun Road.

Because they are quite hard and split easily in thick slabs, certain of these argillites have been used as flagstone pavement in parts of the Park. The more observant visitor will note the presence of mud cracks and ripple marks on some of these slabs. Most of these *mud cracks* were developed when ancient silt or clay deposits dried out and underwent shrinkage; such cracks are roughly polygonal in shape and form a rock surface that resembles a section cut through a large honeycomb. The mud cracks were preserved when a second layer of sediment was deposited on the first. If the deposits later became lithified, the outlines of the cracks may have been accurately preserved for hundreds of millions of years. Such cracks are usually exposed to view when the rock is split along the bedding plane between the two deposits. Certain of these mud cracks look much as they did at the time they were first formed and provide evidence that the original deposit was subjected to alternate periods of flooding and drying.

Ripple marks consist of a series of undulating wavelike markings which may be developed on shallow bottoms by either waves or currents. Ripple marks, like mud cracks, commonly furnish the geologist with clues as to the conditions that existed at the time that the formation was deposited. Thus, if ripple marks have asymmetrical sides they were probably

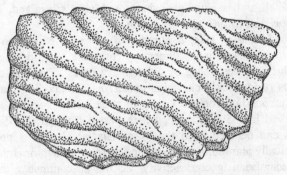

Ripple marks occur at a number of places in Glacier National Park. They help the geologist to reconstruct climatic and geographic conditions of the geologic past.

formed by currents, the long slope on the ripple mark being the direction from which the current came. On the other hand, symmetrical ripple marks with sharp or slightly rounded ridges separated by gently rounded troughs were probably formed by the oscillatory movement of sea water along a coast. Both types of ripple marks occur in the rocks of Glacier National Park and provide much information about the geologic history of the area. Good examples of mud cracks and ripple marks can be seen on the **Hidden Lake** self-guiding trail.

But sedimentation was not the only geologic process in operation in the area at this time—some of the material deposited in the Precambrian sea shows evidence of igneous activity. One of these formations, the Siyeh Limestone, contains a *sill*—a tabular body of igneous rock which was intruded between beds of Siyeh Limestone. This sill, which is about one hundred feet thick wherever it occurs, is composed of *diorite,* a granular dark-gray, crystalline igneous rock. As the hot molten magma was squeezed between the layers of limestone, the rocks on either side of it were baked and recrystallized to form marble, a metamorphic rock (see p. 20). The lighter colored marbles which lie above and below the dark diorite serve to accentuate further the sill, which is seen as a distinct band across the **Garden Wall** and is equally obvious on **Mount Cleveland, Mount Wilbur,** and **Little Chief Mountain.**

In addition to sills, a number of *dikes* have been intruded into the Precambrian sedimentary formations. Unlike the sills, which lie between beds of sedimentary rocks, dikes are tabular bodies of igneous rock which cut *across* the bedding planes of the surrounding rocks. Several of these vertical seams can be observed in the Park; one, which passes through the **Pinnacle Wall,** can be seen from the Many Glacier Hotel. Its presence is typically marked by a snow-filled, vertical trough or chute.

Precambrian igneous activity was not all confined within the earth—in places lava was extruded onto the ocean floor where it hardened into rounded masses or "pillows." These lumpy, ellipsoidal, pillow structures are typical of submarine lava flows wherever they are found. The lava, called the Purcell Lava, is exposed at **Granite Park** and in the vicinity of **Boulder Pass.**

Although there is no record of terrestrial life during Precambrian time, there is evidence that life was present in the sea. This is known from the presence of massive beds of fossil algae which occur in the Siyeh Limestone. These tiny one-celled plants secreted calcium carbonate and built great masses or "reefs" of organic limestone. Their rounded surfaces sometimes resemble rosettes or a head of lettuce; they can be seen exposed along the Going-to-the-Sun Road and on the face of the **Pinnacle Wall** overlooking **Iceberg Lake.**

Missing Pages

Although much is known of the Precambrian history of the Park area, the happenings of Late Precambrian and Paleozoic time are obscured by great gaps in the geologic record. However, it is probably safe to assume that Late Precambrian sediments were deposited in the area but later removed by erosion when the sea withdrew and the Park area was uplifted. Following this period of erosion, when thousands of feet of sediments may have been removed, there was another invasion of the sea at which time additional sediments were deposited on the old eroded Precambrian surface. This encroachment of the sea probably began during the Early Paleozoic and deposition may have continued throughout part of Paleozoic time. Unfortunately, Paleozoic rocks have not been found in Glacier National Park but formations of this age are exposed in certain mountains south of the Park. However, geologic evidence suggests that the Precambrian formations of the Park were once overlain by similar Paleozoic strata which have since been stripped away by erosion.

The Sea Returns

The next clearly decipherable event in the geologic history of Glacier National Park began about 120 million years ago during the Cretaceous Period; it started with the deposition of thousands of feet of sand and limy mud in the old geosyncline. These sediments, some of which contain the remains of relatively advanced forms of invertebrate life, completely buried the rocks that had been deposited during earlier geologic time.

The Crust Buckles

As the Cretaceous Period neared its close, the sediments that had been deposited in the old geosynclinal trough were subjected to pressures of unbelievable magnitude. These forces, which acted very slowly and came primarily from the southwest, greatly compressed and elevated the geosynclinal deposits. Thus were the sediments of an ancient sea gradually transformed into the great Rocky Mountain system. These mountain-building or *orogenic* movements were so severe that the Precambrian and Cretaceous strata were strongly wrinkled, folded, and fractured.

Although the deformation described above took place near the end of the Mesozoic Era, the newly created mountain system was subjected to continued stress during Early Cenozoic time. As pressure continued from the west the arched layers of rocks were overturned, buckled, and finally broken. So great were these forces that rock layers from the west were shoved over on top of the rocks on the eastern side of the fracture—it has been estimated that this great rock mass was thrust as much as forty miles to the east. This type of *fault,* or displacement of the strata, is called an *overthrust*—a fault wherein the rocks on one side of the fault are thrust over and above the other side. Faults of this type bring about a reversal in the normal order of rock layers by thrusting older strata over and above younger ones. Thus in certain places in the Parks, Precambrian rocks hundreds of millions of years old are found lying on top of Cretaceous beds that are only about sixty million years old.

The overthrust in Glacier National Park has been named the *Lewis Overthrust* and is one of the best-known features of this type in the world. The effects of the overthrust are apparent in many parts of the Park, but they are especially evident on the east face of **Chief Mountain** where Precambrian formations can be seen resting on younger Cretaceous strata. Actually **Chief Mountain** is but an erosional remnant of the great thrust sheet that was driven eastward during the Lewis Overthrust. Technically it is known as an *outlier*—a veritable "island" of older rocks surrounded by much younger strata. Good views of Chief Mountain can be had from the Chief Mountain International Road between Babb, Montana, and the International Boundary (see map).

Evidence of the *fault plane* (the surface along which the thrusting occurred) can also be seen on the side of **Summit**

Mountain. This mountain, located about three miles north of U. S. Highway 2, can be seen from near **Marias Pass.** The fault plane appears as a nearly horizontal line beneath the vertical cliff of Precambrian strata which overlies the grayish Cretaceous shales.

Erosion Begins

Paradoxically the very forces that created the great overthrust block also produced conditions that acted to destroy it. The uplift of the area produced new, vigorous streams and rejuvenated older ones and these streams dissected the overthrust sheet,

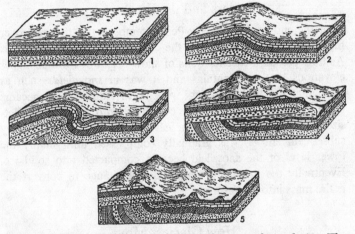

How Big Chief and Little Chief mountains were formed. (1) First sediments were deposited in a geosyncline that covered much of the present Rocky Mountain region. (2) Compressive forces from the west folded these sediments during the Laramide Revolution. The fold overturned toward the east. (3) The fold developed into a thrust fault with displacement toward the east. (4) The thrust reached its maximum extent, and Precambrian rocks then rested upon those of Cretaceous age. (5) Erosion removed most of the overthrust mass, leaving two remnants, or outliers, that are now called Big Chief and Little Chief mountains. From *Historical Geology,* Second Edition, by Russell C. Hussey, © 1947 McGraw-Hill Book Company. Used By Permission of McGraw-Hill Book Company.

carving sheer cliffs and deep canyons. Finally, by Middle Tertiary time, most of the overriding beds had been completely removed by erosion, leaving only outliers such as Chief Mountain described above.

But despite their importance, it was neither orogenic movements nor stream erosion that shaped the face of Glacier National Park. Rather, it was the work of ice which created the spectacular glaciated topography from which the Park derives its name. This phase of landscape development began about one million years ago with the advent of Pleistocene time—the Great Ice Age.

Glaciers and the Work of Ice

Near the end of the Tertiary Period temperatures dropped over much of North America and the climate became much colder. As time passed, concentrations of snow began to form at higher elevations in the mountains and it was in *snowfields* such as these that the Park's glaciers were born. As the snow became more compact it was slowly converted into granular, pellet-sized ice particles called *firn* or *névé*. This material, covered by subsequent snows, was gradually compressed until finally the lower level of the snowfield became compacted into solid ice. Eventually the firn underwent sufficient change to convert the entire mass into *glacier ice*.

How Glaciers Move

When enough glacier ice had accumulated, the ice, reacting to the force of gravity, slowly started to move downslope. This downward glacial movement followed the narrow, V-shaped, stream-cut valleys which laced the mountainsides and it was here that the ice did its work.

Visitors to the Park's glaciers commonly ask: "How fast does a glacier move?" This question is not easy to answer for there are many variables that must be considered. Indeed, to the casual observer a glacier appears to be a stationary mass of ice. But glaciers *do* move and their rate of forward progress may

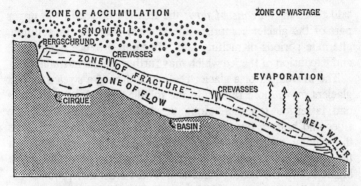

A glacier is marked by a zone of accumulation and a zone of wastage. Within a glacier, ice may lie either in the zone of fracture or deeper, in the zone of flow. A valley glacier originates in a basin, the *cirque*, and is separated from the headwall of the cirque by a large crevasse, the *bergschrund*. (Reproduced from L. Don Leet and Sheldon Judson, *Physical Geology*, third edition, 1965, by permission of Prentice-Hall, Inc., Englewood Cliffs, New Jersey.)

range from less than one inch to as much as one hundred feet a day (the latter occurring only under rather extreme conditions). For example, the late Dr. James L. Dyson, who made detailed studies of the Park's glaciers,* noted that the smallest glaciers may move as little as six to eight feet a year, and the larger ones as much as twenty-five to thirty feet a year.

Factors which affect the rate of movement include (1) the size of the glacier (the thicker it is the faster the movement), (2) the slope and topography of the land, (3) the temperature of the area (the glacier moves faster as the temperature rises), and (4) the amount of water in the ice.

The exact nature of glacial movement is not thoroughly understood for it appears to move as the force of gravity and pressure from the weight of accumulated ice causes the ice in the lower levels of the glacier to become plastic and subject to slow flowage. Ice in this lowermost portion of the glacier is

* James L. Dyson, *Glaciers and Glaciation in Glacier National Park*, West Glacier, Montana: Glacier Natural History Association, revised edition, 1962. This well-written, easy-to-understand publication provides an excellent introduction to glaciation in the Park.

said to be in the *zone of flow;* the less plastic ice in the upper part of the glacier comprises the *zone of fracture*. In addition, alternate periods of melting and refreezing produce contraction and expansion of the ice which may further glacial movement.

The movement of a glacier resembles that of a stream in that glaciers move more rapidly in the middle than along their sides and faster on top than along the bottom; friction along the sides and bottom retards movement there. As a glacier follows the winding path of its valley and passes over irregularities in the valley floor, tension will produce cracks in the rigid, more brittle ice in the zone of fracture. These fractures produce fissures called *crevasses* which may be hundreds of feet deep. These are especially numerous in midsummer when the upper surface of the glacier is weakened by streams of melt water and accumulations of slush. Hikers to **Sperry** and **Grinnell Glaciers** normally get an opportunity to look down into the gaping jaws of a crevasse, but this should be done *only* when in the company of an experienced guide or Ranger-Naturalist. Crevasses are often concealed by thin crusts of snow which break at the slightest weight, consequently they are a constant hazard to persons moving over the glacier's surface.

How long does a glacier continue to move? Most glaciers will flow until they reach an area where air causes the ice to melt as fast as the glacier advances; the *ice front* then becomes stationary and begins to melt because of the warmer temperatures. If the ice wastes more rapidly than it advances, the glacier will eventually retreat. Unfortunately most of the Park's glaciers appear to be retreating and many have disappeared completely.

In other parts of the world, glaciers may continue to move until they reach the sea; there large pieces of the glacier may break off to form *icebergs*. These float away and will eventually melt when they reach warmer waters. **Iceberg Lake** (see map) contains similar pieces of ice and snow which break off the snowbanks that surround part of the lake. However, the small glacier which originally produced icebergs and which gave the lake its name has vanished.

How Glaciers Shaped the Landscape

So great is the erosive power of glaciers that few objects will deter them as they relentlessly plow down their valleys or override mountains to blanket an entire continent. Glacial erosion is accomplished by (1) *plucking* or *quarrying* as the ice dislodges and picks up protruding fragments of bedrock, (2) *abrasion,* when the ice-plucked blocks and other rock debris scratch and polish the bedrock over which the glacier moves, and (3) *plowing* or *sledding,* as loose material is slowly pushed ahead of the glacier or as rock material falls from the valley walls and comes to rest on top of the ice.

Valley glaciers, such as those that once filled the canyons of Glacier National Park, are the most effective agents of glacial erosion and there is much evidence of their work in the Park. Glaciers of this type typically have their origin in *cirques*—great semicircular depressions which have developed as glacial erosion deepened and enlarged the head of a mountain valley. Cirque development occurs at the head of the glacier as water enters crevices between the cirque wall and the glacier and later freezes around blocks of rock; when the glacier moves, these blocks are plucked from the wall. Cirques, many of which have become filled with water, thus creating *cirque lakes,* can be seen in many parts of the Park. These include **Ellen Wilson, Iceberg, Cracker, Ptarmigan, Upper Two Medicine, Hidden,** and **Avalanche Lakes.**

In some places cirques have developed close together and on opposite sides of a mountain. As each cirque developed, glacial erosion reduced the ridges separating them to a sharp, narrow, jagged ridge called an *arête.* Although there are many such ridges in the Park, the **Garden Wall** and **Ptarmigan Wall** are probably the most spectacular and best known. When cirques on opposite sides of a ridge pluck their way backward they may meet to form a notch or low place called a *pass,* or, more technically, a *col;* **Logan, Piegan,** and **Gunsight Passes** have been developed in this manner.

Another conspicuous feature produced by glacial erosion is a peak called a *matterhorn* or *glacial horn*. These lofty, steep-sided, pyramidal peaks are formed throughout the Glacier National Park area and include peaks such as **Kinnerly Peak, Mount Wilbur,** and **Reynolds, Clements, Split,** and **Sinopah Mountains,** and **Little Matterhorn.**

Valley glaciers have greatly affected the valleys through which they have passed; the glacial ice has ground away the rocks of the valley walls and planed down the floor of the valley. Thus, young stream-cut valleys traversed by glaciers will be widened and deepened, transforming their typical V-shaped profile into the shape of a U as at **McDonald** and **St. Mary Valleys.** In addition, the valley walls and floor are typically scratched, grooved, and polished by rock debris carried in the glacial ice. Markings thus produced are called *glacial striae* if they are scratches; deeper marks are known as *glacial grooves.* Although common throughout the area, these features of glacial abrasion are especially noticeable in a well-exposed road cut half a mile above the **Loop** on **Going-to-the-Sun Road.**

A main glacial valley, known also as a *glacial trough,* is commonly more deeply eroded than are the tributary or side valleys leading into it. When the glaciers melt, the lower ends of the tributary valleys may be left suspended in midair above the main valley floor; valleys of this type are called *hanging valleys.* Most of the major valleys in the Park have their hanging valleys, some of which have waterfalls at their mouths. Falls which have developed in this fashion include **Grinnell, Bird Woman, Florence,** and **Virginia Falls.**

In addition to their ability to erode, glaciers are also capable of transporting and depositing great quantities of earth materials. These rock fragments, called the *load* of the glacier, range in size from finely pulverized rock (known as *rock flour*) to great boulders. Thus, the glacier's load consists of rocks and soil randomly intermingled irrespective of size, weight, or composition. When the ice melts, this debris is deposited, thus forming a variety of deposits which are collectively designated as *glacial drift.* There are two types of drift: (1) *till,* unstratified drift which has not been sorted by water action but has been deposited directly by the ice; and (2) *outwash,* or stratified drift, consisting of sediments that have been sorted and deposited in definite layers by the action of glacial melt water.

Finer glacial sediments, such as rock flour, are rather commonly deposited by melt water from the glaciers. Indeed, finely powdered rock flour dissolved in the water is responsible for the milky or cloudy appearance of the streams which issue from the glaciers. Although the water in these streams has been called *"glacier milk,"* all "glacier milk" does not have the appearance of milk; the color may vary considerably and is governed by the bedrock over which the glacier has passed and from which the pulverized rock has been derived. This in turn determines the color of the melt water and eventually the color of the lakes into which the "glacier milk" flows.

Most glacial sediments are deposited as the glacier melts and deposits its load; this type of material, which is not bedded as is most sedimentary rock, is called *till.* Deposits of till often form topographic features called *moraines*—ridges or mounds of boulders, gravel, sand, and clay deposited by a glacier. There are several types of moraines, each being named with respect to their relation to the glacier responsible for it. *End moraines* are mounds of till which accumulate at the *terminus,* or end, of a glacier; these mark the former or present position of the ice front. *Recessional moraines* are deposits of till left at various points as a glacier recedes or is temporarily stable. *Lateral moraines* are ridges of material formed on each side of a glacial valley; they consist of material that has been eroded from the walls of the valley or that has fallen from the valley sides onto the glacier's surface to be carried along by the ice. Moraines are common throughout much of the Park; the **Going-to-the-Sun Road** crosses several and they are prominently associated with **Grinnell, Sperry,** and other glaciers.

Glacial sediments have played a significant role in the development of many of the Park's lakes. For example, **Lake Josephine** is partially contained by a moraine, and **Lake McDonald** has been dammed by outwash deposits of stratified gravel deposited by the melt water of glaciers that once extended into McDonald Valley.

Thus, through the combined geologic processes of glacial erosion and deposition, ice has shaped the face of Glacier National Park. Will the shrinking glaciers—now reduced to about fifty in number—eventually disappear as those before them? Or will the remaining glaciers maintain their precarious foothold on the mountainsides until the return of some future Ice Age? Time alone will tell. In the meantime, countless visitors will visit this Park each year and enjoy the breath-taking scenery of this ice-sculptured mountain refuge.

Plants and Animals of Glacier National Park

Because this is the only region south of Canada with a high-mountain subarctic climate, the fauna and flora of Glacier are of unusual interest. Moreover, within the Park there are four great life zones, each of which supports its own characteristic species of plants and animals. Biologists estimate that more than 1000 species of trees and wildflowers, approximately 270 species of birds and mammals, and 22 types of fishes are found in the varied environments within Glacier National Park.

Glacier is well known for its colorful arrays of mountain flowers, some of which are in bloom from early spring until late autumn. Among the more colorful alpine flowers are the yellow glacier lilies (which are common in the mountain meadows), the false hellebore, scarlet painted cup, purple pasque-flower, red heather, brown-eyed Susan, yellow arnica, and blue gentian. One of the most common flowers in the Park is the bear grass. This lily, which has grasslike leaves and small creamy-white flowers, is so spectacular that it has been designated the official Park flower.

Typical shrubs include huckleberry, bearberry, creeping juniper, shrubby cinquefoil, thimbleberry, and serviceberry. Conifers (cone-bearing trees) are the dominant trees in most of the Park; among them are lodgepole, ponderosa, limber, and western white pine, Engelmann spruce, Douglas and subalpine fir, western red cedar, and subalpine larch (which is the official Park tree). The more common "non-evergreen" trees include quaking aspen, black cottonwood, northwestern paper birch, willows, and western thorn apple.

The animals of Glacier are also interesting and varied and can frequently be seen along Park roads and trails. Larger mammals that may be observed are American moose, wapiti (or American elk), mule and white-tailed deer, the bighorn (Rocky Mountain sheep), black bear, and mountain goats. The latter are apt to be seen at higher elevations of the Park and are one of the area's more interesting animal attractions. These creatures, which are not true goats but a goat-antelope related to the chamois of the Alps, have amazing climbing ability and easily negotiate the sheer sides and precipitous cliffs of the more rugged mountain peaks.

Ground and pine squirrel, chipmunk, marmot, porcupine, snowshoe hare, and beaver are among the more typical small mammals. In addition, many interesting birds live in the Park; look for hawks, eagles, water ouzels, pileated woodpeckers, ptarmigan, larks, finches, and sparrows.

Tips for Tourists

The visitor to Glacier National Park will find a number of activities awaiting him. This is essentially an "outdoorman's" Park for there are many well-marked trails and interesting places to hike and camp. However, the visitor with less time to spend can also get a good look at the Park in a relatively short time. Because the Park is large and its activities varied, one should consult the Naturalist Program published by the Glacier Natural History Association and available at entrance stations and visitor centers throughout the Park. Or you may request one by mail from Superintendent, Glacier National Park. This helpful bulletin gives the times, locations, and types of naturalist activities available in the Park. In addition, the very helpful publication, *Guide to Glacier National Park,* is highly recommended. This detailed guide book (written by Dr. George C. Ruhle, formerly Chief Park Naturalist in the Park) is available from the Glacier Natural History Association, West Glacier, Montana 59936.

There are unlimited opportunities for the camera enthusiast at Glacier National Park; mountain panoramas, wild animals, and a host of wildflowers are but a few likely subjects. During the summer excellent scenic views can be made early in the morning (between 5 and 6 A.M.); colors are bright and haze is at a minimum at this time.

The scenic overlooks along Going-to-the-Sun Road are excellent places for picture-taking. If you have telephoto lens perhaps you can photograph a mountain sheep or goat on the rocky crags. Wildflowers are most numerous and varied in the valley floors and alpine meadows; try close-up views of these. Pictures taken after midmorning will be clearer if a haze filter is used; the light here can be deceiving so rely on your exposure meter.

Glacier National Park at a Glance

Address: Superintendent, West Glacier, Montana 59936.

Area: 1,013,129 acres.

Major Attractions: Outstanding glaciated topography of Rocky Mountains, with numerous glaciers, lakes, and other geologic attractions; fossil algae; mountain flowers and wildlife.

Season: Approximately June 15–September 15 (Going-to-the-Sun Road opens on or about May 30 and closes about October 21).

Accommodations: Cabins, campgrounds, hotels, motels, trailer sites. *For reservations contact:* Glacier Park, Inc., Glacier Park Lodge, East Glacier Park, Montana 59434 (summer); 1735 East Fort Lowell, Box 4340, Tucson, Arizona 85717 (winter).

Activities: Boating, boat rides, camping, fishing, guided tours, hiking, horseback riding, mountain climbing, nature walks, picnicking, scenic drives, swimming, water sports, and cross-country skiing in winter.

Services: Boating facilities, boat rentals, food service, gift shops, guide service, health service, laundry, nursery, post office, religious services, service stations, telegraph, telephone, and transportation.

Interpretive Program: Campfire programs, guided hikes,

visitor center exhibits, nature trails, nature walks, roadside exhibits, self-guiding trails, and trailside exhibits.

Natural Features: Canyons, erosional features, fossils, geologic formations, glaciation, glaciers, lakes, mountains, rivers, rocks, unusual birds, unusual plants, volcanic features, waterfalls, wilderness area, and wildlife.

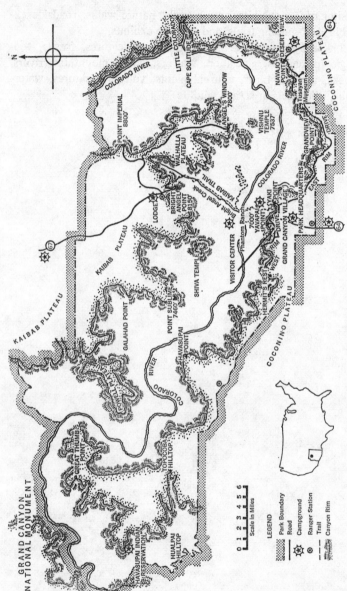

Grand Canyon National Park, Arizona.

12. Grand Canyon National Park, Arizona

King of the Canyons

Few geologic features rival the Grand Canyon in size, beauty, and spectacle. This immense chasm—it is about one mile deep, 9 miles wide, and 217 miles long—is an area of outstanding scenic and scientific attraction.

There are many reasons why hundreds of thousands of visitors throng to this Park each year. The lay person comes to marvel at the immensity and beauty of the canyon, for where else can one peer over the edge of a flat-lying plateau and look down on the tops of mountains? But the canyon holds even greater attraction for the scientist because it is a treasure house of biological and geological information. Geologists have long studied the Grand Canyon for it enables them to see a "slice" of the earth's crust that is without parallel. Not only can the structure and composition of this crustal segment be observed, equally important is the fact that more than two billion years of history have been recorded in the canyon's colorful walls. And, because of the great range in elevation— from nine thousand feet on the North Rim to two thousand feet at river level—the Canyon is of considerable interest to botanists and zoologists. Here, at a single latitude, can be found plants and animals ranging from desert-dwelling reptiles to high-altitude evergreen trees.

Earth History In Review

In the process of excavating the Grand Canyon, the forces of erosion have laid bare a succession of strata which represent a veritable manuscript in stone; included are rocks ranging from Precambrian to Recent in age—a time span of perhaps two billion years. Yet, geologically speaking, the Grand Canyon is a relatively young feature for the carving of this awesome gorge began only about seven million years ago.

Because of the vast amount of geologic time involved and the complexity of the geologic record, space will permit only a condensed review of the geologic history of the Grand Canyon region. However, even a superficial knowledge of the canyon's rocks and their geologic significance will enhance the visitor's understanding of this unusual natural feature.

Chapter 1.—Precambrian Rocks

The most ancient and perhaps most distinctive rocks in the canyon originated during Precambrian time. The oldest of these, the *Vishnu Schist,* is Archeozoic in age and can be seen in the inner gorge. These highly metamorphosed deposits consist primarily of mica and quartz schists believed to have originally been silty shales and mudstones. Present also are *quartzites* (metamorphosed sandstones), some of which still retain traces of stratification, providing additional evidence of the rocks from which the metamorphics were formed. As a result of their great antiquity and the many changes that they have undergone, the Vishnu rocks have been severely deformed and in places intruded by numerous igneous dikes.

Immediately overlying the Vishnu Schist is the Grand Canyon Series of Proterozoic age. Consisting essentially of conglomerates, sandstones, shales, and limestones, these rocks have not been as greatly metamorphosed as have the underlying Vishnu formations. There is evidence that this part of Precambrian history was a time of considerable volcanic activity for thick basaltic lava flows occur near the middle of the section of sedimentary rocks.

Following the deposition of the Grand Canyon Series, the region was elevated and its surface fractured into massive fault blocks which produced lofty fault-block mountains (see p. 26). This great crustal disturbance was followed by a long period of erosion during which the mountains were eroded down to their very roots. It is on this erosional surface, or *unconformity,* that the flat-lying Cambrian sedimentary rocks were deposited.

Chapter 2.—The Paleozoic Rocks

Following the uplift and erosion of the Grand Canyon Series, there was resumed deposition in the Grand Canyon region. It was during this time—the Cambrian Period—that the rocks comprising the *Tonto Group* were formed. Composed mostly of shales, sandstones, and limestones, the Tonto Group has been divided into three formations: the Tapeats Sandstone, Bright Angel Shale, and the Muav Limestone.

The *Tapeats Sandstone* lies directly upon highly distorted crystalline Archeozoic rocks. This brown, rather coarse sandstone contains lens-shaped bodies of conglomerates and was perhaps deposited by an encroaching sea. Directly above the Tapeats lies the *Bright Angel Shale,* a greenish buff marine shale which contains fossil shells, seaweed, and trilobites—small three-lobed creatures distantly related to the horseshoe crab. The gray to buff *Muav Limestone,* also of marine origin, lies above the Bright Angel Shale. Collectively having a thickness of as much as 965 feet, rocks of the Tonto Group are well exposed on the Tonto Platform.

Unfortunately Chapter II of the Grand Canyon's "book" of earth history has some missing pages for Silurian and Ordovician rocks do not occur in the canyon. Why should there be a gap of more than one hundred million years in the geologic record? Perhaps these rocks were removed by erosion, or, more probably, the Grand Canyon region may have been above sea level during this part of geologic time and no sediments were deposited in the area. Information is also scant for another period of the Paleozoic—the Devonian is represented only by the *Temple Butte Limestone*. Although fossils are rare, the remains of Devonian corals, snails, and fish have been found in this thin-bedded, lavender, sandy limestone.

It is not until Mississippian time that Paleozoic geologic events again become clear; the formation representing this part of the geologic story is the *Redwall Limestone*. Occurring near the middle of the canyon wall and composed of massively bedded layers of limestone, this is one of the most striking features in the canyon. Interestingly enough, the crystalline limestone comprising this formation is actually bluish gray in color. Its distinctive red hue is the result of stains produced by iron oxides which have washed down from the overlying Supai Formation. The composition of the Redwall Limestone and the fossils which it contains suggest that these deposits accumulated on the bottom of a rather shallow widespread Mississippian sea.

The *Supai Formation,* which overlies the Redwall, is of considerable interest because it contains rocks which might possibly represent two distinct periods of geologic time. The lower part of the Supai is composed of reddish sandstones and shales which some geologists believe to be Pennsylvanian in age. The upper portion consists of nonmarine red shales interbedded with ledges of reddish brown sandstones. These beds, which resemble the overlying Hermit Shale, are considered to be Permian in age. The Supai sandstones contain interesting tracks which are believed to have been made by early amphibians or reptiles. Examples of these tracks may be seen at the Visitor Center, the Yavapai Museum, and along the Kaibab Trail.

The latter part of the Paleozoic "chapter" is not difficult to read; above the Supai Formation there begins a thick succession of Permian strata. The youngest of these, the *Hermit Shale,* is composed of brick-red sandy shales interspersed with layers of siltstones and sandstones. Clearly nonmarine in origin, these deposits contain fossil insects, worm trails, and vertebrate tracks; plant remains include many species of ferns and some primitive coniferous plants.

The *Coconino Sandstone* occurs above the Hermit Shale and is believed to have been deposited as dune sand in a Permian desert. Forming the most precipitous cliffs in the upper canyon wall, this *eolian* (wind-deposited), cross-bedded (see p. 478), pale buff sandstone contains footprints of small creatures believed to be ancestral reptiles or amphibians.

Following the formation of the Coconino Sandstone, there was a marked change in Permian geography as marine deposition resumed in the Grand Canyon region. This transition is evidenced by the *Toroweap Formation,* a series of gray limestones and red to yellow sandstones which are of undisputed marine origin. (The Toroweap was originally included in the overlying Kaibab Formation and is not shown on older maps.)

The most recently formed Paleozoic rocks in the Grand Canyon are those of the *Kaibab Limestone.* Consisting of massive beds of buff to gray crystalline limestone, the Kaibab contains numerous fossils of marine snails, sponges, and corals; shark teeth have also been found in these rocks. The Kaibab Limestone forms the surface of the Coconino and Kaibab Plateaus which border the Grand Canyon; it also forms the steep cliffs which mark the rim of the canyon.

Chapter 3.—The Mesozoic Rocks

Although it is possible that a thick section of Mesozoic rocks once covered the surface adjacent to the canyon, only remnants of Triassic rocks are now present. These are the red, sandy shales and sandstones of the *Moenkopi Formation* which occur in Cedar Mountain near the southeast corner of the Park (see map). But despite the missing Mesozoic strata in the canyon proper, the presence of erosional remnants in the vicinity of Grand Canyon suggests that the area was once blanketed with a covering of Mesozoic sedimentary rocks.

Chapter 4.—Cenozoic Time:
Carving of the Canyon

During late Cenozoic time, the Grand Canyon region was again subjected to crustal disturbance; these movements resulted in the uparching or doming of the rocks to form the area now called the *Colorado Plateau*. This extensive, high tableland encompasses about 45,000 square miles of northwestern New Mexico, southwestern Colorado, southern Utah, and northern Arizona; it includes the Coconino Plateau which forms the South Rim of the canyon and the Kaibab Plateau which marks the canyon's North Rim. The uplift, which probably began near the end of the Tertiary Period, continued into Pleistocene and Recent time and raised the region thousands of feet, thereby forming the present dome-shaped plateau.

Before the region was elevated, the Colorado River and its tributaries flowed rather sluggishly across a flat lowland of slight elevation. Then as the Colorado Plateau was raised, *stream gradients* (the difference in elevation between the source and mouth of the stream) were increased causing the streams to flow more swiftly. This re-energizing of the Colorado River—technically called *rejuvenation*—greatly increased the river's ability to erode and carry sediments. Thus, approximately seven million years ago, the Colorado River began rasping away the rocks of the Colorado Plateau as it continues to do today.

The Mighty Colorado. The visitor seeing the Grand Canyon for the first time may find it difficult to believe that this spectacular gorge began as a simple gully. But to the *geomorphologist,* the geologist who studies the origin and development of landscapes, the Grand Canyon is but a gully magnified many times over. This is evident because the shape of the canyon, the nature of its tributaries, and the character of its walls indicates that it was deepened by the downcutting of a stream and widened by various types of weathering and mass wasting (p. 162).

Standing on the rim of the canyon and looking into its yawning depths, you may well wonder if the apparently small stream in the hazy distance could really have carved this colossal gorge. But before doubting the geologic efficiency of the Colorado River perhaps we should review its "qualifications." The second longest river in the United States (about two thousand miles), the Colorado drains an area of some 246,000 square miles covering one twelfth of the United States and a small part of Mexico. At the gauging station in the bottom of the canyon, the river averages three hundred feet in width and ranges from twelve to forty-five feet in depth. Flowing at speeds of from two and a half to as much as twenty miles per hour, during a 24-hour period this turbulent stream may carry past a given point an average of almost one million tons of silt, sand, gravel, and boulders.

Most of this rock debris, called the *load* of the stream, is acquired by erosion of the sides and bottom of its channel. Some of this material, such as salt and minerals, is transported in a dissolved state (in solution); still more, silt and fine sand, for example, is carried suspended in the water. Those particles which will not dissolve in water and are too heavy to be carried in suspension will slide or roll along the stream bed. As it flows through the canyon, the Colorado uses its load to further erode the rocks over which it passes. Each rock fragment then becomes a cutting tool for *abrasion* as loose rock fragments moved by the water slowly wear away the bed and banks of the stream. Eventually the abraded rock particles become smooth and rounded, and the stream bed is gradually worn down to a lower level. The river also erodes by *hydraulic action* as loose rock fragments are lifted and moved by the force of the river's current. This process is similar to the effect produced when soil is churned up and washed away when water from a garden hose is sprayed on loose earth.

Today the Colorado River has had its load reduced because of the presence of the Glen Canyon Dam upsteam from the Grand Canyon. Rock debris now enters the Colorado by means of tributaries downstream from the dam.

While the river is employed in deepening the canyon, other

geologic processes are actively engaged in widening it. The arid climate of the Grand Canyon region is one which facilitates *mechanical weathering,* the process whereby natural agents reduce rock to small fragments without changing its mineral composition. In such areas, large daily temperature changes bring about alternate expansion and contraction of the rocks which may produce fractures. During the winter, water in these fissures may freeze and expand, further splitting the rock. This process, called *frost wedging,* is capable of prying off great blocks of rock from the canyon walls.

Mass wasting, the erosional process that occurs when large masses of rock material move downslope by gravity, has also been instrumental in shaping the canyon. Some of these movements, such as landslides, have been rapid and moved great quantities of rock. But most have been imperceptibly slow as masses of *talus* (accumulations of rock debris) on steeper slopes have inched slowly downhill because of their own weight.

In addition, some of the more soluble limestones have been dissolved by water flowing over them; pitting caused by solution is evident on certain of the limestone cliffs. Other cliffs have literally been "sandblasted" as strong winds have steadily blown sand against their faces.

Even the most casual observer will soon note that not all of the rocks have been equally affected by erosion. Indeed, it is the nature of this *differential erosion* that gives the canyon its rugged sculptured appearance. Because of their resistance to erosion, some of the formations such as the Redwall Limestone, Coconino Sandstone, and the Kaibab Limestone form massive, vertical cliffs. Other strata, for example, the softer Hermit and Bright Angel Shales, erode more easily and produce gentle slopes or platforms. This variation in the rate of erosion has played a large part in the formation of the unusual architectural forms within the canyon.

Plants and Animals of Grand Canyon National Park

As the crow flies, the North and South Rims are only ten miles apart, yet the fauna and flora of these two areas differ considerably. There are two major reasons for this: (1) the great physical barrier formed by the canyon, and (2) the fact that the North Rim is from one thousand to two thousand feet higher than its southern counterpart.

There is an even greater variation in elevation between the North Rim (nine thousand feet) and the floor of the inner gorge which is only two thousand feet above sea level. This difference is clearly reflected in the disparate faunas and floras of the two areas. For example, the species which inhabit the warm, dry bottom of the canyon include desert plants such as mesquite, cactus, yucca, and agave; typical animals are rattlesnakes and spiny lizards, forms similar to those which normally inhabit northern Mexico. These typical desert-dwelling organisms differ greatly from those which populate the high North Rim—the latter are not unlike the fauna and flora of southern Canada. Thus the beautiful blue spruce (which occurs naturally only at higher elevations) is restricted to the North Rim. There is also variation between the animals of the two rims; certain animals, such as the Kaibab squirrel, are confined to the North Rim while the Abert squirrel is found on the South Rim.

Some plants and animals inhabit both rims of the canyon. These include ponderosa pine, piñon, and juniper trees, mule deer, chipmunk, skunk, Steller's jay, mountain chickadee, and nuthatch.

Tips for Tourists

The most popular activity at this National Park might properly be described as "canyon watching," for each hour of the day brings new color and a different perspective to the canyon. The visitor who takes time to view this incessantly changing kaleidoscope of color will truly appreciate the vastness and matchless beauty of this great abyss. In addition, there are a host of other activities especially planned to further one's enjoyment and understanding of the canyon.

A pleasant nonstrenuous hike can be made on the **West Rim Trail** which leads from El Tovar Hotel to Maricopa Point.

The trail (one mile long) follows the rim and passes many fine vantage points from which to view the canyon. The more adventurous (and physically able) may elect to take one of the foot trails to explore the depths of the canyon. These include the **Bright Angel Trail** which leads from the South Rim to the Colorado River (a round-trip distance of sixteen miles). This twisting trail is rather demanding and even experienced hikers should plan to stay overnight on the canyon floor and make the return trip the following day.

The **South Kaibab Trail** takes you from Yaki Point (see map) to the river in the bottom of the canyon. This rugged seven-mile trail is quite steep, challenging, and especially difficult when coming out of the canyon; those wishing to take a shorter hike can stop at Cedar Ridge which is one and a half miles down the trail. The informative guide leaflet available for this trail points out many of the geologic formations which have been described earlier. Fossil tracks and footprints, fossil plants, a fault line, and other interesting geologic phenomena are also explained.

Hikers who prefer to take the South Kaibab Trail into the canyon and then ascend by means of less steep Bright Angel Trail can reach the latter by means of the 2½-mile **River Trail** which connects the two.

In the bottom of the canyon at the river, the South Kaibab Trail joins the **North Kaibab Trail** which leads to the North Rim (see p. 163).

Hiking in and out of the canyon can be a very strenuous task: it should not be attempted by other than experienced hikers in good physical condition. Even veteran hikers find canyon trails a challenge; unlike mountain climbing when most effort is expended when the hiker is fresh, the strenuous 5000-foot pull out of the canyon comes near the end of the journey. Anyone, regardless of hiking ability, should allow plenty of time to reach the canyon rim before dark, dress for maximum protection from heat, and carry an ample supply of water (a gallon per day per person is recommended). For additional information about canyon hiking, prospective hikers should obtain the "Hiker Information Bulletin" from one of the ranger stations.

To reach the bottom of the canyon with a minimum of effort, take one of the mule trips conducted by a Park concessioner. These trips (limited to persons weighing no more than two hundred pounds and over twelve years of age) are made on specially trained mules and under the supervision of experienced guides. The shortest ride, the **Plateau Point Trip,** requires about seven hours and goes to the Tonto Plateau some thirty-two hundred feet into the canyon. From there one can get a good look at the muddy Colorado River as it roars through the inner gorge.

The **River Trip** goes to the very bottom of the canyon; there you can stand on the riverbank and appreciate fully the power of the mighty Colorado. This trip is the same as the Plateau Point Trip as far as Indian Gardens. From this point the mules follow Bright Angel Trail down Garden and Pipe Creeks to the river's edge. Although requiring more time than the trip to Plateau Point, this ride is well worth the additional time and expense involved.

For those who have time the two-day **Phantom Ranch Trip** is highly recommended. Upon reaching the canyon floor, members on this tour follow the riverbank, cross the famous suspension bridge, and shortly thereafter arrive at Phantom Ranch. Here the group dismounts to enjoy a relaxing swim, a hot meal, and a well-earned night's rest. The return trip to Yaki Point is made by a shorter but equally interesting route arriving at the rim about 1:30 P.M.

For an interesting review of the human history of this region and the exploration of the Colorado River visit **Kolb's Studio.** Especially interesting is the lecture and motion picture of the boat trip down the Colorado (admission charged). In addition, colorful Indian ceremonial dances are performed each afternoon outside the **Hopi House** (an authentic reproduction of one of the terraced dwellings of the Hopi Indians of this region) in Grand Canyon Village (no admission charge).

Persons wishing to "get away from it all" can find no better place than the **Havasupai Indian Reservation** near the western boundary of the Park (see map). The first leg of this rugged trip (about 205 miles) may be covered by automobile; the last segment (8 miles) must be made by foot or horseback.

The Havasupai Tribe live in a lush river valley surrounded by towering canyon walls. While there you can simply enjoy the quiet simplicity of the secluded area or visit some of the more remote spots in the valley by foot or horseback. It is possible to camp or stay in the lodge or dormitory operated by the Havasupai Tribe, but you must prepare your own food. Before visiting this area contact Tourist Manager, Havasupai Tourist Enterprise, Supai, Arizona, for additional information and reservations.

The Grand Canyon is uniquely photogenic, for color and lighting vary according to the time of day and year. In the early morning there is a minimum of haze and dust; in the late afternoon long shadows add perspective and intensify the canyon's color. In general, avoid shooting between 10 A.M. and 2 P.M.; during this period the position of the sun precludes the side and back lights which add depth and drama to canyon photographs. Sunsets are particularly effective, especially when taken at periodic intervals to show the changing colors. Lighting is deceptive here so use a light meter if possible. A haze filter is a must for most pictures. Finally, do not limit your picture taking to any one part of the canyon; effective photographs can be made at all observation points along the rim as well as in the interior of the canyon.

Unlike the South Rim which may be visited year-round, the North Rim is open only during the summer season (mid-May to mid-October). There is a marked contrast between the climate, vegetation, and wildlife of these two areas and to really know Grand Canyon National Park one should visit both of them. In the winter Park roads are closed with snow, but during the summer months the cool North Rim is a welcome relief to the more arid climate of the south side of the canyon. And, although activities are not as varied, the visitor to the North Rim will find much to keep him occupied.

Grand Canyon National Park at a Glance

Address: Superintendent, Box 129, Grand Canyon, Arizona 86023.

Area: 673,575 acres.

Major Attractions: Most spectacular part of the Colorado River's greatest canyon, which is 217 miles long and 4 to 18 miles wide; exposure of rocks representing a vast amount of geologic time.

Season: North Rim closed mid-October to mid-May; South Rim open all year.

Accommodations: Cabins, campgrounds, group campsites, hotels, motels, and trailer sites. *For reservations contact: South Rim*—Fred Harvey, Grand Canyon, Arizona 86023; *North Rim* —Utah Parks Company, P.O. Box 400, Cedar City, Utah 84720.

Activities: Boating, camping, fishing (Arizona license required), guided tours, hiking, horseback riding, nature walks, picnicking, scenic drives, mule rides, and airplane rides.

Services: Food service, gift shop, health service, laundry, post office, public showers, religious services, service station, telegraph, telephone, transportation, picnic tables, rest rooms, and general store.

Interpretive Program: Campfire programs, museums, nature trails, roadside exhibits, self-guiding trails, and trailside exhibits.

Natural Features: Canyons, deserts, erosional features, forests, fossils, geologic formations, rivers, rocks and minerals, wilderness area, wildlife, and Indian ruins.

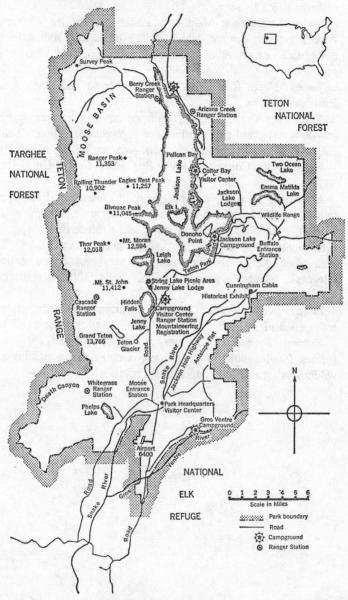

Grand Teton National Park, Wyoming.

13. Grand Teton National Park, Wyoming

Mountains Without Foothills

Located in northwestern Wyoming—a scant seven miles from Yellowstone National Park—is Grand Teton National Park. Here, the shining peaks of the Teton Range rise with startling abruptness over the lake-studded valley called Jackson Hole.

Unlike the folded mountains of Shenandoah and Rocky Mountain National Parks and the Cascade Range which is of volcanic origin, the Teton Range is a *fault-block mountain range;* these majestic peaks have literally been thrust upward along great fracture zones near the eastern front of the range. Then attacked by wind, rain, running water—and finally great valley glaciers—this massive uplifted fault block was gradually carved into the rugged spires that characterize the present Teton Range.

The beauty of the Tetons is greatly enhanced by the absence of foothills; thus, when viewed from the east, there is nothing to obstruct one's view of this spectacular alpine vista. Why are there no foothills present? How are fault-block mountains formed? To answer these questions we must, as always, turn to the geologic record of this unusual area.

Birth of the Tetons

Although the mountain-building movements that created the Teton Range occurred within relatively recent geologic time, the rocks comprising the heart of the range date back to the beginning of earth history. These ancient igneous and meta-morphic rocks, which consist primarily of granite, schist, gneiss, and quartzite, were formed during Precambrian time and are more than 2600 million years old. Certain of the ancient crystalline rocks have been intruded by masses of molten igneous rock which later solidified to form *dikes* (see p. 64). Tabular structures, typically seen as long black streaks on the mountain walls, the dikes are composed of *diabase,* a basaltic type of igneous rock. The dikes, three of which can be seen on the flanks of **Mount Moran** and **Grand** and **Middle Teton,** range from about 20 to 125 feet in width. In places the diabase forming the dike is more resistant than the rock it has intruded, thus the diabase stands out in relief as on the face of Mount Moran. However, the dike on the Grand Teton has weathered away to form a long narrow depression between the more resistant surrounding rocks. This depression, and one of similar origin on Middle Teton, is commonly followed by mountaineers climbing these peaks.

The Precambrian rocks described above served as a founda-tion for the Paleozoic formations which were deposited in the area much later. Consisting primarily of sedimentary rocks such as limestone, sandstone, and shale, these strata have a total thickness of approximately three thousand feet. The rocks and the fossils which they contain indicate that these deposits accumulated on the floors of ancient seas which occupied this area about 230 million years ago. Despite the fact that most of the sedimentary formations have long since been removed by erosion, Paleozoic strata can still be found atop Mount Moran. The presence of sedimentary beds accounts for the relatively flat top of this mountain whose outline is quite unlike the sharp, jagged profile which typifies most of the Teton peaks.

More recently, during the Tertiary Period, the Paleozoic sedimentary strata and certain exposed parts of the Pre-cambrian crystalline rocks were covered by lava and other volcanic products. Although these extrusives are considerably younger than the underlying Precambrian and Paleozoic rocks, they were probably extruded tens of millions of years ago and

North Cascades—one of our newer National Parks—is noted for its rugged, ice-carved terrain. NATIONAL PARK SERVICE PHOTO BY M. WOODBRIDGE WILLIAMS

Even in isolated areas such as the Chisos Mountain "Basin," in Big Bend National Park, concessionaire-operated services provide for the visitor's needs. JENNIE A. MATTHEWS

Obsidian Cliff—a massive bluff of volcanic glass—was one of the wonders that Jim Bridger found in Yellowstone National Park. NATIONAL PARK SERVICE

Spruce Tree House, in Mesa Verde National Park, is situated in a large alcove that has developed in the massive Cliff House Sandstone of Cretaceous age. JENNIE A. MATTHEWS

The geysers in Yellowstone National Park are located in rather well-defined areas called geyser basins. Here, Castle Geyser, in the Upper Basin, is seen in the violent phase of its eruption. NATIONAL PARK SERVICE

The ceiling of the Temple Room in Wind Cave National Park is adorned with boxwork—an intricate network of delicate calcite veins. NATIONAL PARK SERVICE PHOTO BY JACK E. BOUCHER

Thor's Hammer—a Bryce Canyon landmark—is a feature of the Navajo Loop Trail. NATIONAL PARK SERVICE

Visitors to Rocky Mountain National Park pause at Rock Cut (elevation 12,110 feet), on Trail Ridge Road. Gorge Lakes—of glacial origin—are in the canyon in the middleground. The peaks in the background are located along the Continental Divide, and heavily wooded Forest Canyon lies in the valley two-thousand feet below. NATIONAL PARK SERVICE

Acadia visitors cruise past the east face of Great Head, a typical rocky promontory. Note the sea cave (arrow) that has been eroded in the face of the steep cliff. NATIONAL PARK SERVICE PHOTO BY JACK E. BOUCHER

The "Narrows" of the Virgin River, in Zion National Park, have proved to be excellent examples of stream erosion in massive sandstone layers. At the most narrow segment of the gorge a width of approximately eight feet can be measured. The walls of the gorge sweep vertically upward for hundreds of feet, sometimes completely obscuring the sky. UTAH TOURIST AND PUBLICITY COUNCIL PHOTO BY ALEXIS KELNER

Haleakala Crater—one of the world's largest—is the feature attraction in Haleakala National Park. Here the rim of the crater forms a backdrop for the cinder cones and the solidified lava that flowed toward the lower right part of the photograph. JENNIE A. MATTHEWS

This photo of Yosemite's "Incomparable Valley" was taken from the vicinity of Artist Point. The massive cliff at the left is El Capitan, Half Dome looms in the middle background, and the Cathedral Rocks and Bridalveil Fall are on the right. U. S. GEOLOGICAL SURVEY PHOTO BY F. E. MATTHES

Petrified Forest National Park, in northeastern Arizona, contains one of the world's largest concentrations of fossil wood. Note the eroded clays in the background which give rise to the so-called badlands topography. FRED HARVEY

Wyoming's magnificent Teton Range forms the backdrop for a corral in Grand Teton National Park. GRAND TETON LODGE COMPANY

hence are considerably older than the glacial deposits of the Park. These volcanics, which are well exposed at lower elevations on the west flank of the range, extend northward where they are continuous with exposures of similar rocks in Yellowstone National Park. Thus, prior to the uplift of the Teton fault block this general area was covered by a thick section of Paleozoic sedimentary rocks which was overlain in most places by the younger Tertiary volcanics. Deeply buried beneath these lay the ancient Precambrian igneous and metamorphic basement rocks.

As you observe such lofty peaks as Grand Teton (13,776 feet above sea level) you will no doubt wonder how and when those once-buried Precambrian rocks attained such heights. Once again, the record in the rocks provides reliable evidence which enables the geologist to postulate rather accurately both "how" and "when" the Tetons were elevated.

About sixty million years ago near the end of the Cretaceous Period, there was great crustal unrest in the Rocky Mountain region; and the great thickness of rocks which had been accumulating in the ancient western geosyncline (p. 364) was gradually being lifted above sea level. This uplift, which did not occur as a single, violent upthrust but as complex series of earth movements, lasted until the middle or latter part of the Tertiary Period; it was near the end of this mountain-building disturbance—relatively recently, geologically speaking—that the Teton range was created. Finally, for reasons not yet thoroughly understood, insurmountable stresses developed within the earth and the crust was sheared to form the famous *Teton Fault*. This long, generally north-south fracture permitted two massive segments of the crust to be greatly displaced with respect to each other. The rocks nearest the fault were most severely effected; here a great fault block more than forty miles long and about thirteen miles wide was thrust perhaps as much as twenty thousand feet upward. As the west block was displaced upward and tilted westward, the block east of the fault zone was depressed forming a valley known as a *fault trough*. It is this valley that is today called Jackson Hole. It should be noted that the Teton Fault was not the only major earth fracture accompanying the elevation of the Teton Range. Geologic field studies indicate the presence of numerous secondary faults in the Teton block and the depressed block which underlies Jackson Hole.

Sculpturing of the Landscape

The latest chapter in the geologic history of the Park began during Pleistocene time and the advent of the Ice Age. As glaciers formed in snowfields (see p. 144) high in the Teton Range, the ice began its inexorable journey downslope. These white rivers of ice ground steadily down the narrow stream-cut gorges which marked the face of the Teton fault block, and those that reached the valley joined forces to form piedmont glaciers. Certain of the piedmont glaciers coalesced to form a gigantic apron of ice which occupied the site of the present-day Jackson Lake.

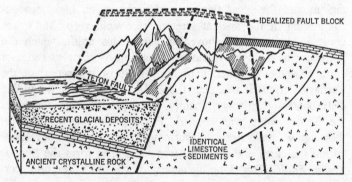

The fault structure of the Teton Range near Jenny Lake. Reproduced from *Jenny Lake Nature Trail Guide Pamphlet* by permission of Grand Teton Natural History Association.

Then, as the climate slowly became warmer, the ice began its retreat. Today only small glaciers occur in the Park; these are confined to the shadowed and more protected portions of the range. But though the glaciers have all but vanished, their geologic handiwork is quite in evidence. The gorgelike, stream-carved, V-shaped canyons have been modified into the typical U-shaped glacial trough (p. 148), and ice-carved, sharp-pointed, faceted peaks crown the once-flat surface of the Teton fault block. Present also are cirques, hanging valleys, glacial lakes, glacial polish and striae, and other glacial phenomena such as those discussed earlier.

Yet the work of the glaciers was not wholly confined to erosion; their work in the valley was essentially constructional. As the glaciers melted they deposited their sediments as *moraines* (ridges of glacial debris deposited by the glaciers) and *outwash plains*—plains formed by deposition of materials washed from the edges of the glaciers (see p. 149). Most of the glacial sediments were deposited in Jackson Hole, for this marked the terminus of many of the glaciers and it is here that the depositional record is most complete.

Among the more noticeable glacial features seen in the valley are unusual basin-shaped depressions in the outwash plain. These pits, called *kettles,* mark the place where blocks of ice (left by melting glaciers) were buried in outwash. When the ice melted, the overlying sediments slumped downward to form these unusual concavities. (These pits are known locally as "potholes," but they are more correctly called kettles; the former term correctly refers to an erosional feature found in certain stream beds.)

Although many of the glacial features mentioned above can

Sequence in the formation of a kettle. A block of stagnant ice is almost buried by outwash in A. The eventual melting of the ice produces a depression, as shown in B. In some instances, outwash may completely bury the ice block. Some kettles are formed in till. Reproduced from L. Don Leet and Sheldon Judson, *Physical Geology*, third edition, 1965, by permission of Prentice-Hall, Inc., Englewood Cliffs, New Jersey.

be seen in other National Parks, they are not usually as closely spaced and spectacularly exhibited as in the craggy profile of the Tetons and the expansive floor of Jackson Hole. With so much encompassed in such a small area, it is not surprising that Indians, trappers, and settlers were attracted to the Jackson Hole area early in the nineteenth century. Nor have the Tetons lost their charm—two and a half million visitors throng to this remarkably scenic area each year.

Plants and Animals of Grand Teton National Park

The biological attractions of Grand Teton almost equal its geological features. Long famous for its wildlife, Indians and trappers first descended upon Jackson Hole more than one hundred and fifty years ago. Today, thanks to protective measures supplied by Park Rangers and other U. S. Government personnel, the fauna and flora of this area remain much as they were during the past.

Among the Park's more interesting creatures are a number of large mammals including moose, elk, bear, and mule deer. The moose, the largest member of the deer family, is rather commonly seen by the more observant visitor—especially along the Moose-Wilson Road and in marshy areas near the beaver ponds west of Jackson Lake Lodge. The wapiti, or American elk, also feed in the Jackson Hole area and may be seen on occasion. Smaller mammals include beaver, chipmunk, ground squirrel, marmot, pika, marten, and rabbits.

Among the more than two hundred species of birds occupying the varied habitats available in the Park are herons, Canada geese, bald eagles, ospreys, and the rare trumpeter swans; these are most likely to be seen in the bottom woodlands of the Snake River. In the valley and at lower elevations in the mountains are a number of smaller forms such as mountain chickadee, gray jay, magpie, water ouzel, western tanager, and mountain bluebird. Hikers in the Teton high country may spot the black rosy finch.

Like the Park's fauna, Grand Teton plants vary considerably according to the elevation at which they are found. The conifers, or evergreen trees, grow around the lakes and well up on the mountain slopes; principal among these are Douglas and subalpine fir, Engelmann's and blue spruce, and lodgepole, limber, and whitebark pines. In the lower reaches along the stream banks and the river bottom are aspen, willows, and cottonwood trees. Typical shrubs include huckleberry, serviceberry, silverberry, creeping mahonia, and sagebrush.

During the summer months the valley and lower slopes of the mountains are often bedecked with a profusion of wildflowers. These displays, which vary in species from one month to the next, include pentstemon, larkspur, blue lupine, twinflower, wild geranium, painted cup, scarlet gilia, balsamroot, and pinedrops. Flowers which are more likely to be seen at higher elevations in the Park are alpine forget-me-nots, white Colorado columbine, and the lamb's-tongue or fawn lily. The latter grow near the edges of snow patches and thus have been called the glacier or snow lily.

Tips for Tourists

The visitor to Wyoming's Jackson Hole country will find a full schedule of outdoors activities awaiting his pleasure.

There are three interesting museums in the Park: the **Fur Trade Museum** at **Moose Visitor Center** where unusual exhibits relate the story of the early fur trade and "mountain men" of Jackson Hole and the Rocky Mountain West; **Jenny Lake Museum** features exhibits of Teton geology and the story of mountaineering in the Park; and the **Colter Bay Visitor Center Museum** where excellent exhibits tell of the geology, fauna and flora, and early human history of the Park. Visitors entering the Park from the north (from Yellowstone should make the Colter Bay Visitor Center their first stop for there is a fifteen-minute illustrated orientation program which will better prepare you to enjoy this scenic and historic area.

An especially interesting geological side trip can be made

down **Signal Mountain Road.** This is a five-mile scenic drive up the 7730-foot peak from which the road derives its name. Observation turnouts along the way and at the mountain's summit provide unmatched vistas of the Tetons and the valley. Persons wishing to see evidence of another well-known geologic event should drive to the **Gros Ventre Slide,** which is located not far from the eastern boundary of the Park. To reach the Gros Ventre River road go to Kelly Post Office and store (see map), turn left (north) and proceed about 1.1 miles, at which point the dirt River Road takes off to the east. Turn right and follow this winding road about four miles to the slide area. (*Bonney's Guide,* which may be purchased throughout the Park, has a detailed road log and an interesting account of this area.) The Gros Ventre Slide is a classic example of the rapid movement of earth materials by landslide and is featured in many geology textbooks. This landslide occurred in 1925, at which time a great mass of rock broke away from the side of Sheep Mountain and skidded into the river valley where it accumulated as a mass of broken rock and soil. The slide material formed a natural dam which stopped the flow of the Gros Ventre River and created Slide Lake. In 1927 this great earthen dam ruptured as a result of high water. The lower Gros Ventre valley was flooded, and the community of Kelly was washed away but has since been rebuilt.

The Park visitor who wishes to relive some of the history of Jackson Hole will want to visit **Menor's Ferry** and the **Chapel of the Transfiguration,** which was built in 1925 and is still used for Episcopal Church services. The building is constructed of native pine and has rustic pews hewn from quaking aspen; a large plate-glass window behind the altar provides an inspiring view of the majestic Tetons.

You may also wish to drive to the **Jackson Hole Wildlife Range** which is located inside the east entrance of the Park northeast of Highway 287. There are an information station, exhibits of wildlife, and pastures in which elk and buffalo can often be seen.

No other National Park has mountaineering facilities to compare with those of Grand Teton; small peaks await the neophyte and the Grand Teton loftily awaits the approach of the more experienced climber. There is even instruction for the rank beginner—the Exum School of Mountaineering quickly trains the novice in the proper climbing techniques and safety precautions; practice climbs are made within a matter of hours. Park regulations pertaining to mountain climbing are strictly enforced to insure the safety of the climber. They state: *"All climbers are required to register at Mountaineering Headquarters at Jenny Lake before starting to ascend any peak and must report their return from each expedition. Solo climbing is not permitted."*

Concessioner-operated cruises are conducted on Jackson and Jenny Lakes; inquire at an information station for schedules and rates. For the more adventurous there is an exciting float trip down the Snake River. This six-hour voyage, which begins near Buffalo Fork and ends at Moose, is made in a large rubber raft (holding eighteen persons steered by boatmen operating huge paddles at bow and stern.

The glistening, snow-flecked rock spires of the Teton are subjects worthy of any photographer. As in most mountain situations the light here is deceiving and the uninitiated tend to overexpose consistently. Careful use of a light meter will help remedy this but two or three different exposures of each subject is the best "insurance." A medium yellow filter will give just the right emphasis to clouds if you are using black and white film; a haze filter is almost a "must" if you are shooting in color. Don't limit your photography to any particular time of day—the mountain mood changes almost hourly according to time and the nature of the weather. You can photograph bison or elk at the Wildlife Range—a telephoto lens will be helpful here, as you should remain at a safe distance from these wild animals. Finally, do not overlook the opportunity to capture the Teton's reflection in the blue waters of the lakes; exceptional photographs can be made on clear days when the water is calm.

Grand Teton National Park at a Glance

Address: Superintendent, Box 67, Moose, Wyoming 83012.

Area: 310,350 acres.

Major Attractions: Series of peaks comprising the most impressive part of the Teton Range; once noted landmark of Indians and "Mountain Men." Includes part of Jackson Hole; near winter feeding grounds of largest American elk herd.

Season: Year-round.

Accommodations: Cabins, campgrounds, hotels, trailer village, and motels. *For hotel and motel reservations contact:* Grand Teton Lodge Co., Jackson, Wyoming 83001 (summer); 209 Post Street, San Francisco, California 94108 (winter).

Activities: Boating, boat rides, camping, fishing (license required), guided tours, hiking, horseback riding, mountain climbing, nature walks, picnicking, scenic drives, and swimming.

Services: Boating facilities, boat rentals, food service, gift shop, guide service, health service, laundry, post office, public showers, religious services, service station, telephone, transportation, picnic tables, rest rooms, and general store.

Interpretive Program: Campfire programs, museums, roadside exhibits, conducted hikes, self-guiding trails, and trailside exhibits.

Natural Features: Erosional features, forests, geologic formations, glaciation, glaciers, hot springs, lakes, mountains, rivers, rocks and minerals, unusual birds, unusual plants, large mammals, waterfalls, and wilderness area.

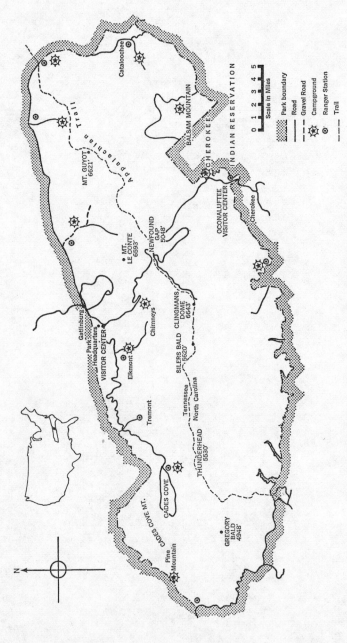

Great Smoky Mountains National Park, North Carolina-Tennessee.

14. Great Smoky Mountains National Park, North Carolina-Tennessee

The "Smoking" Mountains

Piled up astride the Tennessee-North Carolina border, the Great Smoky Mountains are among the highest and most beautiful in the eastern United States. They are not, it is true, as lofty or imposing as the craggy Tetons or the rugged Sierra Nevada; however, the gently rounded summits of the Smokies possess the dignity of old age, for they were formed many millions of years before the younger western ranges mentioned above. Even so, the Great Smoky Mountains can hardly be classified as hills—even by westerners—for despite their great age and the effects of long-continued erosion, these ancient mountains still boast sixteen peaks over six thousand feet high. The tallest of these, Clingmans Dome in the heart of Great Smoky Mountains National Park, rises 6643 feet above the Appalachian Highlands. It should be noted that the Great Smoky Mountains are not an isolated mountain mass; rather, they mark the southern culmination of the mighty Appalachian Range.

To the early settlers these were the "smoky mountains," a name given them because of the soft blue, velvety haze that normally covers the valleys and frequently drapes the sides of the mountain peaks. This haze, which imparts a characteristic smoky appearance to the entire range, is caused by the geographic location and heavy vegetation of the mountains. The air circulation is such that warm air flows into the mountains from the south and west and as it moves up to cooler elevations the moisture given off by the plants condenses to form lingering, wispy clouds.

Origin of the Smokies

Unlike that of many of our National Parks, the geology of Great Smoky Mountains National Park is rather subtly expressed. Indeed, many of the Park's rocks are hidden from view, but if we look beneath the luxuriant forest vegetation and fertile soil, climb the higher mountain peaks and explore the stream beds, we will discover ancient rocks that relate a long and fascinating geologic story.

Ancient Seas. These rocks reveal a story that began more than 600 million years ago near the end of Precambrian time. During this part of earth history the Great Smoky Mountains region differed greatly from that of today. There were no mountains here then; instead, the area now occupied by the Smokies was probably flooded by an extensive shallow sea. Here—over a period of many millions of years—there accumulated a great mass of sediments. These sediments, which in places attained a thickness of more than twenty thousand feet, consisted largely of rock particles believed to have been washed down from an eroding land mass which lay to the east.

As time passed, the ancient mud, sand, and gravel were converted into a great sequence of sedimentary rocks (p. 19) that is now called the *Ocoee Series*. Unfortunately, no fossils have been found in these Precambrian rocks, for they formed at a time when life was not abundant. But fossils have been found in Paleozoic rocks to the northwest in the Tennessee Valley and some of these rocks, those of the *Chilhowee Group* of Cambrian age, extend into the Smokies. These formations consist primarily of sandstone and shale and in places contain vertical tubelike structures believed to have been bored by Cambrian worms. Elsewhere there are outcrops of rocks of Ordovician age. These formations, exposures of which can be seen at Cades Cove, Tuckaleechee, Wears Cove, and Whiteoak Sink, consist primarily of limestone and shale. Some of the limestones exposed in Cades Cove contain the fossilized remains of invertebrate animals which dwelt in the ancient Ordovician seas.

Elevation of the Mountains. The accumulation of sediments in the late Precambrian and Paleozoic seas was but the first chapter in the geologic story of the Great Smoky Mountains. The next episode began more than 200 million years ago, probably near the end of the Pennsylvanian Period or the beginning of Permian time. The record of this portion of earth history is best seen in the igneous and metamorphic rocks of the Park, for these rocks differ greatly from those of sedimentary origin. For example, some of the rocks are riddled with igneous dikes and sills (p. 16), places where molten rock materials were injected into the surrounding sedimentary rocks. Present also are metamorphic rocks that have been derived from shales that were originally deposited as mud on the floor of those early prehistoric seas. Both the igneous intrusions and the hard, slaty metamorphic rocks indicate a period of great crustal disturbance, a time when powerful earth movements strongly deformed the crust of the earth in the southern Appalachian region. So intense were these movements that the rocks which had formed in the ancient, trough-like seas, were uplifted thousands of feet. It was thus that the sites of ancient seas were converted into a chain of towering mountains.

Evidence of the severity of this disturbance, called the *Appalachian Revolution,* can be seen throughout the Park today. Many of the rocks were wrinkled and folded into a series of wavelike structures called *anticlines* and *synclines*. The anticlines represent places where the rocks have been arched upward; synclines are troughlike or downfolded rock structures. The observant visitor is likely to see structures of both types at certain locations where Park roads have been carved out of the mountainside. At other places (for example, Newfound Gap), the rocks have actually been turned on end. As time passed and internal pressures increased, some of the rocks were subjected to such severe stress that they buckled and broke, producing a type of rock displacement called a *fault* (p. 24).

In parts of the Smokies these pressures were so great that faulted rock layers from the southeast were pushed north and west for as much as thirty-five miles. This type of fault, called an *overthrust fault,* is produced when the rocks on one side of the fracture are thrust over and above the other side of the fault. Faulting of this kind brings about a reversal in the normal sequence of rock strata by thrusting older layers over and above younger ones. This is why in certain places in the Park (for example, at Cades and Wear Coves), Precambrian rocks of the Ocoee Series which are about 600 million years old, are found resting on Ordovician rocks that are only about 400 million years old. This faulting, which occurred during the Appalachian Revolution, took place along a great fracture that has been named the *Great Smoky Overthrust.*

But how, you may ask, can we see exposures of Ordovician rocks in the cove, or valley, areas if the Precambrian rocks have been thrust over them? The rocks in the coves have been attacked by erosion, and the overlying Precambrian strata have been stripped away, thus revealing the underlying Ordovician strata. Erosion has, in effect, cut a "window" in the overlying Precambrian thrust sheet, thereby permitting us to look through it and see the underlying Ordovician shales and limestone. This type of geologic phenomenon is called a *fenster,* a name derived, most appropriately, from the German word for "window." The present-day climate of this region is such that the soluble Precambrian limestones are poorly resistant to erosion and the "windows" are slowly being enlarged, thereby giving rise to some of the flat-lying coves or valleys that we see today.

Shaping of the Mountains. For all their mighty glory, the newly uplifted Smokies—like all mountains—were destined to be but transitory creations on the face of the earth. For no sooner had uplift started than the forces of erosion began to gnaw at the mountain crests. Wind, rain, ice, and snow—each of these geologic agents played an important part in carving the mountains that we see today. And so it was that over the past 200 million years the Smokies gradually assumed their present form. Although the record is not preserved, it is believed that the newly formed mountains were subjected to prolonged erosion throughout all of Mesozoic time. Moreover, the wearing away of the early Smokies must have continued during the Cenozoic Era, for there is evidence that the whole region has been worn down nearly to a plain at several times in the past.

Perhaps you are wondering why—after these millions of years of erosion—we still have these mountains today. One reason is because the rocks of the Great Smoky Mountains have more successfully resisted erosion than those in surrounding areas. But even more significant is the fact that the southern Appalachian region has been subjected to intermittent periods of vertical uplift. These periodic elevations produced broad, gentle upwarps that raised the area above the base level of the sea. And, as uplift occurred, the streams in this region were rejuvenated and they renewed their downcutting activity with increased vigor. The present valleys and ridges of the Great Smoky Mountains, therefore, were not caused by sudden upheaval during the Appalachian Revolution. Rather, they are the product of post-Paleozoic erosion and dissection of the southern Appalachian Highlands.

The Smokies and the Ice Age. Unlike many of the world's mountain ranges, the Great Smoky Mountains seem never to have felt the direct effects of glaciation. These mountains were situated hundreds of miles south of the Pleistocene ice sheets and no evidence of glaciation has yet been discovered in them. The Smokies were, nevertheless, indirectly affected by the climatic extremes of the ice ages.

Because of the proximity of the northern icefields, there can be little doubt that the climate of the southern Appalachians during the Pleistocene was much colder than it is today. What is more, there is evidence to suggest that the colder temperatures of the Ice Age made possible the operation of certain geologic processes that are no longer at work in the warmer modern-day climate of the southern Appalachians.

One reason for this assumption is the presence of immense, angular boulders that are located well up on the sides of some of the mountain slopes. These monstrous rocks, which appear to have been broken from ledges of rock located nearer the mountain summits, were probably dislodged by frost-wedging. They resemble similar concentrations of boulders which are presently forming above timberline in arctic and subalpine regions in other parts of the world. Today there appears to be no such boulder-forming process operating in the Smokies. Instead, the boulders are no longer moving downslope but are grown over with plants and are slowly being reduced to soil. Can we not assume, then, that the Great Smoky Mountains of Pleistocene time may have had barren summits above timber line as do so many of our mountains in colder climates? And, since the climate was much colder, might we not also postulate the presence of snowfields and wide expanses of barren rock on the ancient mountaintops? These timberless areas would have been susceptible to a type of frost action that is currently producing similar boulders in colder and higher regions elsewhere.

Massive boulders can also be found farther down the mountain slopes and in some localities they litter the valley floors. These bulky stones—some as much as thirty feet in diameter—have been transported many miles from their original source. Boulders of this size are much too large to have been moved by streams; they were probably steadily inched forward by alternate thawing and freezing, and the swelling of the water-saturated clays that surrounded them.

Effects of the Recent. The termination of the Ice Age marked the advent of the Recent, that epoch of geologic time in which we are living today. Unlike their earlier—and sometimes violent—history, the post-Ice Age story of the Great Smoky Mountains was largely one of peaceful change. During this time the climate gradually assumed its present character, an important transition that made possible the development of the modern plant and animal life so abundant in the Smokies today.

Life in Great Smoky Mountains National Park

Although the geology of the Park is relatively subdued, this is certainly not true of its plant life. Indeed, more than anything else, it is the luxuriant vegetation that is the hallmark of this 800-square-mile woodland preserve. For example, Great Smoky Mountains National Park boasts the largest virgin forest of hardwoods in North America. Here, too, are 130 species of native trees—almost as many kinds of native trees as are found in the whole of Europe. Add to these the more than thirty-five hundred other plant species which thrive in this area and you can readily see why some botanists have called the Smokies the "cradle of vegetation" for eastern America.

This lush flora is not surprising when we consider the fertile soil and favorable climatic conditions of the Great Smoky Mountains. At some points in the Park the annual rainfall exceeds eighty inches, thus producing an environment that is almost like that of a rain forest. The moisture-laden climate has also had a pronounced effect on the geology of this area. Rock weathering is accelerated in regions with humid climates and the relatively rapid chemical decomposition of the bedrock has produced unusually fine soils. These rich soils further facilitate the growth of the dense vegetation so typical of this area.

One of the more noteworthy elements of the flora are the well-developed spruce and fir forests that blanket the higher elevations in the eastern part of the Park. These trees are typical of the Canadian life zone which reaches its southernmost extension in the Smokies. However, the plants reach their greatest development in certain of the coves in the lower reaches of the Park. Trees that characterize the coves area include eastern hemlock, basswood, yellow poplar (or "tulip tree"), yellow buckeye, sugar maple, American beech, cucumber tree, and northern red oak.

In addition to their great number and variety of species, the plants of this Park are famous for their large size. Trees which have been known to attain record proportions include eastern hemlock, Fraser magnolia, yellow poplar, silver bell, cucumber tree, and yellow buckeye. As examples of the unusual size of some of the virgin trees, consider the following: a yellow poplar with a trunk circumference of more than fourteen feet, a Canada hemlock whose trunk measures more than nineteen feet in circumference, and a yellow birch or "tulip tree" with a record girth of twenty-four feet. Moreover, certain plant species that are normally relatively small also attain amazing dimensions in the Smokies. One mountain laurel was found to have a seventeen-inch-thick trunk and a height of forty feet, and rhododendrons have been known to grow twenty-five feet tall.

Among the more unique aspects of the Great Smoky Mountains' plant community are the treeless areas which mark the summits of certain of the mountain peaks. These areas do not, as one might suspect, indicate elevations above timber line, for neighboring mountaintops that are as high or higher may be heavily forested. Why, then, are certain of the mountains "bald"? The origin of these *grass balds*—and this is the name that has been given these treeless grassy patches—is not understood. One interpretation is that they are cut-over areas that were used by the Indians or pioneers to pasture livestock. But the Cherokees have their own explanation of these high meadowlands. According to Indian legends, these clearings were made so that their ancestors could watch the movements of a hostile tribe that might attack them. Another Indian story is that the early Cherokees cleared these areas as an observation point from which to warn of a monster that was stealing the Indian children. You can visit one of these grass balds by taking the Andrews Bald Trail that leaves from Forney Ridge Parking Area on Clingmans Dome.

Not all of the balds are grassy. In places there are treeless areas that are covered with dense tangles of shrubs. These areas, called *heath balds* or *slicks,* resemble grass balds from a distance, but closer examination reveals them to be an almost solid mass of mountain laurel and rhododendron. Some of these "slicks" are virtually impenetrable unless one chops his way through dense undergrowth which in places may be as much as ten feet high. The popular hike to Alum Cave Bluffs visits a typical heath bald.

Wildflowers, such as trillium, violet, and phacelia, are profuse during April and May; and this is also the season for dogwood, silver bell, and serviceberry. May and June bring mountain laurel and flame azalea, while rhododendron appears from mid-June through July.

About mid-October the hardwood leaves begin to "turn" and the summery colors of the flowering plants are gradually replaced by the warm hues of autumn. Then the mountainsides are cloaked in a multicolored mantle of reds, browns, golds, and yellows.

The heavily forested Great Smoky Mountains also support a varied fauna. However, with the exception of the white-tailed deer, which are most numerous in the Cades Cove area, and black bear which occur throughout the Park, most of the Park's animals are small. These include raccoon, opossum, fox, weasel, woodchuck (or groundhog), skunk, and several types of squirrel. The densely wooded Smokies are also a haven for birds, and about two hundred species have been recorded within the Park's boundaries. Among these are the song sparrow, wood thrush, brown thrasher, cardinal, Carolina chickadee, tufted titmouse, rose-breasted grosbeak, Carolina juncos, and various woodpeckers and warblers. Wild turkey are also commonly seen in the vicinity of Cades Cove.

Tips for Tourists

In terms of visitors per year, Great Smoky Mountains is our most visited National Park. This is due, in part, to the area's many natural and historic attractions, but equally significant is the Park's proximity to large centers of population. Unfortunately, most of these visitors seldom leave their automobiles, except perhaps momentarily at a scenic overlook along the highway. And more is the pity, for the heartland of the Great Smokies is off the beaten path and deep in the forest glades, and the visitor who lingers here will not be disappointed; for the Park's well-rounded program of activities provides something for virtually every taste.

There are interesting exhibits on display in both of the Park's Visitor Centers. The **Sugarlands Visitor Center** on the Tennessee side of the Park has exhibits and a slide show that are designed to provide a preview of the area's many attractions. Visitors to the North Carolina side will want to stop at the **Oconaluftee Visitor Center** where the main theme of the displays revolves around the human history of this part of the Smokies. And while you are there, don't fail to tour the Pioneer Homestead, a reconstructed outdoor display that portrays the way in which the early settlers lived. Inquire at both of these Centers about the summer program of naturalist activities that is offered in the Park.

Great Smoky Mountains National Park is a popular place with the fisherman, for more than seventy kinds of fishes are known to inhabit the more than six hundred miles of streams in this mountain area. These include rainbow and brook trout, the latter of which are found only in the cooler waters of the higher elevations in the Smokies. You will need a fishing license (Tennessee or North Carolina depending upon where you are fishing) and you should consult a Park Ranger about the latest fishing regulations.

Because of the riot of color attendant almost every season in the Great Smokies, this Park offers remarkable photographic opportunities for the cameraman who uses color film. However, the typical "smokiness" of these mountains make a haze filter almost a necessity for distant shots of the mountains.

Unique backgrounds are available at the Cades Cove "open air" museum and also at the Pioneer Farmstead. If you are lucky you might even get a chance to "shoot"—on film, of course—some of the white-tailed deer that live in Cades Cove.

Great Smoky Mountains National Park at a Glance

Address: Superintendent, Gatlinburg, Tennessee 37738.

Area: 516,626 acres.

Major Attractions: Loftiest range east of the Black Hills and one of the oldest uplands on earth; diversified and luxuriant plant life, often of extraordinary size.

Season: Year-round.

Accommodations: Campgrounds and limited group campsites; motels and hotels also available in nearby towns.

Activities: Camping, fishing (license required), guided tours, hiking, horseback riding, nature walks, picnicking, and scenic drives.

Interpretive Program: Campfire programs, museums, nature trails, roadside exhibits, and self-guiding trails.

Natural Features: Forests, mountains, rivers, unusual plants, waterfalls, wilderness area, and wildlife.

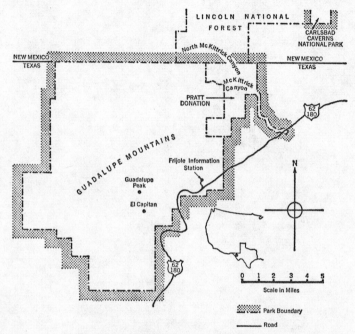

Guadalupe Mountains National Park, Texas.

15. Guadalupe Mountains National Park, Texas

The Top of Texas

The Guadalupe Mountains are located in south-central New Mexico and trans-Pecos Texas. A sprawling, V-shaped range, the Guadalupes extend southeast to northwest for fifty-five miles and are from twenty to thirty miles wide. The two arms of the "V" extend northwestward and northeastward into New Mexico, but the apex of this triangular wedge of mountains juts into Culbertson County, Texas. Here the point of the "V" is marked by El Capitan, a massive, cliff-flanked mountain that rises abruptly above the surrounding plains. Directly north of El Capitan lies 8571-foot Guadalupe Peak—the highest point in Texas.

The Capitan Reef

Although Guadalupe Mountains National Park contains a number of interesting natural features, its major attraction is a 4000-foot section of marine limestones of Permian age. This is the massive *Capitan reef,* the world's largest known fossil reef.

Because it is so well exposed and easily reached, geologists from all over the world have come to study this great thickness of fossiliferous carbonate rocks. They see here the classic example of how reefs formed in the geologic past and they can relate its structure to certain reefs that are forming in present-day seas.

The average visitor to the hot, arid plains of trans-Pecos Texas finds it hard to believe that this area was once the site of a great inland sea. But this is precisely what one would have seen here during the Permian Period, some 240 million years ago. During this chapter in earth history, west Texas and southeastern New Mexico were covered by an ancient ocean. This body of water, called the Delaware Basin, was roughly oval in shape and covered an area of approximately 10,000 square miles. A channel to the open sea was located to the southwest of the basin and this brought in fresh water to replace that lost by evaporation.

The carbonate (limy) rocks and the kinds of marine fossils found in the Guadalupes suggest that the water in the Delaware Basin was warm and clear. The reef-forming organisms of today thrive in this type of environment. This must have been equally true in the geologic past, for near the end of Middle Permian time, large reefs began to form in the shallower water along the margin of the basin. There were a number of these reefs, but the largest—the Capitan reef—was a massive horseshoe-shaped structure composed largely of lime-secreting algae, sponges, and a variety of shellfish.

The Capitan reef apparently extended from 350 to 400 miles around the margin of the basin. The water behind the reef was shallow, but in front of it to the south the sea was almost two thousand feet deep.

The structure of the reefs and their associated rocks indicate that the reefs grew both horizontally and vertically. But most of the growth was horizontal and toward the sea. As the reef grew seaward, pieces of reef material were often broken loose by storm waves. These rolled down to form slopes of fragmental debris that graded downward into the deeper-water muddy sediments. At other times there were submarine mud slides along the reef front. These produced masses of mud and rock fragments that flowed rapidly onto the basin floor, and spread over the nearer part of its bottom. It was on this foundation that the reef continued to grow seaward.

By Late Permian time the reef had grown to such an extent that it formed a barrier between the open basin and the shallow waters of the back-reef lagoon. This *barrier reef* prevented sea water from entering the lagoon and as the water in the lagoon evaporated it became increasingly salty. As time passed this water became concentrated to form brine from which much rock salt and gypsum was precipitated. In places more than two thousand feet of *evaporites*—as these salt deposits are called—were deposited in the great dead seas of the Permian Period.

The final development of the Guadalupes took place much later in geologic history during Late Cretaceous or Early Tertiary time. It was then that the area was uplifted and tilted slightly to the northeast. The magnificent exposures of the reef complex that we see today were exposed as a result of faulting associated with this uplift.

In places the carbonate rocks of the Guadalupes are honeycombed with caves. Most of these have been dissolved out of the relatively soluble Capitan Limestone. This is the same geologic formation in which Carlsbad Caverns (p. 103) has developed. No major caves occur within the Park's boundaries, but some might be discovered at a later date.

Plants and Animals of Guadalupe Mountains National Park

Today the Guadalupe Mountains form a highland that rises from an elevation of 3100 feet in the Pecos Valley near Carlsbad, New Mexico, to 8751 feet at Guadalupe Peak sixty miles to the southwest. Running water has carved numerous twisting, gorgelike canyons in the south end of the range, and here can be found a unique assemblage of plants and animals.

The fauna and flora vary considerably according to altitude. At lower elevations the vegetation is not unlike that of northern Mexico. Drought-resistant plants such as creosote bush, lechuguilla, and walking-stick cholla characterize this flora.

The vegetation changes considerably at higher elevations in the Guadalupes. The upper canyons support a variety of broadleaf trees and a scattering of ponderosa pines. But the most remarkable botanical province is in the highland along the rim of the Guadalupe escarpment. The forests here are comprised of ponderosa and limber pines interspersed with Douglas fir. A few groves of quaking aspen also grow in the upper parts of the mountains.

The animal life of the Park is almost as varied as the plants. Mule deer are common in the highland canyons and at the base of the escarpment. The original species of elk that inhabited the range has been extinct for many years, but a herd of elk that was introduced in 1925 appears to be thriving. Bighorn sheep, once common throughout the Guadalupes, may still be present in greatly reduced numbers. A small group has been reported in the vicinity of Guadalupe Peak. Although not commonly seen, a few mountain lions and black bears inhabit the more rugged and isolated canyons.

Included among the smaller mammals are ringtails, skunks, raccoons, bobcats, rabbits, rock and ground squirrels, porcupines, gray foxes, and coyotes. In addition, more than two hundred birds have been observed within the area.

Guadalupe Mountains National Park at a Glance

Because this is a relatively new National Park—it was authorized October 15, 1966—the area has not yet been developed and is not open to the general public. However, advanced planning is well under way and facilities for visitors should be ready in the not too distant future. An information station is open during the summer, and plans call for a visitor center, scenic drives, and campgrounds. In addition, there will be interpretive exhibits to explain the Permian reef complex and the biological features of the area. At present, Guadalupe Mountains National Park is under the supervision of the Superintendent of Carlsbad Caverns National Park, Box 1598, Carlsbad, New Mexico 88220.

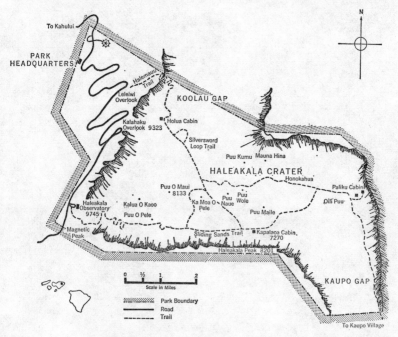

N

To Kahului

PARK
HEADQUARTERS

Halemauu Trail

KOOLAU GAP

Leleiwi
Overlook

Kalahaku
Overlook 9323

Holua Cabin

Silversword
Loop Trail

Puu Kumu Mauna Hina

HALEAKALA CRATER

Honokahua

Puu O Maui
8133

Puu
Wole

Puu
Naue

Ka Moa O
Pele

Paliku Cabin

Oili Puu

Haleakala
Observatory
9745

Kalua O Kaoo

Puu O Pele

Puu Maile

Magnetic
Peak

Sliding Sands Trail

Kapalaoa Cabin
7270

Haleakala Peak 8201

KAUPO GAP

| 0 | ½ | 1 | | 2 |
Scale in Miles

Park Boundary
Road
Trail

To Kaupo Village

Haleakala National Park, Hawaii.

16. Haleakala National Park, Hawaii

House of the Sun

High above the Pacific Ocean on the island of Maui is one of the world's largest volcanic craters. This great depression, the major attraction of Haleakala National Park in the Hawaiian Islands, is seven and a half miles long, two and a half miles wide, and twenty-one miles in circumference. The floor of this huge pit covers an area of nineteen square miles and is located three thousand feet beneath the volcano's summit; its surface is dotted with numerous cinder cones, lava flows, and other volcanic phenomena of considerable interest.

According to island legend, a Polynesian god named Maui once climbed to the top of this mountain to capture the sun in order to force it to move more slowly through the sky. Maui did this at the request of his mother so that she might have more daylight hours in which to complete her work. Thus, to early Hawaiians this great volcanic mountain was *Hale-a-ka-la,* "House of the Sun," the name that is still used today.

Haleakala was originally designated as the Haleakala Section of Hawaii National Park which also included the Mauna Loa-Kilauea Section on the Island of Hawaii. However, Haleakala was established as a separate National Park effective July 1, 1961, and today has a total land area of almost twenty-seven squares miles.

Erosion and Explosion

Like the other islands in the Hawaiian Archipelago, Maui is the product of volcanic eruptions which are believed to have taken place during Late Tertiary time, some ten to twenty-five million years ago. In fact, the Hawaiian Islands are actually the tops of a great range of mountains which rise some fifteen thousand feet from the ocean floor. Some of these, like Haleakala, rise as much as ten thousand feet above the water or more than twenty-five thousand feet above its base.

The island of Maui is composed of two inactive volcanoes. Haleakala, which forms the eastern part of the island, is thirty-three miles long, twenty-four miles wide at its base, and stands 10,023 feet above sea level. The volcano forming the western part of the island is considerably smaller and is attached to Haleakala by means of an isthmus about seven miles wide.

The great "crater" of Haleakala, like the "craters" of Crater Lake (p. 120) and Mauna Loa (p. 205) is not truly a crater in the geologic sense of the word. However, it was not formed by collapse as were the calderas of Crater Lake and Mauna Loa; rather, this great basin has been produced by the forces of erosion. Most of this erosion has occurred since the volcano became inactive and was concentrated in the area now known as *Koolau* and *Kaupo Gaps* (see map). These two great valleys have been carved out of the summit of Haleakala and join together to form the extensive erosional basin now referred to as Haleakala "crater."

After creation of the basin by erosion, Haleakala experienced renewed volcanic explosions; these resulted in a series of cinder cones and lava flows which can still be seen in the crater. These eruptions originated from a series of vents which extended diagonally across the basin floor. The tallest of these cinder cones, **Puu O Maui** (the "Hill of Maui"), rises one thousand feet above the floor of the crater.

Although it is not definitely known exactly when Haleakala was last active, the latest eruption is believed to have taken place in about 1750. During this eruption, lava issued from two vents and formed the **Keoneoio Flow,** which can be seen above La Perouse Bay in the southwest corner of the island. Because of this relatively recent eruption and the presence of earthquake activity still being recorded in the area, Haleakala is believed to be a *dormant* or "sleeping" volcano. On the other hand, Kilauea in Hawaii Volcanoes National Park (p. 205) is in periodic eruption and is thus classified as an *active* volcano. Volcanoes which are not known to have erupted within historic times are said to be *extinct*. Even so, some so-called "extinct" volcanoes, for example Lassen Peak in California (p. 233), have suddenly become active after long periods of quiescence.

Plants and Animals of Haleakala National Park

Haleakala contains an interesting group of native plants and animals. The most unusual plants of the Park are the rare *silverswords,* large yuccalike plants which are actually members of the sunflower family. These plants, which are almost as famous as the Park in which they grow, are characterized by large, dagger-shaped, silvery leaves and a cluster of many purple-petaled flowers atop a tall spherical stalk. Although the silversword also occurs in Hawaii Volcanoes National Park, it is most typically associated with Haleakala. In addition to silverswords, the flora of Haleakala consists of ferns, rushes, sedges, Hawaiian raspberry, primrose, geraniums, and a host of other species peculiar to this area.

The brightly colored native birds are the most noticeable part of the park fauna. These include numerous native forms such as the *nene* or Hawaiian goose (a form that has completely forsaken the water for life on land); *pueo,* the Hawaiian short-eared owl; *iiwi,* a scarlet-bodied bird with black wings and tails; the *kolea,* or golden plover; and *koae,* the white-tailed tropic-bird. Among the birds that have been introduced to Haleakala there are skylark, finch, sparrow, Chukar partridge, mockingbird, mynah, Japanese white-eye (or mejiro), and California valley quail.

Other animals of this area include rats, mice, mongooses, wild goats and pigs, and a host of native insects.

Tips for Tourists

On your visit to Haleakala National Park you will want to take advantage of the recreational and interpretive facilities that have been provided for you. The **Crater Observatory** overlooks the crater and contains exhibits and orientation displays; this is the logical place to concentrate your visit.

The drive from the **Park Entrance** near **Hosmer Grove** leads to Park Headquarters. Here you can obtain camping and hiking permits, information on crater trips, or assistance in planning your visit. On the road to the Crater Observatory you should stop at **Leleiwi** and **Kalahaku Overlooks** for good views of the crater. Continuing up the road you soon reach **Puu Ulaula** which marks the summit of the crater rim. If the weather is good, you will get a magnificent view from this point and various interpretive devices are provided to help identify the many peaks and islands that can be seen.

The only way to enter the crater is on foot or on horseback. Good, well-marked trails await the hiker, and there are three cabins within the crater for visitor use. Experienced hikers might make the round trip in and out of the crater in one day; however, it is best to allot two or three days for this trip. You can make crater cabin reservations by writing to the Park Superintendent.

In addition to the longer trails within the crater, there are several shorter walks such as those to the top of **White Hill** or along **Halemauu Trail** from the Park road to the rim of the crater (about eight-tenths of a mile). Or you may prefer to hike a short distance down the **Sliding Sands Trail.** Despite the name, the loose material forming this trail is not composed of sand; instead, it is made of fine-grained cinders and ash which were deposited during periods of eruption. Should you decide to descend this trail, do not go too far; you are hiking at a very high altitude and the return trip may prove to be quite exhausting.

You can also explore the crater floor by horseback in the company of an experienced guide. For information inquire at Park Headquarters.

Numerous photographic opportunities await the visitor to Haleakala; the colorful cinder cones, native plants and birds, and various geologic features are especially worthy subjects. Consult your light meter before shooting; the intense sunlight at this elevation sometimes results in overexposure, particularly when using color film. Camera enthusiasts may encounter special problems with the dense banks of fog that frequently fill the crater, hiding the landscape from view. Luckily, these commonly disappear as suddenly as they blow in, so do not give up too quickly!

Haleakala National Park at a Glance

Address: Superintendent, P.O. Box 456, Kahului, Maui, Hawaii 96732.

Area: 27,600 acres.

Major Attractions: Haleakala crater, one of the world's largest, located in the top of a 10,023-foot volcanic mountain; exotic fauna and flora including the rare silversword plant.

Season: Year-round.

Accommodations: Cabins and campgrounds. For reservations on crater cabins contact Superintendent, P.O. Box 456, Kahului, Maui, Hawaii 96732.

Activities: Camping, hiking, horseback riding, nature walks, picnicking, and scenic drives.

Services: Telephone, saddle horses, picnic tables, and rest rooms.

Interpretive Program: Nature trails, roadside exhibits, self-guiding trails.

Natural Features: Erosional features, rocks and minerals, unusual birds, unusual plants, and volcanic features.

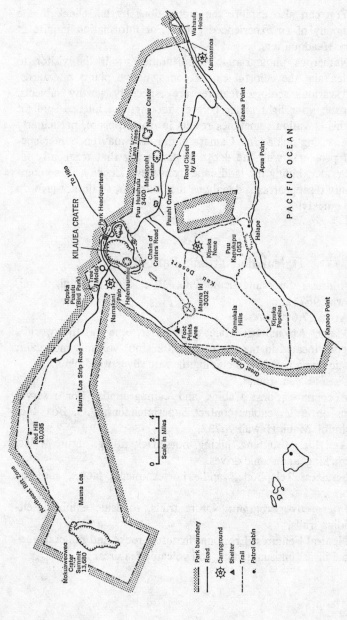

Hawaii Volcanoes National Park, Hawaii.

PACIFIC OCEAN

Wahaula Heiau
Kamoamoa
Napau Crater
Lava Trees
Kaena Point
Puu Huluhulu 3400
Makaopuhi Crater
Road Closed by Lava
Pauahi Crater
Apua Point
To Hilo
Park Headquarters
KILAUEA CRATER
Chain of Craters Road
Tree Molds
Kipuka Puaulu (Bird Park)
Namakani Paio
Halemaumau
Kipuka Nene
Puu Kapukapu 1053
Halape
Kau Desert
Mauna Iki 3032
Kamakaia Hills
Kipuka Pepeiau
Kapaoo Point
Foot Prints Area
Great Crack
Mauna Loa Strip Road
Red Hill 10,035
Northeast Rift Zone
Mauna Loa
Mokuaweoweo Crater Summit 13,680

Scale in Miles
0 2 4

········ Park boundary
——— Road
✹ Campground
▲ Shelter
– – – Trail
● Patrol Cabin

17. Hawaii Volcanoes National Park, Hawaii

Hawaii—Islands of Fire

As mentioned earlier, the Hawaiian Islands consist almost exclusively of mountains of lava which issued from great fissures in the ocean floor. Hawaii, the largest and easternmost island of the Hawaiian chain, is composed of five large volcanic cones —two of these, Mauna Loa and Kilauea, are classified as active volcanoes.

Mauna Loa

Mauna Loa is believed to be the world's largest active volcano. It is a *shield volcano* or *lava dome* and is composed almost completely of numerous lava flows deposited one upon the other. This great mountain rises approximately 13,680 feet above sea level and more than 30,000 feet above the bottom of the sea; at its summit there is a large depression which is referred to as **Mokuaweoweo Crater.** However, Mokuaweoweo is not a true crater; rather, it is a *caldera* similar to that in which Crater Lake has formed (p. 120). This great caldera was created as the summit of Mauna Loa collapsed when the supporting magma was withdrawn. Today the caldera is about three miles long, one and a half miles wide, and almost six hundred feet deep; its floor is covered with lava flows of several different ages.

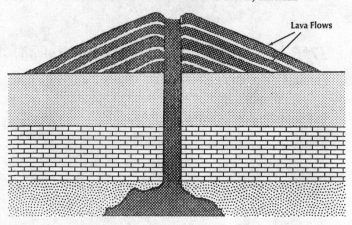

Lava Flows

The Hawaiian volcanoes, such as Mauna Loa and Kilauea, are called shield volcanoes. They are composed of countless layers of lava superimposed upon one another.

Periodically, Mauna Loa's periods of quiescence (which have ranged from a few months to as much as twenty-three years) are interrupted by volcanic activity. These eruptions may occur as *central* or *summit eruptions,* at which time volcanic material emanates from the top of the volcano, or as *flank eruptions* during which lava pours from *fissures* or *rift zones* (great cracks) located at lower levels on the side of the mountain. During these eruptions which up to 1950—the last Mauna Loa eruption—have occurred on an average of every three and a half years, great quantities of lava flow down Mauna Loa's flanks, and frothy pumice, cinders, and ashes are blown over the countryside. Indeed, lava flows from Mauna Loa cover more than two thousand square miles of the surface of the island.

During the great eruption which began on June 1, 1950, Mauna Loa produced more than 600 million cubic yards of highly fluid lava; the major lava flows associated with this activity moved at an average speed of about six miles per hour and reached the sea in less than three hours. Upon entering the water, the hot lava generated vast clouds of steam, some of which rose thousands of feet into the air. Luckily for the islanders, Mauna Loa, like certain other volcanoes, frequently warns of impending eruption, thus enabling scientists to notify the island residents of possible forthcoming activity. For this reason, any change in the level of the caldera floor or increasing earthquake activity in the area is carefully noted as a possible sign of future eruption.

The part of the Park in which Mauna Loa is situated is not as well developed as the area around Kilauea Crater. However, you can visit **Mokuaweoweo Crater** by hiking the 18.2-mile **Mauna Loa Trail** which starts at the end of the **Mauna Loa Strip Road.**

Kilauea

Although Mauna Loa is considerably larger, Kilauea is by far the most active Hawaiian volcano. The mountain itself is a typical shield volcano composed of innumerable lava flows; its summit contains a broad shallow caldera two and a half miles long, two miles wide, and about 450 feet deep. This large depression is partially enclosed by a nearly perpendicular wall composed of steep cliffs called *fault scarps;* these sharp breaks were formed by collapse of the surrounding caldera rim.

On the floor of this great caldera is **Halemaumau,** the most active vent on Kilauea and the traditional home of Pele, Polynesian goddess of volcanoes. From time to time, Halemaumau has contained an active lava lake which occasionally "boiled over" and overflowed onto the surrounding caldera floor. Within the caldera of Kilauea there are numerous reminders of past volcanic activity; materials ejected during eruptions of 1790, 1919, 1924, and 1954 can be clearly distinguished. Most of Kilauea's eruptions have been quiet and nonexplosive; however, once in 1790 and again in 1924 volcanic explosions (probably due to underground accumulations of steam) rocked Kilauea and the surrounding area. Debris thrown out by these explosions can still be seen on the caldera floor.

Most of the lava which occurs on the floor of **Kilauea Crater** is called *pahoehoe* (pa-ho-ay-ho-ay). This type of lava is characterized by a smooth, satiny or glassy surface; it may resemble a series of parallel strands of twisted rope. At other places in the Park (for example, along the **Mauna Loa Strip Road**) you will notice great jumbled masses of blocky lava; this is called *aa* (ah-ah). Oddly enough, both of these lavas are identical in composition; the great variation in their physical appearance is caused by differences in the state of enclosed gas at the time the lava solidified as well as by the temperature and state of crystallization present at solidification.

Probably the best-known and most spectacular eruption of Kilauea began on November 14, 1959. At this time a fissure opened on the southwest wall of **Kilauea Iki,** a pit crater located adjacent to the eastern margin of Kilauea Crater (see map). The eruption was marked by great lava fountains which played along a line about twelve hundred feet long; one fountain gradually grew in height until it was throwing lava about nineteen hundred feet into the air. Volcanic materials ejected by this great fountain built a cone-shaped hill more than one hundred and fifty feet high and buried part of the Crater Rim Road. Between the periods of high fountaining there were periods of inactivity of varying lengths of time.

A great deal has been written about the Hawaiian volcanoes and their eruptions during historic time; much of this information is summarized in *Volcanoes of the National Parks of Hawaii* (available from Hawaii Natural History Association, Hawaii Volcanoes National Park, Hawaii 96718). This very excellent publication should be read by anyone who visits either Haleakala or Hawaii Volcanoes National Park.

Plants and Animals of Hawaii Volcanoes National Park

Hawaii Volcanoes National Park is well known for its varied vegetation and unusual fauna of native birds. There is a marked contrast in the flora, which ranges from lush tropical plant life on the windward side of the island to the sparse vegetation of the **Kau Desert** south and west of Kilauea. **Crater**

Rim Drive enters both of these areas, thus accentuating the great difference between plant life in these two diverse environments.

The lush tropical vegetation of the Park is dominated by the ohia tree with its scarlet to yellow blossoms called *lehua*. These relatives of the eucalyptus are the most common native trees in the islands and are the symbol of the island of Hawaii. Also present are ferns, sandalwood, koa, and mamani; all are plants of unusual beauty. For visitors who are interested in the plant life of the island there are two self-guiding nature trails which feature this unusual flora.

Native wildlife consists largely of birds; many of the same species that inhabit Haleakala National Park of Maui are found here at Hawaii Volcanoes National Park. Notable among these are the *koae* (or white-tailed tropic-bird), the American golden plover or *kolea,* the *iiwi* and *apapane,* which are commonly seen in the ohia trees, the *elepaio,* one of the fly-catcher family, *io,* the Hawaiian hawk, and the Hawaiian short-eared owl or *puco*. Extremely rare is the *nene* or Hawaiian goose, which in recent years has been re-established in Haleakala and Hawaii Volcanoes National Park after coming perilously close to extinction. In addition to the native birds, there are introduced forms such as finch, house sparrow, Chinese and blue pheasants, quail, partridge, cardinal, rice bird, Japanese hill robin, mynah, and skylark.

Among the animals living in the Park are wild pigs and goats, and mongoose; all have been introduced by man.

Tips for Tourists

Although you may not be so fortunate as to visit the Park during an actual eruption, Hawaii Volcanoes' varied attractions will make any visit to the area well worth while.

A visit to the **Thomas A. Jaggar Memorial Museum** in Kilauea Visitor Center at Park Headquarters is a good place to acquire basic information of the Park and its volcanic features. There you can hear a talk on volcanism and see colored motion pictures of recent eruptions; also the volcanic story is further explained by means of exhibits, paintings, maps, and relief models.

The **Hawaiian Volcano Observatory,** which is operated by the United States Geological Survey, is located about two miles from Park Headquarters; geologists stationed here have kept a careful record of Kilauea's activity since 1912. The laboratory is operated for observational and research purposes and is not open to the public; however, a special *seismograph* (a machine used to record the intensity of earth vibrations or earthquakes) display can be seen through a public viewing window. A seismograph is also available at Kilauea Visitor Center. Records obtained from the Observatory's seismographs and *tiltmeters* (instruments which measure variations in the tilting of the surface of the caldera floor and the summit area) are used to predict eruption potential and aid in the understanding of how volcanoes behave and why. Scientists assigned to the Observatory also keep a detailed record of all volcanic eruptions, collect samples of lava and volcanic gas, and gather descriptive and temperature data on the various volcanic materials which emanate from the volcano.

Hawaii Volcanoes National Park offers numerous opportunities for picture taking; geological, biological, and scenic subjects are virtually unlimited. A wide-angle lens will be most helpful in certain areas of limited size (for example, to show all of Halemaumau) and a haze filter will be useful when shooting color film.

When hiking or riding in the Park, please stay on established trails; many of the flows are supported only by a thin crust and hidden cracks and lava tubes are an ever-present source of danger. In addition, be especially cautious when approaching the rim of Halemaumau. The lip of this fiery pit is weak in places. For your safety and protection a path and observation platform have been provided. Please use them.

Hawaii Volcanoes National Park at a Glance

Address: Superintendent, Hawaii Volcanoes National Park, Hawaii 96718.

Area: 220,344 acres.

Major Attractions: Two active volcanoes, Mauna Loa and Kilauea, lava flows, craters, rare birds and plants, seascapes, and ancient Hawaiian ruins.

Season: Year-round.

Accommodations: Cabins, campgrounds, and hotel. *For reservations contact:* Kilauea Volcano House, Hawaii Volcanoes National Parks, Hawaii 96718.

Activities: Camping, fishing, hiking, horseback riding, mountain climbing, picnicking, scenic drives, and during eruptions volcano watching.

Services: Food service, gift shop, post office, telephone, picnic tables, and rest rooms.

Interpretive Program: Museum, nature trails, roadside exhibits, self-guiding trails, trailside exhibits, and talks and movies on volcanism.

Natural Features: Lava tubes, desert, forest, geologic formations, marine life, mountains, rocks and minerals, seashore, unusual birds and plants, volcanoes, volcanic features, and wilderness areas.

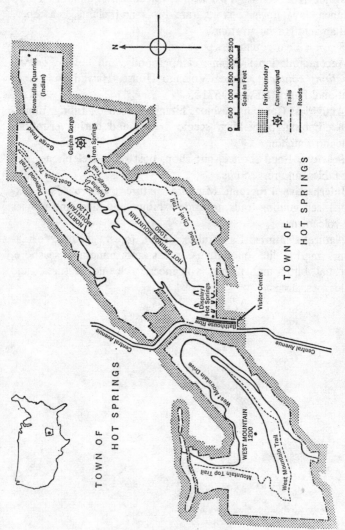

Hot Springs National Park, Arkansas.

18. Hot Springs National Park, Arkansas

Fountain of Youth

Unlike most of the National Parks, Hot Springs National Park is located in the midst of a bustling city. Yet within its 1032 acres are Hot Springs Mountain and most of North Mountain. Three other mountains, Sugar Loaf, West and Indian Mountains, are partially located within Park boundaries. But the major attraction in this Park is not mountains; rather, it is the famous natural hot waters which relieve the sick and help to keep the physically fit in condition.

Historians believe that the springs were first seen by white men in 1541, at which time Hernando de Soto, a Spanish explorer and discoverer of the Mississippi River, explored the Hot Springs region. But de Soto and his party were not the first men to view the springs. According to tradition, the Hot Springs area was a favorite camping ground of the Indians long before 1541. There is also some indication that the Indians valued the water for its therapeutic properties and that fierce battles were fought to control the land on which the springs were located.

With the coming of the white man in about 1800, the region was settled and word of the healing hot waters soon spread throughout the area. Early records indicate that people were bathing in the springs as early as 1804. Finally, in order to protect the springs from destruction and undue exploitation, the U. S. Congress in 1832 set aside the hot springs and an adjacent four square miles of land as Hot Springs Reservation. In 1921 Congress decreed that the name should be changed to Hot Springs National Park.

Nature's Hot Water Heater

Probably the most commonly asked question at Hot Springs is: "Why is the water hot?" And a good question this is, for it is one that has puzzled geologists since the turn of the century. In an attempt to determine the origin of the hot water, a number of geologic studies have been made. These have provided answers to some of the problems about the rocks and geologic structure of the Park area, but the exact mechanism of the natural "hot water heater" is still in doubt.

It is known, however, that all of the forty-seven hot springs are located along a fault line located in the valley between West and Hot Springs Mountains. Confined to a relatively small area (about twenty acres), the springs occur principally along the outcrop of the Hot Springs Sandstone. This formation is about two hundred feet thick in this area and was deposited in an ancient Mississippian sea which covered this part of Arkansas about 300 million years ago. Much later in geologic history (probably during the Pleistocene Epoch), the sandstone and adjacent formations were subjected to lateral compression which lifted the Park area and produced great folds in the rocks. These same forces severely fractured the rock strata, thereby establishing a series of openings through which the heated groundwater could reach the surface.

But despite the fact that the geologic history of the area has been well established, geologists still do not agree on the source of heat of the spring water. Although a number of theories have been suggested, most scientists support either the *meteoric water* theory or the *juvenile* or *magmatic* water theory. The former suggests that the water originates as *meteoric water* (water derived from the atmosphere such as rainfall), which is soaked up by porous rocks cropping out in the valley floor between Sugar Loaf and West Mountains. Then, as the rainwater seeps downward through the highly fractured rock, it is believed to be heated as it passes close to a mass of buried hot rock. After being heated, the water flows to the surface through fractures in the tilted sandstone strata along the southwestern base of Hot Springs Mountain.

According to the other major theory, the waters are of *juvenile* or *magmatic* origin. This type of water has not previously existed as atmospheric or surface water; rather, it is derived from heated rocks within the earth's interior. Thus, it has been suggested that the heated water has been discharged from a slowly cooling mass of igneous rock buried relatively near the earth's surface and that the water rises to the surface by means of fractures or fissures which extend into the deep interior of the earth. As in the case of the meteoric origin, there are a number of objections to the juvenile or magmatic theory.

In addition to the theories mentioned above, it has also been suggested that the heat may have been derived from (1) radioactive minerals, (2) great pressures due to overlying rock burden, (3) chemical reactions taking place near the water, and (4) friction as great masses of rock slid past each other during deformation of the strata.

Even though the origin of the springs and their heated water may be subject to controversy, the performance of the springs leaves nothing to doubt—they continue to produce almost a million gallons of water per day with an average temperature of more than 143° F.

Plants and Animals of Hot Springs National Park

The Park area is typical of the densely wooded Ouachita Mountains and the rocky hills support an interesting flora. Oaks, pines, and hickories dominate the forested areas, and masses of huckleberries and other shrubs form a thick undergrowth. Wildflowers are varied and plentiful and bloom throughout the year; serviceberry, redbud, dogwood, aster, goldenrod, bitterweed, and coreopsis are among the more decorative forms. In the damper areas, ferns, sedges, mosses, and lichens are common.

Among the more interesting plants in the Park are the tiny blue-green algae which occur in certain of the hot springs. Algae, a primitive type of plant, are common in many cool lakes, ponds, swamps, and streams where they are commonly referred to as "pond scum." However, species of blue-green algae which inhabit hot waters require special adaptation of structure to survive in high temperature environments and are thus not nearly so common. Another unusual characteristic of these minute plants is their "rock-building" ability. These hot-water algae help convert calcium and silica salts in the hot waters into insoluble rock deposits called *tufa*. Deposits of tufa can be seen in outcrops at the Display Springs behind Bath-house Row.

Animals found within Park boundaries include such common small mammals as squirrels, rabbits, opossums, foxes, raccoons, and small rodents. Reptiles and amphibians include a number of species of snakes, lizards, turtles, frogs, and salamanders. There are, in addition, over two hundred species of birds which have been seen in the Park.

Tips for Tourists

Hot Springs National Park is not only different in that it is located in the heart of a city; it provides a variety of activities not to be found in any other National Park. Because the city of Hot Springs is a resort and convention center, it has such recreational facilities as movies, nightclubs, horse-racing, golfing, sailing, and fishing. The climate is favorable, and outdoor recreation is possible almost year-round.

One of the more popular activities is to take the baths which are given in the government-approved concessioner-operated bathhouses along **Bathhouse Row** and elsewhere in the city. Charges vary according to the type of equipment and accommodations provided, but the water is the same in all. Rates include charges for services of attendants and are set by the National Park Service. The water from the springs is generally believed to possess definite beneficial therapeutic effects and seems to be especially helpful to persons with arthritic ailments or in a general run-down condition. (Examination by a physician is recommended before bathing in the thermal waters.)

In addition to the activities mentioned above, some of the more "typical" National Park activities are also available. For example, there is a modern Visitor Center in downtown Hot Springs at the corner of Central and Reserve Avenues. Here there are displays of rocks, minerals, and fossils, exhibits on the mechanism of the springs and the geologic history of the area, and information on the plants and animals in the Park.

Some interesting geological features can be seen on the **Promenade Nature Trail,** which is located above and behind Bathhouse Row. This pleasant one-mile trail begins at the Visitor Center on Reserve Street and follows the Grand Promenade to Fountain Street.

Numbered markers and the trail leaflet (obtainable at the Visitor Center) point out and explain the various plants and geological formations along the trail. You can cover the entire distance in about one hour, and benches along the way permit opportunities for rest and relaxation.

In addition to an introduction to the plants and wildlife of the Park, interesting geological features to be seen on the promenade include unusual *quartz veins* which appear as white streaks in the sandstone and good exposures of *conglomerate,* or *"pudding stone."* Conglomerate is a sedimentary rock that is composed of other rock fragments held together by a natural cementing agent. The cement in this conglomerate is probably derived from minerals deposited by the hot springs water. Also of interest is a fine exposure of *tufa.* As mentioned earlier (p. 216), this sedimentary rock was formed by exposure of the thermal waters to the air and/or by the action of algae helping to convert dissolved calcium and silica salts from the water into solid rock. Needless to say, this process requires a considerable period of time and the rock outcrop along the trail took a long time to produce.

There are also stops at the **Heat Exchange Building,** where thermal water from the reservoir is cooled before going to the bathhouses, and two **Display Springs** which have been left open for public display. The other forty-five springs have been sealed to help prevent contamination of the water and loss of radon gas. Message repeaters at both the Heat Exchange Building and the Display Springs will tell you the story of these two features. In addition, printed copies of the narration may be obtained at the Visitor Center.

Hot Springs National Park at a Glance

Address: Superintendent, Box 1219, Hot Springs National Park, Arkansas 71901.

Area: 3535 acres.

Major Attractions: Hot springs used in the treatment of certain ailments; deposits of calcareous tufa around hot springs.

Season: Year-round.

Accommodations: Campgrounds and hotel and motel accommodations in Hot Springs, Arkansas. Contact Hot Springs Chamber of Commerce for a list of accommodations available.

Activities: Camping, guided tours, hiking, nature walks, picnicking, and scenic drives.

Services: Picnic tables, rest rooms, and bathhouses.

Interpretive Programs: Campfire programs, museum, nature walks and hikes, and self-guiding trails.

Natural Features: Geologic formations, forests, hot springs, rocks and minerals, and mountains.

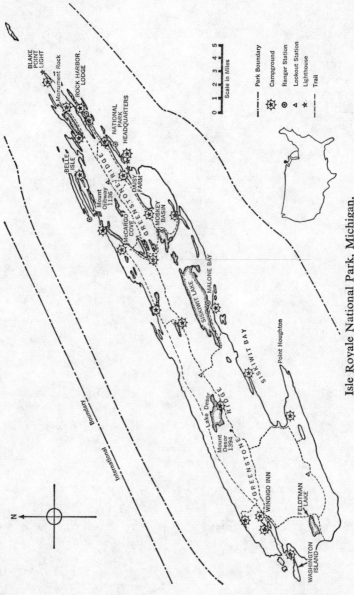

Isle Royale National Park, Michigan.

19. Isle Royale National Park, Michigan

The Royal Island

The largest island in the largest freshwater lake in the world is the site of Isle Royale, one of our more unique National Parks. Because of its geographic location—in the northwestern part of Lake Superior, forty-five miles from the nearest Michigan shore and fifteen miles from the Canadian mainland—Isle Royale differs from other National Parks in a number of ways. It is, for example, our most inaccessible National Park: it can be reached only by boat or seaplane. It is also a true "wilderness" area, for there are no roads or wheeled vehicles in the Park; visitors must explore the island on foot, by boat, or not at all. Also unusual is the fact that there is more water than land within the Park. The Park's boundaries, which extend four and a half miles in all directions from the island's shores, enclose an expanse of more than six hundred square miles of water. The land area of the island, on the other hand, consists of only about two hundred and ten square miles. What is more, Isle Royale is actually closer to Canada than it is to the mainland of Michigan, the state of which it is a part.

Yet despite its geographic isolation, the island has long attracted human visitors. There is evidence that the area was visited by prehistoric Indians almost four thousand years ago, and French explorers—who named the island in honor of King Louis XIV—claimed Isle Royale as part of the Canadian territory as early as 1669. Much later, in the early 1800s, man was again attracted to the "Royal Island," this time to establish commercial fisheries. But probably the most exciting period in Isle Royale's history took place between 1844 and 1855 and again in the 1870s when prospectors and miners swarmed over the island in search of copper, which is found in certain of the island's rocks (p. 224).

Although the copper mines have been inactive for many decades, man still heeds the call of this water-locked wilderness. Here, in a superlative natural setting of unspoiled forests and peaceful harbors, one can relax in solitude and serenity far from the cares of everyday living.

Origin of the Island

Visitors to Isle Royale commonly ask: "Why is the island here?" But—because Isle Royale is located in Lake Superior—before this question can be answered we must first ask another question: "Why is Lake Superior here?" Fortunately, thanks to detailed geologic studies of the Lake Superior region, we can answer both these questions with some degree of accuracy, for the rocks of the island contain many clues to the geologic history of this area.

The Geologic Setting

To understand clearly the origin of Lake Superior and Isle Royale, we should first know something about the general geologic features of the region in which they are located. Both of these areas lie almost wholly within the *Canadian Shield,* a vast two-million-square-mile expanse of Precambrian rocks. Each continent displays at least one of these so-called *shields,* which are stable areas that since Precambrian time have never been deeply covered. Most of the shields were sites of extensive mountain building during Precambrian time but have subsequently become stable and now form large interior plains on our continents. Because its rocks contain fabulously rich deposits of iron, nickel, copper, silver, and gold, the Canadian Shield is the most studied of all of the shields. It is also the world's largest area of exposed Precambrian rocks.

The Rocks

The rocks which comprise the surface of the Canadian Shield represent the youngest division of rocks formed during the Proterozoic Era, or Late Precambrian time. They are called the *Keweenawan Series,* and derive their name from the Keweenaw Peninsula on the south side of Lake Superior in northern Michigan.

The geologic formations which make up the Keweenawan Series are believed to have an aggregate thickness of almost ten miles. Of these, about fifteen thousand feet consist of *red beds:* iron-stained sandstone, siltstones, shales, and conglomerates. The presence of ripple marks, apparently made by stream currents, and mud cracks, which indicate periodic drying of the sediments, indicates that the red beds were probably deposited in an ancient delta. What is more, the reddish color of many of the sediments suggests that the sediments have been oxidized, perhaps after exposure to the atmosphere. The conglomerate is well exposed on the southeast side of the island near the old Rock Harbor Lighthouse and Middle Island Passage. Indeed, one of the inlets in this area has been named Conglomerate Bay. There are good exposures of the sandstone at Houghton Point and along the southwestern end of the island almost to Grace Harbor. The sandstone and conglomerate are also prominent on and comprise most of Feldtman Ridge.

Geologic cross section of Lake Superior showing the Lake Superior Basin and Isle Royale (on left). Reproduced from *Geology of the Great Lakes* by permission of Jack L. Hough and University of Illinois Press.

Interbedded with and lying above the sedimentary red beds, is a thick sequence of volcanic rocks. Measuring as much as thirty-five thousand feet in thickness, these rocks consist of a succession of lava flows which are noticeable everywhere except at the southwestern end of the island. The flows, whose volume has been estimated at twenty-four thousand cubic miles, poured repeatedly from great *fissures,* or cracks, in the earth's crust. In places the surfaces of certain flows are still rough and slaggy and many of the rocks have cavities or *vesicles* left by escaping gas. Some of the gas cavities have become filled with minerals which were deposited there long after the gas-charged lava had cooled. Igneous rock of this type is said to be *amygdaloidal,* and the mineral-filled vesicles are called *amygdules.* You can see amygdaloidal basalts at a number of points in the Park including the Minong Ridge and Amygdaloid Island.

In the great copper-producing district of the Keweenaw Peninsula the remarkably rich deposits of metallic copper are largely associated with the amygdaloidal basalts. Copper does, however, also occur as a filling around pebbles in certain of the Keweenawan conglomerates.

Most geologists believe that this copper is a *secondary* mineral deposit; that is, it is younger than the rocks in which it is found. After the lava and conglomerates were formed, the copper is thought to have been deposited by hot, mineral-bearing solutions which invaded these relatively porous rocks. It is interesting to note that Isle Royale's copper deposits were worked by the Indians long before the advent of white men. This is known because the large number of "Indian pits" which occur on the island are the ancient worked-out prospecting pits of this prehistoric race. Archeologists believe that the Indians removed the copper, which occurs in a pure or *native* state, by the simple process of pounding it free from the lava *matrix* (surrounding rock). This was accomplished by pounding the lava with rounded beach cobbles called hammer stones. Shallow pits, the visible remains of these ancient mines, are common at **McCargo Cove** and certain other localities on the island.

The Lake Superior Basin

The Keweenawan sedimentary and volcanic rocks were originally deposited in a horizontal position. But as the great floods of lava poured out upon the surface, their enormous weight caused the crustal rocks of the Lake Superior region to sag downward, thus producing an immense downwarped area called a *syncline* (p. 26). This vast structural trough is called the **Lake Superior Basin;** the south rim of the basin forms the Keweenaw Peninsula and where the north side of the basin emerges from Lake Superior we have Isle Royale. The structural relation of the syncline to Isle Royale can be seen in the geologic cross section on p. 223. You will notice that this is an *asymmetrical syncline,* because the rocks on the Isle Royale (north) side of the basin dip underground at an angle of about 15 degrees, whereas those on the south (Keweenaw Peninsula) side dip in at an angle of 35 degrees or more.

Following the subsidence of the Lake Superior Basin, the syncline was apparently filled by rocks that were younger and less resistant than the older rocks around the rim. Still later there was faulting on the southern side of the basin near Keweenaw Bay and these fractures further modified this portion of the syncline. But the final deepening of the basin was accomplished in fairly recent geologic time by ice and running water. Recent studies of the Great Lakes area indicate that during the Pleistocene Epoch the site of Lake Superior was a large river valley and that glacial ice followed this course. The basin as we now know it was probably excavated by tongues of glacial ice that scoured out the soft sediments in the bottom of the syncline, but had relatively little effect on the more durable Middle Keweenawan rocks that make up the sides of the basin.

Today this great structural trough is occupied by Lake Superior, the largest body of fresh water in the world. Isle Royale —because of the structural deformation of the crustal rocks— rises more than seven hundred feet above Lake Superior's surface.

The Landscape

Although the face of Isle Royale has been modified by events of fairly recent geologic time, its basic surface features are related to the lava flows of Precambrian time. As mentioned earlier, the island is composed of a series of ancient lava flows with interbedded layers of sandstone, siltstone, and conglomerate. These rock layers crop out in elongated, northeast-south-west trending ridges, many of which extend from one end of the island to another, a distance of forty-five miles. Such prominent landmarks as Feldtman and Greenstone Ridges mark the elongate outcrop pattern of the Keweenawan sandstones and lavas respectively. Greenstone Ridge, the highest and longest of these, literally forms the "backbone" of Isle Royale. Extending from Washington Harbor to Blake Point, the ridge rises to more than thirteen hundred feet above sea level—seven hundred feet above Lake Superior. These elevated areas tend to be rather barren of vegetation and their northwest-facing slopes are usually much steeper than those which face southeast.

In between the ridges there is a series of valleys which also trend northeast-southwest. The depressed areas tend to have more vegetation than the ridges and contain lakes, streams, swamps, and—around the margin of the island—embayments of Lake Superior.

A short distance from shore, portions of submerged ridges break the surface of the lake to form lesser islands and islets such as **Long, Raspberry, Passage,** and **Amygdaloid Islands.** Conversely, most of the harbors and channels such as **Rock Harbor** and **Amygdaloid Channel** are located in depressions between certain of the nearshore ridges.

This alternating series of ridges and valleys has given the island a "corrugated" or "washboard" type topography which has been produced by uneven erosion of the uptilted ends of the rocks. The ridge and valley topography has been further accentuated by glacial erosion, for during Pleistocene time, Isle Royale was overridden by great continental glaciers (p. 65).

The ice probably passed over the island several different times, but only the last period of glaciation is known with certainty. This final ice invasion probably occurred about one million years ago and it left behind unmistakable signs of its presence. Some of the island's rocks were ground smooth by the ice; others were grooved and scratched. These markings, which can be seen at a number of places (p. 227), indicate that the glacier moved essentially parallel to the trend of the ridges and valleys. As the ice ground over the island, the depressions between the ridges were further deepened by glacial quarrying as rock material was picked up by the steadily flowing ice. Thus was formed the aligned series of basins now occupied by the scores of lakes which dot the face of Isle Royale.

When the last of the glaciers melted, Isle Royale lay beneath the waters of a huge lake which developed in front of the melting ice sheet. The lake formed partly because glaciation had changed the system of outlets that drained Lake Superior and the other Great Lakes, and partly because the earth's crust, which had been depressed by the great burden of the overlying ice sheet, had not yet adjusted upward. But as time passed, the land surface gradually adjusted upward in response to the removal of its icy burden and the great glacial lake began to drain through a new series of outlets. Thus began the gradual recession of the ancient lake and the eventual emergence of Isle Royale.

Lake Superior finally reached its present level by means of several stages, each of which left behind a record in the form of abandoned beaches, sea cliffs, and sea caves. The remnants of these prehistoric shorelines can be seen at a number of places in the Park. One of the more prominent of these is **Monument Rock,** a seventy-foot monolith located near the northeastern tip of the island. This great pinnacle of rock was carved by the erosive action of waves and winter ice when Lake Superior stood at a higher stage.

It should be noted that the present shoreline of Isle Royale is not necessarily its last, and new ones may be formed at some future time. This is possible because Isle Royale is continuing to emerge from Lake Superior at a rate of approximately fifteen inches per century.

Plants and Animals of Isle Royale National Park

Because of its geographic location and the surface relief of the island, Isle Royale supports an interesting and diversified flora. Most of the island is forested, and along the cool, moist shoreline you will see such trees as white spruce, balsam fir, paper birch, jack pine, and quaking aspen. Some of the more common trees in the interior and higher parts of the island are northern red oak, white pine, sugar and red maple, yellow birch, and bigtooth aspen. Among the many smaller plants that grow on the island are bunchberry, raspberry, blueberry, and a host of wildflowers including orchids, devil's-club, and the Indian pipe or "ghostflower."

The wildlife of Isle Royale is of interest because it consists only of those species that have been able to negotiate the great stretches of water that separate the island from the Canadian shore, fifteen miles away, and the Michigan mainland which is forty-five miles away. This has posed no serious problem for the birds, and more than two hundred species have been observed in the Park. These include eighty species of woodpeckers, bald eagle, osprey, a variety of songbirds, and the herring gull.

Smaller mammals which have apparently been able to swim, float, or drift to Isle Royale include beaver, mink, weasel, muskrat, snowshoe hare, and red squirrel. There are also a few larger mammals on the island and these include red fox, timber wolves, and moose. The moose, which are one of the Park's main animal attractions, are believed to have reached Isle Royale by means of an "ice bridge" when Lake Superior froze across to the Canada shore in the early 1900s. Wolves were apparently unknown on the island prior to about 1949; it is believed that they also reached the island over the ice. Look for moose along stream banks and on the shores of inland lakes and harbors. You might also see them near the lodges and campgrounds, for they are often unafraid of man. But remember that moose are wild animals and should be treated with caution. The wolves are seldom seen by the average visitor.

Tips for Tourists

Unlike most of our National Parks, you will not be able to see Isle Royale from the comfort of an automobile. You may, of course, skirt the island in your own boat or see it from the deck of one of the popular sightseeing vessels, but the visitor who wishes to see the "real" Isle Royale must take to the island's trails. This is easy to do, for more than one hundred and twenty miles of footpaths lead through the forests, over ridge and valley, and along the shores of the island's lakes.

During the visitor season, Ranger-Naturalists may conduct nature walks to several localities within the Park. These include the four-mile hike to **Suzy's Cave,** a half-mile walk to **Rock Harbor Lighthouse,** and the hike on **Mott Island,** which covers two miles. There is also a conducted walk to **Mount Franklin** (four miles) and to **Lookout Louise,** a two-mile hike. Shorter trips are offered daily to introduce visitors to the **Windigo** and **Rock Harbor** areas. Each of these guided trips has its own special flavor and will introduce you to a different facet of the Park. The hike to **Lookout Louise** has interesting geologic attractions, including wave-cut Monument Rock, ancient beaches, and a prehistoric copper mining site. And from Lookout Louise you will get one of the most beautiful views in the Park. Ancient shoreline and wave-erosion features are also seen on the Suzy's Cave nature walk.

The conducted tour to **Mount Franklin** will acquaint you with the "washboard" topography of Isle Royale, for the trail leads over a succession of valleys and ridges as it crosses from one side of the island to the other. Along the way you will see a swamp, evidence of beaver activity, and be treated to an excellent view of the north side of the island. Mount Franklin, at an elevation of 1074 feet above sea level, is the highest point on northeastern Isle Royale. It was named in honor of Benjamin Franklin, who, after the Revolutionary War, convinced the British that Isle Royale should be placed within United States boundaries.

Many of the conducted tours require a boat ride for which there is a fee that must be paid to the concessioner who operates the boats. There is, of course, no charge made for the services of the Ranger-Naturalists who accompany such cruises. You should get your ticket at the Lodge office before trip departure time, which will be posted on bulletin boards throughout the Park.

Because of its lack of roads and wheeled vehicles, travel

within Isle Royale National Park must be accomplished by boat or on foot. Consequently, most Park visitors make good use of the many foot trails which lead to points of interest on the island. Although a few of the Park's trails are briefly described below, prospective hikers on Isle Royale will do well to purchase a copy of *Wilderness Trails: A Guide to the Trails in Isle Royale National Park* by R. M. Linn and R. G. Johnsson. This useful booklet contains location maps and information about points of interest along twenty-four of the Park's more popular footpaths. The publication is available in the Park, or it may be purchased by mail from Isle Royale Natural History Association, Isle Royale National Park, Houghton, Michigan 49931. Starting points on some of the trails can be reached only by boat, but others are located within walking distance of the lodges or major campgrounds.

Although no special photographic problems will be encountered on Isle Royale, most beach and lake scenes will have more light available than you might think; better use a light meter on these. A light meter and a tripod will also be helpful for scenes taken in the forest. By all means do not forget about lake sunsets—these can be used to create unusually dramatic color photographs.

Isle Royale National Park at a Glance

Address: Superintendent, 87 N. Ripley Street, Houghton, Michigan 49931. For additional information see *Wilderness Trails: A Guide to the Trails in Isle Royale National Park,* by R. M. Linn and R. G. Johnsson. This publication is available in the Park, or it may be purchased by mail from Isle Royale Natural History Association, Isle Royale National Park, Houghton, Michigan 49931.

Area: 539,341 acres.

Major Attractions: North woods forest and water wilderness; largest island in Lake Superior; large moose herds; basalt lava formations; ancient shorelines; Indian and historic copper mining.

Season: Late June to Labor Day.

Accommodations: Cabins, campgrounds, and lodge. *For reservations contact:* National Park Concessions, Inc., Isle Royale National Park, Michigan 55617. Winter address: National Park Concessions, Inc., Mammoth Cave, Kentucky 42259.

Activities: Boating, boat rides, camping, fishing, guided tours, hiking, nature walks, and picnicking.

Services: Picnic tables, shelter huts, general store, rest rooms, boat docks, boat rentals, restaurant, gasoline (for boats), and laundry.

Interpretive Programs: Campfire programs, self-guiding tours, museums, nature trails, trailside exhibits.

Natural Features: Erosional features, forests, geologic formations, lakes, rocks and minerals, swamps, unusual birds, unusual plants, wilderness area, wildlife, Indian ruins.

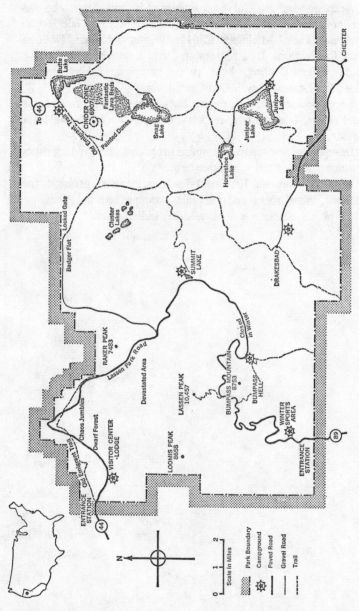

Lassen Volcanic National Park, California.

20. Lassen Volcanic National Park, California

Lassen Peak—Sleeping Giant

Lassen Volcanic National Park, site of the most recent volcanic eruption in the continental United States, is located in northeastern California. The Park, which is an area of considerable geologic interest, bears the name of Peter Lassen, a Danish immigrant who came to the United States in 1830 and settled in California. Like the other pioneers of his day, Peter Lassen thought the peak was an extinct volcano and that its fires had long been cold. But it fires were only banked. On May 30, 1914, it began a period of volcanic activity that was to last for nearly seven years. The volcano awoke from its long slumber with an unexpected series of explosions which belched forth large quantities of dust, rocks, and volcanic gas. These explosions blasted out a new crater near the top of the mountain and subsequent eruptions threw out prodigious quantities of rock, mud, and steam.

Needless to say, this unusual event in northern California caused considerable excitement across the nation and many reporters, photographers, and scientists came to study and record this great phenomenon. Their efforts were rewarded by fiery outbursts which lasted from a few minutes to as much as several hours. Because the area surrounding the mountain was largely uninhabited, there was no loss of life and very little property damage. However, much timber was destroyed by great torrents of thick mud and clouds of volcanic gas which thundered down the mountainside.

During its first year of renewed activity, Lassen Peak experienced some hundred and fifty explosions, none of which were particularly violent. Then, on the night of May 19, 1915 —almost a year after the initial eruption—Lassen's crater became filled with thick glowing lava which poured through two notches in the mountain's summit. Part of this molten rock coursed one thousand feet down the southwest side of the mountain before coming to a stop; molten rock also spilled through the notch on the northeast side of the crater. Heat from the lava melted the snow on this side of the mountain, thus giving rise to great mudflows which plunged into the valleys of Lost Creek and Hat Creek. Geologic evidence indicates that the mud, which was as much as twenty feet deep at the base of Lassen Peak, transported twenty-ton boulders as far as five or six miles.

But Lassen was not through. Still more "fireworks" were in the offing, for just three days later, on May 22, Lassen experienced one of its most violent outbursts. At this time, small mudflows occurred on the north and west flanks of the mountain and a giant mushroom-shaped cloud of volcanic fumes and ash rose more than five miles above the crater. This great cloud was visible from many parts of northern California and showered volcanic ash on areas as distant as Nevada. However, the most spectacular part of the eruption was what has come to be called "The Great Hot Blast"—a tremendous mass of superheated steam-charged volcanic fragments which roared down the mountainside with hurricane speed destroying everything in its path. Technically speaking, such destructive clouds are called *nuées ardentes* (French for fiery clouds), great incandescent masses of gas-charged ash or lava which are discharged with tornadic force. The **Devastated Area** is the result of the 1915 *nuée ardente* and the great mudflows; it is estimated that five million board feet of lumber were destroyed by the fiery products of Lassen's crater. So great was the force from this blast that trees as much as three miles away were uprooted, broken off like matchsticks, and thrown down with their tops all pointing away from Lassen Peak; the bark was stripped from their trunks on the side facing the volcano. Today the Devastated Area is slowly being covered with new tree growth and other vegetation; most of the blown-down trees have decayed and are no longer visible.

Although Lassen experienced numerous small explosions until 1921, the violent eruptions of 1915 appear to have marked the climax of the volcano's activity; after that date, activity gradually decreased. Today, steam rising from small vents in the 1915–16 and 1917 craters and the thermal areas such as **Bumpass Hell** (p. 236) and the **Sulfur Works** (p. 236) are the only evidence of active volcanism in the Park.

Mount Tehama, the Mountain that Broke Off

In previous chapters we have become familiar with such volcanic features as Mount Mazama (p. 236), a composite cone or strato-volcano, and the Hawaiian volcanoes which are shield volcanoes or lava domes (p. 236). Lassen Peak, although of volcanic origin, is different from both of the above; it is a *plug dome* which came into being when a great plug of thick relatively cool lava was squeezed out through a vent on the flank of ancient Mount Tehama. This rather steep-sided mass of protruding lava cooled quickly to form a dome-shaped mass over and around the vent.

Where is Mount Tehama today? Like Mount Mazama in Oregon, old Tehama collapsed leaving behind a great *caldera*. But before it caved in, this lofty mountain stood about one thousand feet higher than Lassen Peak and was more than twelve miles in diameter at its base. And, like Mount Rainier, Mount Shasta, Mount Mazama, and certain other peaks in the Cascade Range, Mount Tehama was a composite volcano built of alternating layers of volcanic ash and lava flows. Today, only remnants of the rim of the Mount Tehama caldera can be seen; the largest of these remnants, **Brokeoff Mountain,** is located about three miles southwest of Lassen Peak. Other fragments of the rim include **Mount Conard, Mount Diller,** and **Pilot Pinnacle** (see map). Field studies indicate that the main vent of Mount Tehama was located over the general area of the present Sulfur Works (see below). This assumption is supported by the fact that the beds of lava and ash comprising the supposed rim of the caldera dip away from about this same point; if these volcanic deposits are projected upward they eventually meet to form an imaginary cone whose summit is approximately above the Sulfur Works.

Other plug domes which formed on the flanks of Mount Tehama are **Chaos Crags, Eagle Peak, Bumpass Mountain,** and **Vulcan's Castle;** these lava protrusions appeared on the north and northeastern slopes of Mount Tehama.

In addition to plug domes, there are numerous other features formed from recent volcanic activity within the Park. These include several examples of relatively small shield volcanoes such as **Prospect Peak, Red Mountain,** and **Mount Harkness.** Although similar in construction to the Hawaiian volcanoes, Red Mountain and Mount Harkness no longer show the typical shield-shaped profile; their original outline has been altered on top by the development of cinder cones which have obscured their formed shape. **Cinder Cone,** a remarkably symmetrical, steep-sided mountain of volcanic ash, cinder, and bombs is another fairly recent volcanic feature. It was last active about 1850–51, at which time there was an eruption of blocky lava in the area now called **Fantastic Lava Beds.**

Among the more dramatic evidences of recent volcanism are the *thermal areas* in Lassen Volcanic National Park. The most accessible of these is the **Sulfur Works** located on Lassen Park Road about one and a half miles from the Southwest Entrance. This area, which is believed to be located in the vent system of old Mount Tehama, is characterized by hot springs, mud pots, and *fumaroles*—holes or vents which emit steam or gaseous vapor. Additional areas of thermal activity can be seen at **Boiling Springs Lake,** the **Devil's Kitchen, Little Hot Springs Valley,** and **Bumpass Hell.**

Bumpass Hell is a barren, sulfur-stained, crater-shaped depression which has been dissolved out of hard igneous rock by acid-bearing steam, mud, and water. Among the numerous geologic attractions here are fumaroles (with such appropriate names as **Big Boiler** and the **Steam Engine**) which issue roaring clouds of superheated steam; *mud pots,* shallow pits or basins filled with boiling mud containing very little water and an

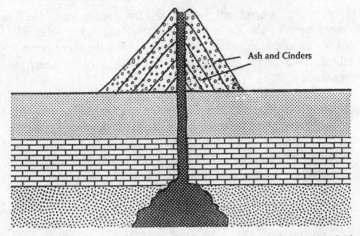

Cinder Cone, in Lassen Volcanic National Park, is a symmetrical, cone-shaped mound of volcanic ash and cinders. It is indicative of an explosive-type volcanic eruption.

abundance of fine-grained mineral matter; *mud volcanoes,* cone-shaped mounds built of mud thrown out of an especially active mud pot; and *solfataras*—fumaroles or volcanic vents which emit sulfurous fumes.

Perhaps you are wondering why certain waters of the Park boil or are converted to steam, while others are quite cold. The thermal areas (Sulfur Works, Bumpass Hell, Cold Boiling Lake, etc.) are located in the southwest corner of the Park—within or nearby the caldera of ancestral Mount Tehama. These areas appear to be situated on a *fault zone* (zone of fracture) in the earth's crust; heat and sulfurous fumes from relatively shallow bodies of molten rock come to the surface along these cracks and groundwater enters the fractures and is carried to depths great enough for it to be heated by the magma.

Plants and Animals of Lassen Volcanic National Park

Lassen Volcanic contains a profuse and varied flora; the most striking plants are those of the lower elevations. Ponderosa pine, white fir, Jeffrey pine, lodgepole pine, sugar pine, western white pine, alder, aspen, and willow are the predominant trees in lower parts of the Park; mountain hemlock, red fir, and white bark pine occur at higher elevations. Among the more prevalent shrubs are Sierra chinquapin, snowbrush, oceanothus, and manzanita.

The mountain meadows of this Park are famous for their beautiful wildflowers, many of which are in bloom from June to late September. Crimson snow plant, "Indian paintbrush" or painted cup, leopard lily, pentstemon, monkey flower, and bleeding heart occur in the grassy meadows; mountain heath, pentstemon, kalmia, and lupine are found in the higher reaches of the Park.

Larger mammals that inhabit the Park include black-tailed and mule deer and the black bear; the latter, however, are not commonly seen. Among the many smaller mammals there are marmot, marten, chickaree, and fox; ground squirrels and chipmunks are also abundant and relatively tame.

Tips for Tourists

Lassen Volcanic is often described as the "outdoorsman's Park" because its 160 square miles are well suited to the activities of fishermen, campers, and back-country hikers. But there is also much to attract the less rugged visitor for most of the Park's features are easily accessible.

The **Loomis Museum** at the **Manzanita Lake Visitor Center** has displays, exhibits, and dioramas depicting all facets of the Park; the geology, biology, archeology, and history of the Lassen Peak area are treated thoroughly. There are pictures of Lassen's eruptions and an interesting display of a seismograph station in operation. In addition, a Naturalist is on hand in summer to answer questions. Maps and special publications dealing with the Park can also be purchased here. The visitor who wants additional information about the geologic history of the Park should obtain a copy of *Geology of Lassen's Landscape* by Paul E. Schulz and published by the Loomis Museum Association, Lassen Volcanic National Park, Mineral, California 96063.

The **Bumpass Hell Trail** is an unusually popular trip to the most spectacular thermal area in the Park and is an easy two-hour walk. Slightly more than a mile long, this beautiful trail winds along the south side of **Bumpass Mountain** past many interesting natural features. For example, along the trail numbered stakes invite your attention to highly polished rocks (*glacial polish*) and a huge *glacial erratic*—a large isolated boulder left perched high on a ridge when an ancient glacier melted. These and other glacial features such as **Lake Helen,** which fills a glacial basin, are evidence of the huge valley glaciers which scoured the slopes of these mountains.

But the main feature of this trail is **Bumpass Hell,** a large thermal area reminiscent of the hot spring and mud pot areas of Yellowstone. In all probability you will smell Bumpass Hell before you actually see it—its sulfurous fumes permeate the atmosphere in the general vicinity of this unusual locality. Discovered by Kendall V. Bumpass in 1864, this area of boiling springs, roaring vents of superheated steam, and bubbling caldrons of mud is easily the most spectacular thermal area of the Park. The foregoing are but a few of the highlights of this unusual trail; you should make this tour if at all possible. However, the visitor is urged to proceed with utmost caution in the thermal areas, *stay on the trails,* and avoid slippery or crusty places; the steam and waters in these areas are *dangerously hot*.

Lassen Volcanic National Park at a Glance

Address: Superintendent, Lassen Volcanic National Park, Mineral, California 96063.

Area: 106,933 acres.

Major Attractions: Lassen Peak, one of the world's largest plug dome volcanoes and most recently active volcano in the continental United States; cinder cones, lava flows, thermal areas, and lakes.

Season: Lassen Park roads normally open from about June 1 to October 31 (opening dependent upon local weather conditions); Southwest and Manzanita Lake Entrances remain open to winter sports area throughout the winter.

Accommodations: Cabins, campgrounds, trailer sites (but no hookups); house trailers are welcome, but must check at ranger station about road conditions. *For reservations contact:* Lassen National Park Company, Manzanita Lake, California 96060.

Activities: Boating, camping, fishing (California license required), guided tours, hiking, horseback riding, mountain climbing, nature walks, picnicking, scenic drives, swimming, winter sports, pack trips, and skiing (in season).

Services: Boat rentals, food service, gift shop, guide service, laundry, post office, public showers, religious services, service station, ski rental, ski tow, ski trails, telephone, transportation, general store, and picnic tables.

Interpretive Program: Campfire programs, museum, nature trails, guided nature walks and hikes, roadside exhibits, and self-guiding trails.

Natural Features: Canyons, erosional features, forests, geologic formations, hot springs, lakes, mountains, mud pots, rivers, rocks and minerals (no collecting), swamps, volcanic features, waterfalls, wildlife and fumaroles.

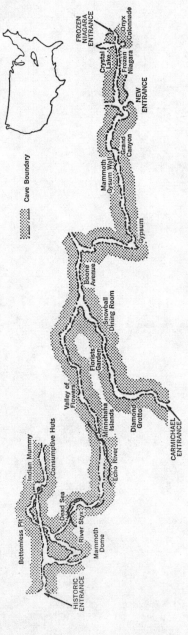

Mammoth Cave National Park, Kentucky.

Some areas, such as the Paradise Ski Area, in Mount Rainier National Park, are the sites of popular winter sports activities. WASHINGTON STATE DEPARTMENT OF COMMERCE AND ECONOMIC DEVELOPMENT

The sheer limestone walls of Santa Elena Canyon—a leading attraction in Big Bend National Park—tower more than fifteen hundred feet above the Rio Grande River (left foreground), which slices through the face of a great fault scarp . Mexico is on the left, and the United States is on the right. NATIONAL PARK CONCESSIONS

During late August or early September the ice caves that open up in Paradise Glacier are the goal of many hikers in Mount Rainier National Park. JENNIE A. MATTHEWS

This anticline is typical of the rock folds that were produced by the mountain-building movements that created the Great Smoky Mountains. U. S. GEOLOGICAL SURVEY PHOTO BY H. E. MALDE

The erosive effects of running water are dramatically displayed in this view down Zion Canyon from Observation Point. NATIONAL PARK SERVICE PHOTO BY GEORGE A. GRANT

These youngsters are examining a portion of one of the world's largest, and possibly oldest, living objects. This section of a giant sequoia tree was removed from Kings Canyon National Park before the trees in this area were under federal protection. AMERICAN MUSEUM OF NATURAL HISTORY

Angel Arch, in Salt Creek Canyon, is one of Canyonlands National Park's more unusual attractions. The naturally carved opening in this graceful arch is estimated to be 150 feet high and 130 feet wide.
WESTERN GATEWAYS MAGAZINE

21. Mammoth Cave National Park, Kentucky

America's Most Historic Cave

Mammoth Cave has long been one of the most famous and popular caves in the United States. It was reputedly discovered in 1799 when a pioneer hunter, Robert Houchins, chased a wounded bear into the cavern entrance. However, other accounts state that the cave had been known earlier because it had been mentioned in certain land transaction records as early as 1798. Be that as it may, the cave has had a long and colorful history and has been used by man for a variety of purposes.

It is known, for example, that the early Indians who lived and hunted in this part of Kentucky used the cave rather extensively. The corridors and chambers of Mammoth Cave have yielded much evidence of these people; pottery, partially burned torches, gourds, sandals, wooden bowls, woven cord, and a host of other artifacts have been collected there. In addition to these objects, the actual remains of some of these Indians have been found by cave guides. The most famous of these is the mummified body of a man believed to have been trapped in the depths of the cave more than two thousand years ago. **Lost John,** as the mummy is now called, is the remains of a man about forty-five years old and five feet three inches tall. He was apparently crushed to death under a six-ton boulder while collecting gypsum from the walls of the cave. Thanks to the even humidity in the part of the cave where he died, his body dried out very slowly and resisted bacterial decay. Today you can see the leathery remains of Lost John as they rest in a glass case near the site of his death.

Much later the cave became famous—and valuable—because of its large deposits of nitrate-containing saltpeter or "peter-dirt." During the War of 1812 approximately 200,000 pounds of nitrates were removed from the cave and used in the manufacture of gunpowder. Opinions differ as to the origin of the nitrate; some believe it was derived from bat *guano* (the droppings of bats which hibernated in the cave). Others assume that the material is of bacterial origin. On the **Historic Trip** you will be able to inspect the remnants of this rather crude but effective mining operation which can still be seen in **The Rotunda.**

Not far from the remains of Lost John, you will see evidence of another fascinating chapter in the history of Mammoth Cave. It was here in 1843 that Dr. John Croghan of Louisville, Kentucky, conducted a rather remarkable but unsuccessful medical experiment. It was Dr. Croghan's belief that the temperature (a constant 54 degrees) and the humidity (87 per cent) of the cave air would be beneficial to those suffering from tuberculosis, or "consumption" as it was then commonly called. Consequently he constructed twelve small cottages in one of the cave's larger corridors and fifteen tubercular patients were moved into these accommodations and remained there for several months before returning to the surface. Unfortunately, one patient died within five months and another died some time later; not long thereafter this ill-fated underground sanitarium was evacuated and the patients returned to the surface. Early records indicate that all of the patients were in much worse condition than before they began this unusual "cure." Ten of the **"consumptive huts"** were constructed of wood and have since been demolished; the two stone cottages still stand and you can see them on the **Historic Trip.**

Origin of the Cave

Mammoth Cave, like the previously described Carlsbad Caverns, has been dissolved out of limestone by the work of subsurface waters. And, as in the case of the famous New Mexico caverns, these limestones were formed on an ancient sea bottom in the distant geologic past. Yet there are distinct differences in the development of the two caves. For example, the rocks from which Mammoth Cave has been dissolved are considerably older than the Carlsbad limestone of Permian age. Moreover, these two great limestone deposits were formed under somewhat different geologic conditions and some of the features in Mammoth Cave originated quite differently from those in Carlsbad Caverns.

In order to understand better the differences between these two great natural wonders, we must briefly reconstruct the geologic history of the Mammoth Cave area. This history dates back about 340 million years to the Mississippian Period of the Paleozoic Era.

During this period of earth history, Kentucky and the states surrounding it were covered by widespread shallow seas. Streams which drained the nearby lands emptied into the Mississippian seas and there deposited their sedimentary load of mud, sand, and gravel. From these sediments came the shales, sandstones, and conglomerates that are seen in the area today. During this same period, limestones were also being deposited in the Kentucky seas. These were formed in part from minerals precipitated from the sea water and partially from the remains of marine plants and animals. With the passage of time, Mississippian deposition continued until approximately twelve hundred feet of limestone, shale and sandstone had been laid down.

Then, early in Pennsylvanian time, the Mammoth Cave area was gradually uplifted and the great inland seas slowly receded. This exposed the Mississippian rocks to the processes of weathering and erosion. Two of these processes, *solution* and *abrasion*, have been important factors in the development of Mammoth Cave.

The Mississippian limestones in which the cave has formed are quite susceptible to solution by groundwater, a process which was discussed earlier in conjunction with the formation of Carlsbad Caverns (p. 104). However, the solvent effect of groundwater in this area is seen not only in the large number of caves and solution cavities which occur in the region but may also be recognized by certain surface features. For example, many of the limestones in the vicinity of the cave contain large *sinks* or *sinkholes*. These range from a few feet to several hundred feet in diameter and were formed as groundwater seeped down into the soluble limestone and gradually dissolved out large solution pockets. There are many such sinkholes in this region and you can see them at several places in and near the Park. For closer observation, there are pullouts, along the Park roads where sinkholes may be seen and where interpretive information is available to the visitor.

Areas which are characterized by sinkholes, disappearing streams, caverns, and other phenomena associated with the work of groundwater are said to display *karst topography. Topography,* a term applied to the collective relief and contour of the land surface, of this type is well developed in the Karst district (a limestone plateau east of the Adriatic Sea) of Yugoslavia. In the United States, regions which display karst topography include portions of Tennessee, Kentucky, and Florida.

As mentioned earlier, continuous solution may cause the walls and ceilings of a cave to collapse, thereby further enlarging the underground passageways. Thus, **The Rotunda,** a large chamber 139 feet wide and with a 40-foot domed ceiling, was formed as great quantities of thinly layered limestone fell from its walls and ceilings over a long period of time.

The chambers and corridors of Mammoth Cave were developed not only by the solvent action of water but also by the abrasive action of sediments carried in running water. This occurred when the water table in this area became low enough to allow the groundwater to flow from the cave, thereby further opening the system and permitting surface streams to enter the cave by means of sinkholes. As they flowed from the surface to the lower levels of the cave, these debris-laden streams scoured and enlarged the solution channels.

The development of this underground drainage system had a profound effect upon the surface drainage of the Mammoth Cave area because large numbers of sinkholes and solution cracks drain the water from the surface very quickly. This produces

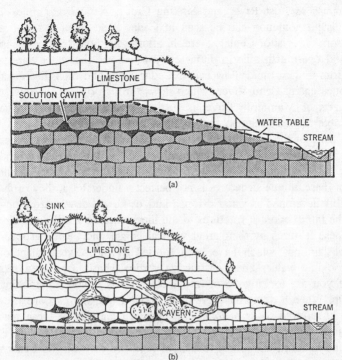

Diagram showing mechanics of cavern formation. (a) Stage 1. Beginning of cavern formation by solution of limestone below the water table to form solution cavities. (b) Stage 2. Solution cavities are drained by lowering of water table either by downcutting of the stream that serves as base level or by change in amount of rainfall. Erosive action of water which enters cavities through sinkholes at the surface enlarges and modifies them to form caverns. Reproduced from *The Geologic Story of Longhorn Cavern*, by William H. Matthews III, by permission of the Bureau of Economic Geology of the University of Texas.

something of a "storm sewer" effect, thus retarding the development of surface streams in the area. Interestingly enough, some of the surface streams that are present may suddenly disappear in a sinkhole only to reappear in another some distance away. This has given rise to such appropriate stream names as Lost River and Sinking Creek. A good example of this phenomenon can be seen at **Cedar Sink** about four miles from the Visitor Center. Here, in a large sinkhole 150 feet deep and one-fourth mile in diameter, you can see the water appear from underground, flow across Cedar Sink, and then disappear once again. **Echo River,** which flows through one of the lowest parts of Mammoth Cave, is further evidence of the effective subterranean stream system in the cavern area.

One of the more striking features formed by the joint forces of solution and mechanical abrasion are the unusual *domes* and *pits* so common in Mammoth Cave. Although the development of these unique structures is not perfectly understood, they probably developed as water dripped and/or flowed downward along the intersection of fractures in the limestone. You will recognize these features by their great vertical height and by the grooves or "fluting" which have been formed by the solvent effect of trickling water. These distinctive vertical shafts are called *pits* if you are looking down into them; *domes* if you are looking up. When a dome has developed above a pit, the two are usually referred to as a *dome-pit.* Typical examples of these features are 95-foot-deep **Silo Pit** and **Roosevelt Dome,** which is one hundred and thirty feet high; both are seen on the **Frozen Niagara Trip** (p. 250). Features such as **Roosevelt, Wilson,** and **Mammoth Domes** are typical "solutional-type" domes and should not be confused with such domelike structures as **The Rotunda,** which was formed largely by ceiling collapse (see p. 246).

Thus, over vast periods of time, the combined forces of abrasion, solution, and cavern collapse have worked together to fashion the more than one hundred and fifty miles of passageways in this great cave.

Deposition Within the Cave

The underground galleries in Mammoth Cave are decorated by accumulations of gypsum and travertine which were deposited by downward percolating mineral-laden groundwater.

Deposition did not begin, however, until the water table had been sufficiently lowered to drain most of the underground chambers.

In this cave, as in Carlsbad Caverns, the more common formations are composed of travertine deposited as flowstone or dripstone (see p. 105). Stalactites, stalagmites, helictites, columns, lily pads, and draperies can be seen in the more scenic chambers and passages. There are, in addition, some rather unusual gypsum deposits which occur in the drier parts of the cave. These features, commonly referred to as "snowballs," "blisters," or "flowers," formed as gypsum-bearing ground-water evaporated in the porous limestones which form the walls and ceilings of the cave. At first the gypsum accumulated as a thin film, but as deposition continued, the gypsum crystals grew larger, forcing the previously formed crystals outward and/or upward. This displacement produced the curved or twisted forms which superficially resemble "flowers," "snowballs," and "needles." You will see good examples of gypsum deposits on the ceilings in **Snowball Room** and similar deposits in **Little Paradise Avenue,** the **Florist's Garden,** and **Specimen** and **Cleveland Avenues.**

Touring the Cave

The visitor to Mammoth Cave has a choice of several cave trips. These range from a rather leisurely ¾-mile trip of one and a half hours to **Frozen Niagara,** to the more strenuous seven-hour **All-Day Trip** which covers the entire length of the cave (seven miles). Check the map of the cave and the descriptions of tours (pp. 250–254) to see which of these trips are in keeping with your physical ability, interests, or time schedule; trip departure times, tickets, and more detailed tour information can be obtained at the information desk in the Visitor Center. You should also attend the short orientation program of colored slides in the Visitor Center auditorium; this will help you determine which of the cave trips are best suited to your needs.

After deciding which trip to take, visit the museum exhibits and read the descriptive material available about the trip you have chosen.

You will probably want to wear a sweater or jacket as the temperature in the cave is a constant 54 degrees and certain parts of the cave are quite damp. Although the trails are good (but occasionally slippery), low-heeled shoes are the safest thing to wear, and rubber soles will add to your safety and comfort. The tour leader assigned to your group is an experienced uniformed member of the National Park Service; your underground trip will be greatly enhanced if you follow his instructions and listen to what he has to say. The tour leader will, of course, be glad to answer any queries about the geology or history of the cave.

Frozen Niagara Trip

A rather short trip, the visit to **Frozen Niagara** requires one and a half hours to complete and covers a distance of about three quarters of a mile; it is electrically lighted all the way. Because one sees a variety of features in a relatively short time, this has long been one of the most popular tours of the cave. During the summer, there are trips to Frozen Niagara about nine times per day to see such features as domes, pits, flowstone, dripstone, travertine dams, and an underground lake.

From the **Visitor Center** you will travel by bus over the scenic hilly road to the **New Entrance,** a sinkhole which marks the starting point of this trip. Descending two hundred and fifty feet by stairway, you will get your first view of Mammoth Cave's famous domes and pits; the 130-foot-high **Roosevelt Dome** and **Silo Pit** (95 feet deep) are especially noteworthy. Before long you enter **Grand Central Station,** a larger chamber formed at the intersection of four corridors. Nearby lies a great pile of rock debris which has fallen from the ceiling of the cave; the guides call this **The Big Break.** Shortly thereafter you will be shown the **Fairy Ceiling or Smooth Ceiling,** an expansive unsupported ceiling measuring two hundred and fifty by eighty-five feet.

Next comes the main feature of the trip—**Frozen Niagara,** a massive flowstone deposit measuring seventy-five feet high and forty-five feet in width. Descending the stairway in front of this "frozen waterfall" you can see green **Crystal Lake** sixty feet below, while lofty **Moonlight Dome** looms high above. As you approach **Onyx Colonnade** notice the varied array of cave formations—stalactites, stalagmites, helictites, and columns, many of which have been named after such imaginary figures as cats, cakes, and bathing beauties. Adjacent **Onyx Chamber** exhibits a variety of "live" formations as evidenced by their wet glistening surfaces. The color of certain of these formations is the result of the oxidation of minerals such as iron and manganese (see p. 107). The many soda straw stalactites in **The Macaroni Factory,** colorful **Rainbow Dome,** and **The Theater Curtain** (a massive travertine column bearing drapelike folds) also contribute to the beauty of **Onyx Chamber.**

The last, and one of the more unusual formations, seen on this trip is **The Great Wall of China,** a travertine deposit composed of *rimstone.* Unlike surface waters which often destroy their dams, underground pools may attempt to dam themselves. When seepage is rapid, water collects into small pools in depressions on the cave floor. As this water evaporates, calcium carbonate is deposited around the rim of the depression; the longer the pool exists, the higher will be the rim or dam formed around its margin. When the water evaporates the dams or terraces are left standing on the cavern floor.

Historic Trip

If you are a history buff this self-guiding trip is for you; it is also short (one and a half miles) and lasts only one and a half hours. Electrically lighted throughout, this trip begins at the **Historic Entrance** and includes such attractions as the saltpeter mining site, Indian mummy, and the stone huts used by Dr. Croghan's tubercular patients. There are interpretive signs to explain the natural features, and Park Guides are also stationed along the trail.

Upon entering the **Historic Entrance** you take a trail that has been followed by cave visitors for more than a century. It is but a short walk from the base of the entrance stairs to **The Rotunda** and the site of the War of 1812 nitrate mining operation. Following the long winding corridor called **Broadway Avenue,** the group soon enters **Booth's Amphitheater.** Here, in 1876, famed Shakespearean actor Edwin C. Booth, mounted a high "stage" of water-worn limestone, and eloquently delivered Hamlet's Soliloquy. **Martha Washington's Statue,** in reality the back-lighted silhouette of a small opening in the cave wall, is also seen in this section of the cave.

Near the **Giant's Coffin,** an immense collapse block fifty feet long, twenty feet wide, sixteen feet tall, and weighing about two thousand tons, you can see the remnant of the ill-fated underground sanitarium. On this trip you will also see gypsum deposits on the ceiling of **Star Chamber** and the mummy of **Lost John.**

Scenic Trip

Although somewhat more demanding than the previously described trips, this 4½-mile walk requires no more than normal physical ability. During the four and a half hours required for the tour there are short rest stops and a longer stop for lunch. Gypsum formations, large rockfalls, lunch in the **Snowball Room,** plus the features seen on the Frozen Niagara Trip, are the major attractions of this tour.

Shortly after leaving **Carmichael Entrance,** one of several man-made entrances into the cave, the tour guide points out a large rockfall known as the **Rocky Mountains.** Next comes one of the drier sections of the cave, where you notice a profusion of gypsum deposits; they adorn the ceiling of **Diamond Grotto,** and are especially numerous and varied in the **Florist's Garden** and along **Cleveland** and **Specimen Avenues.**

After a pleasant lunch in the unusual **Snowball Room** with its remarkable ceiling of gypsum "blisters," you continue the trip via a winding, narrow, high-ceilinged corridor called **Boone's Avenue.** Soon you are walking down **Kentucky Avenue** with its unusual gypsum needles, large rockfalls, and the **Grand Canyon,** a deep gorge that formed after the floor of one passage dropped through the ceiling of another. A short distance later you enter **Grand Central Station** and from this point on the tour is identical to the Frozen Niagara Trip.

"Wild Cave" Tour

This unique tour is designed to provide the visitor a new experience, an adventure few Park visitors realize. The tour lasts three hours, including a brief orientation session. Safety hats, headlamps, and kneepads will be provided. Approximately two and a half hours are spent underground including a walk through large passages with unimproved trails, and a few hundred feet of crawlways.

Persons making the trip must be sixteen or older and in good physical condition. Anyone suffering from claustrophobia, heart ailments, asthma, abnormal blood pressure, or impairment of legs or back should not attempt the tour. Appropriate dress might include sturdy shoes, old clothes, and gloves. A clean outfit should be available, since the clothes worn in the cave may be soiled or torn. The "Wild Cave" Tour is developed for the adventurous Park visitor and is an experience that will long be remembered.

Lantern Tour

Another unusual feature of Mammoth Cave is the Lantern Tour. This unique, three-mile excursion is made by lantern light and is scheduled several times each day. Length of the tour is about three hours, and numerous historic and geologic features will be pointed out by your guide. Lanterns are provided, but rest rooms are *not* available on this trip. Check in the Visitor Center for the latest schedule for this as well as other guided trips through the cave.

Plants and Animals of Mammoth Cave National Park

Life abounds both on the surface of the Park and within the confines of the cave. The most unusual cave-dwellers in the Park are the famous Mammoth Cave blindfish. First discovered in 1838, these peculiar creatures are characterized by eyes which are degenerate or totally absent and the lack of body pigmentation. These sightless creatures have lost their sight as an adaptation to their subterranean existence. Also living in the cave are blind or eyeless crayfish, cave crickets, beetles, snails, worms, spiders, and bats.

Animals living above ground include skunks, foxes, white-tailed deer, rabbits, squirrels, chipmunks, and a wide variety of birds. Among the latter are woodpeckers, bluebirds, robins, thrushes, and hawks, to name but a few of the more than one hundred and seventy species that have been described from this area.

Most of the Park is covered by a dense forest consisting of oak, hickory, maple, and ash, with a scattering of cedars and pine. There are, in addition, redbud, dogwood, mountain laurel, elm, sycamore, and sumac. Among the wildflowers you will find violets, hepatica, yellow adder's-tongue, irises, black-eyed Susans, goldenrod, and cornflowers.

Tips for Tourists

In addition to the various trips through the cave, the visitor to Mammoth Cave National Park has a varied choice of other activities. There is a museum where you can become oriented and learn much about the history, archeology, geology, and natural history of the Park. Take time to enjoy the colored slide show in the auditorium and to browse through the interesting and educational museum displays. Your visit will be richer for it.

From May to October "Miss Green River," a fifty-foot diesel-powered sightseeing cruiser, makes an eight-mile round trip through the more scenic areas along **Green River.** On this hour-long trip you may see deer, beaver, wild turkey, snakes, turtles, and other forms of wildlife. In addition to the several daytime trips, there are also Twilight and Moonlight Cruises. You can obtain details of these trips at the information desk in the Visitor Center.

There are no special photographic tours through Mammoth

Cave and there is little time available for picture-taking on most tours. However, if you are quick with your flash equipment you can still get good pictures—but please, take your used flashbulbs out with you or place them in a waste receptacle rather than drop them along the trail. There is a Photo Shop at Mammoth Cave Hotel and their experienced personnel will be glad to discuss problems of cave photography with you.

Mammoth Cave National Park at a Glance

Address: Superintendent, Box 68, Mammoth Cave National Park, Mammoth Cave, Kentucky 42259.

Area: 51,354 acres.

Major Attractions: A large cavern featuring one hundred and fifty miles of explored passageways, beautifuly decorated chambers, and unusual cave formations of travertine and gypsum. The cave is also famous for its underground boat trip, blindfish, solution pits, high domes, and historical significance.

Accommodations: Cabins, campgrounds, group campsites, hotels, motels, tents, trailer sites. Also available in Cave City and Park City. *For reservations contact:* National Park Concessions Co., Mammoth Cave, Kentucky 42259.

Activities: Boating, boat rides, camping, fishing, guided tours, hiking, nature walks, picnicking, scenic drives, and scenic river cruise.

Services: Boating facilities, food service, gift shop, guide service, kennel, laundry, nursery, post office, public showers, religious services, service station, picnic areas, underground lunch room, and general store.

Interpretive Program: Campfire programs, museum, nature walks, roadside exhibits, self-guiding trails, and underground tours.

Natural Features: Caverns, erosional features, forests, fossils, geologic formations, rivers, rocks and minerals, wildlife, and Indian ruins.

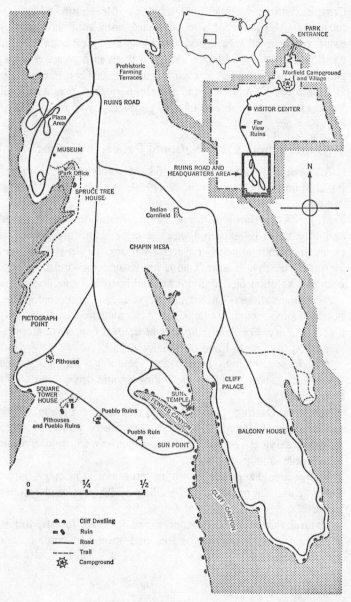

PARK
ENTRANCE

Morfield Campground
and Village

VISITOR CENTER

Far
View
Ruins

Prehistoric
Farming
Terraces

RUINS ROAD

Plaza
Area

MUSEUM

Park Office

SPRUCE TREE
HOUSE

RUINS ROAD AND
HEADQUARTERS AREA

N

Indian
Cornfield

CHAPIN MESA

PICTOGRAPH
POINT

Pithouse

CLIFF
PALACE

SQUARE
TOWER
HOUSE

SUN
TEMPLE

Pueblo Ruins

FEWKES CANYON

Pithouses
and Pueblo Ruins

Pueblo Ruin

BALCONY HOUSE

SUN POINT

CLIFF CANYON

0   ¼   ½

Cliff Dwelling
Ruin
Road
Trail
Campground

Mesa Verde National Park, Colorado.

22. Mesa Verde National Park, Colorado

The Green Table

Looming almost two thousand feet above the southwestern Colorado countryside is canyon-slashed, flat-topped Mesa Verde. This massive erosional remnant is the site of Mesa Verde National Park, the only National Park established for the express purpose of protecting archeological objects.

Bounded on the east by the valley of the Mancos River and on the west by the Montezuma Valley, the Mesa Verde is thought to have been named by Spanish traders who entered this part of Colorado in the middle 1700s. Because of its green, forested top they were prompted to call this expansive tableland *La Mesa Verde,* the Green Table. But although the Spaniards gave the Mesa Verde its name, they were not the first to explore it. This flat-topped mountain was occupied by a race of prehistoric Indians who preceded the Spanish traders by more than seventeen hundred years. The story of this now-vanished race is one of the most intriguing chapters in southwestern archeology, and a study of their ancient culture has raised almost as many questions as it has answered. Although the theme of this book is not archeological, Mesa Verde's early much in order.

Why the Mesa Is There

Despite the great antiquity of its human history, the geologic history of Mesa Verde greatly antedates it. In general, the origin of the mesa is interwoven with the history of the Rocky Mountains, which were elevated near the end of Mesozoic time. The exposed sedimentary rocks of the Mesa Verde and the surrounding area consist of sandstone, shale, and clay strata of Late Cretaceous age. These rocks, which are approximately thirty-six hundred feet thick, were deposited in marine, brackish, or fresh water. The oldest exposed unit in the vicinity of the Park is the *Mancos Shale,* a soft, dark-gray to black marine shale that also contains thin lenses of sandy, yellowish orange limestone. Now found more than a mile above sea level, these rocks were formed from sediments that accumulated as mud on the bottom of a Late Cretaceous sea that extended from Arctic Canada to the Gulf of Mexico. You will see good exposures of the Mancos Shale as you ascend the mesa after leaving the Park Entrance.

During the time that the Mancos sediments were accumulating, the sea bottom slowly subsided, thus permitting the water to maintain an essentially uniform depth. But eventually the sea floor ceased to sink and the Cretaceous sea gradually became more shallow. This was accompanied by an influx of land sediments which washed in from the surrounding shores and gradually filled the ocean basin until the sea was driven from the area. These coarser sediments formed deltaic and beach deposits which today comprise the *Mesaverde Formation.*

The first group of rocks to be deposited in the rapidly filling Mesaverde sea, was the *Point Lookout Sandstone,* a massive cliff-making, marine sandstone. This resistant, coarse-grained rock forms the prominent cliff face of the Mesa Verde and has protected the softer underlying Mancos Shale from erosion. Not all of the mesa's geologic history was written in the sea, for in places there were low, swampy areas which supported dense forests that flourished in the temperate Cretaceous climate. The sediments deposited in these areas gave rise to the strata of the *Menefee Coal,* a thick section of sedimentary rocks consisting of coal, nonporous, black, carbon-bearing shales, and a lesser amount of sandstone. The coal, which is mined commercially in parts of Colorado, contains well-preserved plant fossils at some localities. Look for exposures of Menefee coal and shale in the roadsides as you drive west along the northern escarpment of the Mesa Verde.

The upper, hence youngest, rock unit in the Mesaverde Formation is the *Cliff House Sandstone.* As one might suspect from its name, this is the geologic unit that is most closely associated with the cliff dwellings for which this Park is famous. This pale to dark yellowish orange, cliff-forming sandstone was deposited along the shores of the last sea of Mesaverde time, and beds of this massive sandstone form the rims of the many canyons that have been eroded into the flanks of the Mesa Verde. Here, too—in places where the sandstone contained shale particles that rendered the rock less resistant— prehistoric Indians found ideal niches in which to build their well-constructed cliff houses.

Following Mesaverde time and until the end of the Cretaceous Period, this area underwent alternate and repeated advances and withdrawals of the sea. Finally, near the end of the Cretaceous and during early Tertiary time, the Mesa Verde region was gradually uplifted and the sea was drained away. This final regression of the sea was caused by the great crustal disturbance called the *Laramide Orogeny,* a widespread mountain-building movement that gave rise to the Rocky Mountains and many of the other structural features in the western United States (p. 362).

In some places on the Mesa Verde, there are significant deposits of gravel and well-worn pebbles which provide additional information about the geologic history of the region. For one thing, these pebbles and gravels are unusual in that they are much younger than the Cretaceous rocks on which they are found. What is more, their composition is quite different from the bedrock of the Mesa Verde: they are more like the rocks of the San Juan and La Plata Mountains to the north and east of the mesa. But today the Mesa Verde is isolated from these mountains, surrounded on all sides by deep, broad valleys and canyons. How, then, did these rocks get on top of the mesa?

Although there is no way of being positive, detailed geologic studies of this area have produced evidence that the rocks of the Mesa Verde once were continuous with certain formations exposed in the La Plata and San Juan Mountains. During the Tertiary, perhaps near the end of the period, rock materials derived from the then newly raised mountains were transported southwestward by streams flowing from the uplifted areas. These turbulent mountain streams carried pebbles, cobbles, and boulders which were deposited along the mountainsides and on the valley floors. As time passed and additional sediments accumulated, a geologic feature called a *pediment* was gradually produced. Pediments, which commonly occur between mountain fronts and basins or valley bottoms, are gently inclined, essentially flat erosion surfaces which have been carved from bedrock and are generally covered with stream gravels. These gravels—like those on the Mesa Verde—have been transported to the basins from the surrounding mountains.

Following development of the pediment and at a later date in geologic time, the streams in this part of the Mesa Verde region were rejuvenated and erosion was greatly accelerated until most of the pediment was destroyed by stream erosion. But in some areas the Cretaceous bedrock successfully resisted the forces of erosion. In these places, segments of the ancient pediment have become isolated from the mountainous area and now stand high above the surrounding valleys, and this is the manner in which erosional remnants such as the Mesa Verde were produced. This also explains why rock fragments from the distant San Juan and La Plata Mountains can be found atop the Mesa Verde.

Man Comes to the Mesa Verde

Visitors to Mesa Verde National Park commonly ask: "Why did the Indians settle here, rather than in the valleys?" The Indians cannot, of course, answer this question, for they deserted the area more than seven hundred years ago. But these ancient people did leave evidence that suggests why they chose to inhabit the Mesa Verde and this evidence can be interpreted in the light of what we know of the geology of the area.

Archeological studies indicate that the earliest known inhabitants of the Mesa Verde were a farming people and that they soon recognized the agricultural potential of the fertile soils that had developed on this flat tableland. It is also believed that the presence of the rock shelters and cliff overhangs would have made good places for these early Indians to store their food. This was a most important factor, for without protection for the preservation of produce, the fertile soil and the availability of water would have had no meaning. Then too, the great elevation of the mesa would have provided excellent lookouts from which to spot the approach of enemies.

But probably the mesa's most valuable asset was water. Without water the soil-topped mesa would probably have proved to be uninhabitable. Here, again, the geology of the Mesa Verde is of prime importance, for the rocks of this great erosional remnant serve as an immense natural reservoir. Water derived from rain and snow soaks into the porous Cliff House Sandstone and passes downward until it encounters a thick, nonporous layer of shale within the underlying Menefee Coal beds (p. 259). Unable to penetrate the dense impervious shale, the water then flows along the contact between the shale and sandstone strata and later emerges as springs and seeps where the contact zone is exposed in certain of the canyon heads and cliff walls. Thus, as long as precipitation remained normal, these geologic conditions provided the Indians with a convenient and ample supply of water.

Another geologic factor that favored the habitation of the mesa was the presence of the Cliff House Sandstone. Exposed in many of the mesa's canyons, this massive sandstone formation contains numerous natural rock shelters and overhangs, and the sandstone cliffs provided an ideal setting in which to build the cliff dwellings that were ultimately to become the home of this prehistoric race. The nature of the sandstone is such that its outer layer peels away in great slabs as water and frost gradually bring about its destruction. This type of weathering phenomenon has produced the overhanging sandstone ledges and shallow caves in which the cliff dwellers built their shelters. The Mesa Verde's structure, its soil, and its water—all geologically related—were probably the major factors that influenced the prehistoric Indians to establish their mesa-top and cliffside dwellings.

The Ancient Ones

As mentioned earlier, the Mesa Verde was first inhabited by a group of agricultural Indians. Beginning with a simple culture at about the opening of the Christian Era (A.D. 1), these people attained a relatively high cultural level by the end of the thirteenth century. In studying the cultural advances of this ancient race, archeologists have established four periods of progress, each of which has been given a name. It should be noted, however, that archeological periods are somewhat like geologic periods in that their limits are not always sharply defined and that the dates given for each are only approximate.

The first of these periods, the *Basketmaker Period,* lasted from about A.D. 1 to 450. During this time the mesa dwellers utilized the many shallow cliff caves that have developed along certain of the canyon walls. There is no evidence that these people made pottery, but they had mastered the art of weaving and they fashioned a wide variety of beautifully woven baskets, which were used for a number of purposes including cooking. Bags, belts, sashes, aprons, and other objects were also woven, and these people also made ornaments and tools of materials such as wood, bone, stone, and shell. The Basketmakers grew corn and squash which they stored in their caves for use during the winter, and they hunted mountain sheep, deer, and elk with a primitive dart-throwing stick called the *atlatl.*

During the *Modified Basketmaker Period*—A.D. 450 to 750 —significant advances were made by these early Indians. They began, for example, to make pottery, cultivate beans, and raise turkeys, which were highly prized for their feathers, which were used to make blankets and robes. The Indians of this period also began to use the bow and arrow, a far more effectual weapon than the atlatl. Their greatest step forward, however, was to abandon their primitive caves in favor of roofed dwellings, which they built on the mesa's top. At first these dwellings were little more than shallow, circular pits with flat roofs of poles and mud. Yet, despite their primitive construction, these *pithouses* protected the Indians and their produce from the elements and permitted them to live near their fields. As time passed and the population increased, the Indians gathered together in villages and the houses became more elaborate. Rectangular rooms were added to some of the pithouses, and others were constructed which were square or D-shaped. There is also evidence to indicate that the inhabitants of the Mesa Verde traded with other tribes, for articles foreign to this area have been found in some of the ruins.

The years A.D. 750 to 1100 have been designated as the *Developmental Pueblo Period,* a time of peace, progress, and expansion for the people of the mesa. The beginning of this period was marked by an increasing tendency of the people to cluster their houses together in small communities. The Spanish called their villages *pueblos,* and this is the term applied to the early Indian villages. To distinguish the earlier pueblos from those of the next period (the Great or Classic Pueblo Period), the term "Developmental" was applied. This was a time of architectural and constructional experiment, when new types of houses were constructed of a variety of materials. Moreover, many dwellings of this period were grouped around open courts which contained *kivas,* deep pithouses used for ceremonial purposes.

Advancement is also indicated in the pottery remains of this period, for the potters had greatly improved in their craft since the preceding period. They had also acquired cotton, probably by trade, and this was woven into cloth which was used to supplement the furs, hide, and turkey feathers from which their clothing was made.

The Mesa Verde Pueblo culture reached its climax during the years A.D. 1100 to 1300—the *Great* or *Classic Pueblo Period.* Better-shaped, more elaborately decorated pottery was made, weavers became more skilled, and architectural and constructional techniques were greatly improved. There was also much growth in the people's religious beliefs, and special buildings and numerous large kivas were erected expressly for ceremonial purposes.

But perhaps the most significant development of this period was the marked change in the living habits of the Indians. For reasons not thoroughly understood, the people gradually began to desert the open villages near their farmland and started to congregate in close-knit, more protected village groups. The relocation of the communal areas was accompanied by a change in design and construction of their dwellings. The new houses consisted of terraced structures—some as much as four stories high—with thick, carefully constructed masonry walls. We can only speculate as to why the Indians built such well-fortified dwellings. Some archeologists believe that the peaceful Mesa Verde residents may have become victims of periodic attacks from hostile Indian tribes that had moved into the area. It has also been suggested that there may have been inter-pueblo feuding within the Mesa Verde tribe. But whatever the cause, life on the mesa appears to have undergone a drastic change and there is evidence that the mesa people were steadily declining in numbers as many Indians moved out of the area and settled elsewhere.

The final chapter in the history of the Mesa Verde Indians began about A.D. 1200. It was then that the compact, fortified, mesa-top pueblos were evacuated and the people returned to the caves and rocky ledges in the mesa's canyon walls. In these hidden, easily defended areas they built the thick-walled cliff dwellings for which this Park is famous. Such well-known structures as Spruce Tree House, Cliff Palace, and Square Tower House stand as silent tribute to the craftsmanship of these prehistoric builders. Fortunately some of these dwellings are well preserved and tell us much about their ancient occupants. We know, for example, that the interior walls of many dwellings were smoothly plastered and decorated with colorful designs. These people also produced well-made and highly decorated pottery and cloth, and were accomplished woodworkers. Moreover, the number of kivas and the refinements they display indicate that their religion continued to flourish and achieve new meaning.

But what is *not* known is why the Pueblos vacated their well-fortified villages to return to the cliff-dwelling existence that their ancestors had abandoned hundreds of years earlier. In so doing they left their fields unguarded and at the mercy of marauding tribes and exchanged their relatively comfortable mesa-top pueblos for the narrow confines of the less comfortable and convenient cliff houses. There can be little doubt that a move of this magnitude was precipitated by fear, but the exact nature of the threat to this ancient tribe still puzzles archeologists.

The Desertion of the Mesa

The year A.D. 1276 marked the beginning of the end for the Mesa Verde's early inhabitants. This year marked the start of a twenty-four-year drought, and from this date to the end of the thirteenth century the Mesa Verde region was plagued by a steady decline in precipitation. As crops failed and springs ran dry, the Indians had no choice but to move to areas where living conditions were more favorable. Thus, the cliff dwellers migrated southward and eastward to the Rio Grande drainage in New Mexico, and west to Arizona. It is believed that some of the modern Pueblo Indians who inhabit these areas today may be the descendants of the former inhabitants of the Mesa Verde.

However, the theory of a drought-induced abandonment of the mesa has not satisfied all archeological authorities. The Pueblos, they argue, had survived more severe droughts in other years, so why should they choose this time to leave their hard-earned achievements and start life anew in an unfamiliar area? Although the answer to this puzzle may never be known, archeologists continue to sift through the ruins in search of clues that will shed further light on this thirteenth-century exodus. In the meantime the ruins have been preserved as a lasting monument to the craftsmanship and endeavor of the ancient

Pueblos. Here you can forget, momentarily, modern-day problems and relive a fascinating chapter in the prehistory of the American Southwest.

Plants and Animals of Mesa Verde National Park

Like the geology of the mesa, its flora and fauna are also intricately related to the early habitation of the Mesa Verde. The area's animals provided meat to augment the homegrown squash, corn, and beans and also provided fur, hides, and feathers for clothing and ornamentation. Native shrubs, herbs, and trees were used as food, fuel, medicine, and to roof the ancient dwellings.

The plants and animals of the mesa top represent a mixture of species from the lower, more arid regions to the south with forms from the cooler, high mountains to the north. Most of the mesa is covered by a piñon pine and Utah juniper forest, but at higher elevations there are concentrations of mountain mahogany and scrub oak, as well as fendlera, mock orange, and serviceberry. In some places there are small stands of quaking aspen, ponderosa pine, Douglas fir, and Rocky Mountain juniper. Among the colorful wildflowers that accent the mesa landscape from early spring until fall are lupine, Indian paintbrush, pentstemon, mariposa lily, and sweet pea.

Among the many mammals that live on the mesa are mule deer, black bears, Rocky Mountain bighorns, coyotes, foxes, and bobcats. Smaller species include porcupines, cottontails, chipmunks, and rock squirrels.

In addition, more than one hundred and seventy species of birds have been reported in the Park. These include jays, owls, hawks, woodpeckers, and crows. The reptilian fauna consists primarily of lizards, but a number of snakes, including the prairie rattlesnake, also live on the mesa.

Tips for Tourists

There is much to interest the visitor to the Mesa Verde, but, generally speaking, the activities differ from those in most National Parks. The emphasis here is upon human rather than natural history, and the Park's splendid interpretive program is especially designed to help you appreciate fully your glimpse of life as it was lived by the early Pueblo Indians.

Here, perhaps more than in any of the National Parks, your visit should begin in the Park Museum. If you are to understand the true meaning of what is to be seen in the ancient cliff dwellings and pithouses, you will need to know something of the background and development of the Mesa Verde Indians. In the Museum, there are exhibits and dioramas which depict various phases of pueblo life from the time of the Basketmakers to the Great, or Classic, Pueblo Period, which ended about the year A.D. 1300. Also displayed are replicas of pithouses and cliff dwellings and a large collection of objects that were used by the mesa's early inhabitants. There are also natural history exhibits in a building located south of the Museum and these will acquaint you with the geology and biology of the region.

From early June to Labor Day, members of the Park interpretive staff lead conducted trips through some of the more important cliff dwellings. These include a quarter-mile round trip of **Cliff Palace** which starts in the north end of the ruin. The trail begins at the viewpoint sign in the Cliff Palace parking area on the Ruins Road, and leads down to the ruin. This is the largest and most famous of the cliff dwellings, and as many as four hundred people may have occupied the Palace's two hundred living rooms. There are, in addition to the living areas, twenty-three kivas and numerous small storage chambers.

Ranger-guided trips are also conducted through *Balcony House,* a veritable cliffside fortress. This ruin is noted for its spectacular defensive position and can only be reached by climbing ladders, walking along a narrow ledge, and crawling through a cramped passageway. This well-planned Pueblo "obstacle course" must have been an effective deterrent to would-be raiders of the ancient Indians, but it in no way discourages the thousands of visitors who each year follow the same route to inspect this ancient dwelling. Allot approximately one hour to complete this tour and be prepared to climb a ladder and to walk about a quarter mile. The trips start at the viewpoint sign in the Balcony House parking area on the Ruins Road.

Cliff dwelling trips are subject to change and you should inquire at the Museum for the latest schedule of events. It should also be noted that National Park Service regulations forbid visitors to enter any cliff dwelling except with a Park Ranger on a conducted tour or during visitation periods when a Ranger is on duty. Nor should you disturb, deface, or remove any object from any of the ruins or caves in the Park. The ancient heritage of the Mesa Verde ruins is priceless and every safeguard is taken to assure its preservation.

You will get your best pictures of the cliff dwellings from the canyon rim scenic overlooks. Because most of the caves face west, lighting is best in the afternoon. A telephoto lens is required for the more distant dwellings, and a wide-angle lens will help greatly while photographing within the ruins. A light meter will generally be helpful in determining exposures, especially if you photograph shaded portions of the ruins.

Mesa Verde National Park at a Glance

Address: Superintendent, Mesa Verde National Park, Colorado 81330.

Area: 52,073 acres.

Major Attractions: Most notable and best preserved prehistoric cliff dwellings and other works of early man in the United States; ruins date from late A.D. 500s to late A.D. 1200s.

Season: Year-round; however, concession-operated facilities are normally open from about May 1 to October 15.

Accommodations: Lodge, cabins, campgrounds, and group campsites. *For lodge reservations contact:* Mesa Verde Company, Box 277, Mancos, Colorado 81328.

Activities: Guided tours, scenic drives, museums, picnicking, camping, and horseback riding.

Services: Food service, gift shop, guide service, health service, laundry, post office, public showers, religious services, service station, telegraph, telephone, transportation, picnic tables, rest rooms, and general store.

Interpretive Programs: Guided trips, museums, self-guiding auto trips, roadside and trailside exhibits, and campfire programs.

Natural Features: Canyons, erosional features, geologic formations, and wildlife.

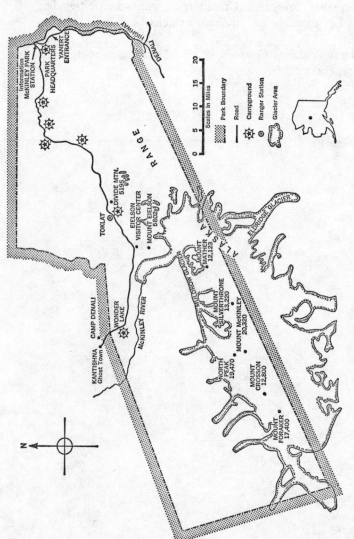

Mount McKinley National Park, Alaska.

23. Mount McKinley National Park, Alaska

"The High One"

Located in south-central Alaska only two hundred and fifty miles south of the Arctic Circle is our nation's second largest National Park. In size, Mount McKinley National Park is second only to Yellowstone National Park, but even Yellowstone cannot match the Alaska Park's stellar attraction, for Mount McKinley is this continent's highest mountain peak. Towering 20,320 feet above sea level, the mountain early attracted the attention of man, for the Indians called it Denali, "The High One." The old Indian name is as fitting today as it was centuries ago, and the name Denali is commonly seen in this part of Alaska.

This majestic mountain has two major peaks, both of which bear the name of the late Sir Winston Churchill, former Prime Minister of England. One of these peaks (formerly called North Peak) rises 19,470 feet above sea level and lies about two miles north of Mount McKinley's ice-sheathed 20,320-foot summit. But despite its dominating presence, Mount McKinley represents a relatively small part of this vast subarctic wilderness area which spreads over 3030 square miles. The terrain within the Park is varied and elevations range from fourteen hundred feet in the valley to "The High One's" lofty summit.

The Alaska Range

One of the youngest and most striking ranges in Alaska, the Alaska Range extends six hundred miles across south-central Alaska in a vast arcuate band. The range is only about thirty miles wide at the Canadian border; however, the arc widens to the west until it is about one hundred and twenty miles wide in the vicinity of Mount McKinley. The crest line of the Alaska Range exceeds eight thousand feet above sea level, and in the center of the arc a cluster of high peaks surround 13,700-foot Mount Hayes.

Mount McKinley National Park is located near the southwestern end of the Alaska Range, and there are a number of high peaks within the Park's boundaries. Southwest of Mount McKinley lie 12,800-foot Mount Crosson and impressive Mount Foraker which rises 17,400 feet above sea level, while Mount Silverthorne (13,220 feet) and Mount Mather (12,123 feet) are located to the northeast (see map). Actually, the great height of Mount McKinley is rather unusual, for less than twenty peaks in the Alaska Range exceed ten thousand feet in elevation. The altitude of the range gradually decreases east of Mount McKinley, until it finally loses its identity as it merges with the Coast Ranges north of the St. Elias Mountains.

Space limitations preclude a detailed discussion of the geology of this complex mountain range, which has been subjected to repeated structural disturbances that have greatly complicated its physical history. Suffice it to say that the range is situated on the site of a great *synclinorium*—a broad regional downwarping of the earth's crust upon which are imposed minor folds. As is usual in this type of geologic structure, the youngest rocks are located near the center of the range and the oldest formations lie on its flanks. Further evidence of crustal deformation can be seen in the great longitudinal faults that cut across the synclinorium. Today these faults are marked by lines of valleys and mountain passes. The mountains are composed of a variety of rock types representing nearly all geologic periods. Many of the rocks have been severely deformed and metamorphosed and there is evidence of considerable igneous activity.

Mighty Rivers of Ice

The Alaska Range is especially spectacular because it rises abruptly from rather low-level surroundings and because it is so far north that the high peaks are covered with snow throughout the year. There are, in addition, large numbers of active glaciers located within the Alaska Range and many of these are found within the boundaries of the Park.

Glaciers of Today

The largest glaciers that now exist in the Park are located on the south side of the mountains in the basin of the Chulitna and Yentna Rivers. These valley glaciers originate high on the southern slope of the Alaska Range, an area that is continually exposed to the moisture-laden winds of the Pacific Ocean. The glaciers that are present on the drier, northern slopes of the range do not receive as much ice-forming moisture, hence they are relatively small. Literally rivers of ice, the valley glaciers follow stream-cut canyons and because they are confined by the valley walls they can only move down the valley. However, the more lofty mountains in the Park have snowfields at their higher elevations and these are the spawning grounds for the few large valley glaciers that flow down the north slope of the Alaska Range. These include the Herron Glacier which originates in the snowfields of Mount Foraker and Peters Glacier which circles around the northwest end of Mount McKinley. Mightiest of all is Muldrow Glacier, a thirty-five-mile river of ice that has its source on the north flank of Mount McKinley. This great valley glacier has flowed to within about one mile of the Park Road, but a rather demanding hike is required to reach its ice. Although most glaciers move very slowly (p. 144), Muldrow Glacier underwent a rather spectacular advance during the fall of 1956 at which time the front of the glacier moved forward almost four miles in less than one year. This accelerated movement, which occurred as the result of a greatly increased supply of ice that thickened the terminal end of the glacier, was rather unexpected for Muldrow had been slowly receding for quite some time.

The glaciers are also responsible for most of the Park's streams. These glacially fed creeks and rivers, most of which consist of water which pours from the snouts, or ends, of the glaciers, are laden with gray silt which is deposited in the river valleys. As they course through the Park, the rivers split into numerous dividing and reuniting channels resembling the strands of a braid. This has resulted in a *braided stream pattern* consisting of many intertwined channels separated from each other by low islands or channel bars. Braided patterns are believed to indicate that the stream has an excessive load and is not capable of carrying on lateral erosion. The braided channels here probably develop as a result of increased deposition due to the rather sudden decrease in stream gradient when the stream leaves the mountains and enters the lowlands at the foot of the Alaska Range. When the water is low, the channel bars make it possible to cross certain streams on foot. But when the water rises, the streams may become raging torrents that are hazardous and difficult to cross.

Glaciers of Yesterday

Although the glaciers of today are among Mount McKinley National Park's most intriguing geologic features, it is the work of past glaciers that has provided us with the Park's most spectacular scenery. During the Great Ice Age of Pleistocene time, much of the Park was covered by glaciers which have left their indelible mark on the landscape. As the glaciers waxed and waned, they sculptured the peaks of the Alaska Range, thereby producing the saw-toothed spires, knife-edge ridges, and broad U-shaped glacial valleys that typify the mountains today. Here and there on the mountainsides one can see great semicircular depressions called *cirques*. These bowl-shaped features were produced by glacial erosion and mark the point of origin of now-vanished valley glaciers. In addition, many lofty, steep-sided, pyramidal peaks attest to the erosive ability of the Pleistocene glaciers. Further evidence of glaciation is provided by *glacial polish*—lustrous rock surfaces which have been ground and polished by the ice—and large and small abrasions called *glacial grooves* and *striations*.

Yet the glaciers did not confine their geologic activities to erosion—their work in the valleys was primarily constructional in nature. As the glaciers melted, they deposited their sediments in ridgelike features called *moraines,* or as broad *outwash plains* formed by the deposition of glacial debris washed from the front of the glaciers. Present also are *kettle lakes,* small, circular bodies of water formed in depressions in the glacial sediments. These pits, called *kettles,* mark the place where blocks of ice left by shrinking glaciers were buried in the glacial rock debris. When the ice melted, the overlying sediments slumped downward to form these unusual concavities.

Elsewhere, the glaciers left behind immense boulders called *glacial erratics.* Many of these great rocks were transported many miles from their original source, hence are likely to be of different composition than the rocks upon which they rest. Some of the erratics have been abandoned high in the mountains and give some indication as to the depth of the ice and snow that once blanketed this part of Alaska.

Life in Mount McKinley National Park

Mount McKinley National Park is justly famous for its unusual plant and animal communities. Here there are types of animals found nowhere else in the National Park System; here, too, are unique plants that have successfully become adapted to the harsh and often inhospitable environment of the subarctic wilderness.

The flora of Mount McKinley National Park is one that must be able to grow in the sparsest of soils and survive the long, freezing winters. Yet despite the rigors of the severe subarctic climate, many different plants live and flourish in this part of Alaska. Below what might be called timber line—up to about three thousand feet above sea level—there is a surprisingly well-developed forest community. This life zone, called the *taiga,* a Russian term used to define the circumpolar forest belt, consists largely of black and white spruce intermingled with balsam poplar, white birch, and aspen. Along stream banks in wetter areas there are occasional stands of alder, aspen, white birch, and willows. In some places the trees appear to be growing at an angle; this occurs in parts of the forest where ground support is weak or where the topsoil has slipped on *permafrost*, the permanently frozen subsoil. Carpet-like growths of lichens and mosses thrive on the forest floor, and such shrubs as alder, currants, blueberry, dwarf birch, and numerous species of willows form the underbrush.

Above timber line lies an expanse of treeless terrain called the *Alpine tundra*. Here the climate approaches that of Siberia and other Arctic regions, and only hardy, specialized plant species can survive the frozen, windswept winter season. Actually, there are two types of tundra within the Park: the *lowland,* or *wet tundra,* characteristic of lower elevations, and the *upland* or *dry tundra* which grows at higher altitudes. The larger plants of the wet tundra which grows at higher altitudes. The larger plants of the wet tundra consist primarily of shrubby species such as blueberry, dwarf birch, and a variety of willows. Growing beneath and among these are thick carpets of mosses and lichens and moisture-loving grasses and sedges. Together these plants form a dense vegetation over and around the many ponds and scattered low hummocks that mark the lowland areas.

The plants of the dry tundra grow at higher elevations where the soils are better drained but where the environment is incredibly harsh. Plants here are dwarfed and stunted, and they typically grow in low-lying, matlike clusters in order to escape the rigors of the bitter Alaska winter. Despite the severity of the tundra environment, plants of the dry tundra do not lack for beauty. Many of them blossom throughout the summer and their delicately colored blooms add much to the beauty of the Park. Colorful wildflowers that accent the mountainsides include the very abundant white-flowered dryas which commonly blankets large areas and the forget-me-not which is Alaska's state flower. Dwarf rhododendron, white heather, dwarf fireweed, asters, blue lupine, and shrubby cinquefoil are other plants that lend color to the Park's scenery.

The animal life in this great wilderness Park is virtually unsurpassed, for few of the world's game preserves provide such unequaled opportunity to observe large wild animals at fairly close range. Moreover, two of these animal species—the Dall sheep and barren-ground caribou—are not found in other areas administered by the National Park Service. The white Dall sheep, a relative of the Rocky Mountain bighorn, have rather slender, curved horns and spend much of their time climbing about the Park's icy cliffs and windswept grassy slopes. Although these animals migrate through the Park with the changing seasons, they may commonly be seen from the Nenana River to Muldrow Glacier on the higher tundra slopes between three thousand and five thousand feet. (Binoculars are most useful in spotting these white mountain climbers.) In addition they are usually visible at several places along the Park Road in the summer. The other species that is unique to Mount McKinley National Park is the barren-ground caribou, a large, heavy antlered deer that migrates through the Park in great numbers. Look for caribou grazing on the grassy hillsides and also in the vicinity of Wonder Lake.

Other large mammals that inhabit Mount McKinley National Park include the huge grizzly bear and the Alaska moose. The former spend much of their time in the open tundra, where they grub about for roots and graze on the tundra grass. The observant visitor may spot grizzlies among the low bushes of the tundra or digging in the gravel channel bars of the Park's braided streams; they are also commonly seen at Sable Pass. The monstrous Alaska moose is the titan of the Park, for bulls weighing as much as fifteen hundred pounds are not uncommon. The male moose is further distinguished by his massive rack of palm-shaped antlers, some of which have a spread of more than five feet. Although these shy creatures normally prefer the protective environment of spruce forests or willow thickets, they occasionally browse the open tundra in search of food.

In addition to the large animals described above, more than thirty species of smaller animals call this Park their home. Among those that you are most likely to see are pika and hoary marmot (especially among the sliderock at the base of cliffs and mountains) and the parka (arctic red) squirrel and the red fox can sometimes be seen on the tundra. Beaver and porcupine are occasionally sighted near streams and ponds in the aspen and willow groves in the valleys. Another common but somewhat unusual small mammal is the snowshoe rabbit, or varying hare. This interesting creature has its own system of natural camouflage that varies with the time of year. During the summer months its dull-brown coat harmonizes well with its natural surroundings, but when snow covers the Park the color of its fur turns almost pure white, thereby allowing the rabbit to blend into the snowy background. The snowshoe rabbit is appropriately named for in the winter its feet are covered with thick pads of hair which serve to keep the animal from sinking too deeply into the snow. Also present—but rarely seen—are the relatively scarce and evasive lynx, timber wolf and wolverine.

Although few birds can survive the Park's long, dark, icy winters, many birds spend the summer in Mount McKinley National Park; and they come from all over the world. There is the long-tailed jaeger which comes to Alaska from its winter habitat on islands near Japan, the golden plover which winters in Hawaii, and the European wheateater which leaves Asia each spring and migrates to Alaska for the summer. The birds vary as much in size as they do in origin, for they range from the tiny kinglet to the mighty golden eagle. Birds that commonly live in the tundra include horned lark, Lapland longspur, wandering tattler, Hudsonian curlew, snow bunting, northern shrike, and ptarmigan. The ptarmigan, like the varying hare, has "built-in" snowshoes; but they are composed of feathers rather than hair. Ptarmigan also change color with the seasons —their feathers are brown and gray in the spring and summer and almost perfectly white in the winter. Another interesting tundra species is the surfbird, a seashore dweller that winters in South America.

Tips for Tourists

The area's varied topography, its diverse plant and animal life, and its true wilderness setting make Mount McKinley National Park a nature lover's paradise. There are trails to hike near Park Headquarters, large game to be photographed, and a fine Visitor Center at Eielson containing displays designed to help you understand better the natural history of this great Park. But probably the major pastime is "mountain watching," for the majestic figure of "The Summit of North America" is the prime attraction and all-dominant feature here. Perhaps "watching *for* the mountain" might better describe the activity of most visitors, for Mount McKinley is shrouded in clouds and fog much of the time. However, there are many vantage points along the Park Road from which—weather permitting—you may spot "The High One."

The interpretive program centers around **Eielson Visitor Center** about sixty-five miles from McKinley Park Entrance Station. Here there are exhibits dealing with the glaciers and glaciation of Mount McKinley and the Alaska Range; there are also displays that relate to the early mountain-climbing history of the mountain. The large picture windows of the Visitor Center provide excellent views of the Park, and a Ranger-Naturalist is on hand to present short talks about the wildlife, geology, glaciers, and early mountain climbs. Books, maps, and similar publications pertaining to the Park are also on sale here. (They may also be purchased at the McKinley Park Entrance Station and at the National Park Service desk at the McKinley Park Hotel.)

The path to the heart of Mount McKinley National Park is the 88-mile gravel road that leads from the Entrance Station to Camp Denali turnoff. Following the scenic intermontane valley north of the Alaska Range, this rough, hilly road winds its way from a low elevation of sixteen hundred feet to almost four thousand feet above sea level near Highway Pass (see map). Pullouts and observation points are located at convenient intervals along the road, and there are interpretive signs which provide information about biological or geological subjects that are pertinent to that particular stop. The road was designed to permit you to travel in reasonable comfort—and at minimum speed—into some of the Park's most scenic areas. The Park Road crosses many rivers, for it parallels the Alaska Range from which the rivers flow into the valley at right angles to the road. Most of the road passes through or near the so-called Outside Range, which are the foothills of the Alaska Range;

these mountains have altitudes of from four thousand to six thousand feet. From the Park Entrance to Eielson Visitor Center, the road alternates between forests and tundra; but after the first seventy miles, the road is confined to the rolling, treeless terrain of the Alpine tundra.

If weather conditions are favorable, you will probably get your first glimpse of McKinley from a point about eight miles west of the Entrance Station, but it will be another fifty miles until really good views come into sight. After that, however, there should be many fine opportunities to see this great snow-capped peak, for the road passes to within twenty-seven miles of Mount McKinley's summit. Seen from within the heart of the Park on a clear day, McKinley appears much closer than it actually is. And, because it rises about seventeen thousand feet above the McKinley River which flows near its base, the great mass of the mountain is even more impressive.

Not located within the Park but closely associated with it is **Camp Denali.** This wilderness retreat is situated two miles from Wonder Lake and is designed for the visitor who wants a taste of life in the Alaska "bush country." This is not a resort in the modern sense of the word: Camp Denali has no juke box, television, or cocktail lounge; nor is there electricity or modern plumbing. But what the camp does offer is an opportunity to experience true wilderness living in one of our nation's "last frontiers." For those with special interests there are "safaris" to photograph wildlife and "wilderness workshops" that explore fully the natural history of the area. For further information on this unique facility, contact Camp Denali: Box D, College, Alaska 99701, until June 1; McKinley Park, Alaska, June 1 to September 10.

A special interpretive program unique to Mount McKinley National Park is the **Sled Dog Demonstration.** Held daily at the National Park Service Dog Kennels, these forty-minute, Naturalist-conducted programs present information about the past and present use of sled dogs in the Park. The various kinds of dogs and equipment are explained and their use is demonstrated by a dog team hitched to a typical freight sled.

Only the weather limits the photographic opportunities in this Park; however, some seasons are better than others for photographing certain subjects. June, for example, is the most favorable time to photograph wildflowers and most of the migratory birds. This is also the best month to sight large herds of caribou. However, moose and grizzly bear can be found throughout most of the summer; the former can often be sighted in spruce-willow areas and along streams, and grizzlies may be spotted in the Sable Pass area. Look for Dall sheep from the Nenana River to the Muldrow Glacier; they inhabit the higher tundra slopes between three thousand and five thousand feet above sea level. A telephoto lens will greatly increase your chance of good wildlife photographs and patience and sharp eyes are equally helpful.

As for Mount McKinley, there are many fine vantage points from which to photograph this imposing mountain. Weather conditions are the controlling factors here; and although there is no definite way to predict it, in recent years there have been more clear days in June than in July and August.

Mount McKinley National Park at a Glance

Address: Superintendent, Box 2252, Anchorage, Alaska 99051.

Area: 1,939,493 acres.

Major Attractions: Mount McKinley, highest mountain (20,320 feet) in North America; large glaciers of the Alaska Range; caribou, Dall sheep, moose, grizzly bears, wolves, and other spectacular wildlife.

Season: Normally open June 1 to September 10.

Accommodations: Campgrounds and hotels. *For reservations contact:* Manager, McKinley Park Hotel, McKinley Park, Alaska 99755 (May 15–September 30) or Mt. McKinley National Park Co., 2522 N. Campbell Avenue, Tucson, Arizona 85701 (October 1–May 15). Limited overnight accommodations and meals on the American plan are available at Camp Denali just north of Wonder Lake (for details and address, see p. 280).

Activities: Boating, camping, fishing, guided tours, hiking, mountain climbing, nature walks, picnicking, and scenic drives.

Services: Food service, gift shop, guided tours, post office, religious services, service station, telegraph and telephone (emergency only), transportation, picnic tables, and general store.

Interpretive Program: Illustrated lectures, nature trails, self-guiding trail, roadside exhibits, and sled dog demonstration.

Natural Features: Geologic formations, glaciation, glaciers, mountains, rivers, rocks and minerals, unusual birds, unusual plants, wilderness area, and wildlife.

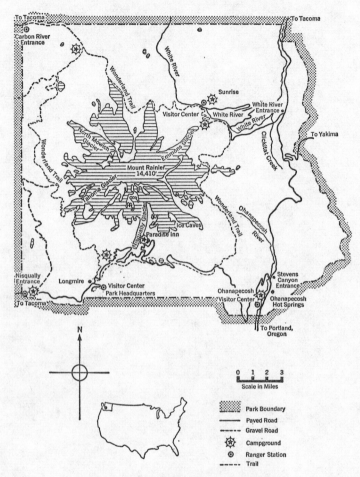

Mount Rainier National Park, Washington.

Mountain Glacier Wonderland

Mount Rainier—one of North America's loftiest and most beautiful mountains—is the prime attraction in the National Park that bears its name. But mighty Mount Rainier does not account for all of the Park's beauty and charm. Sparkling lakes, flowered alpine meadows, plunging streams, and—above all—the icy tentacles of the mountain glaciers are vital supporting elements in this mountain glacier wonderland. There are, moreover, two additional physical factors that serve to accentuate the prominence of Mount Rainier. First, this great peak stands as an isolated cone towering more than nine thousand feet above the ridges of the surrounding Cascade Range. Second, from its furrowed base to its broad, rounded summit, Mount Rainier is shrouded with a perpetual blanket of ice and snow that hides all but its more precipitous cliffs and peaks.

The "American Alps"

Mount Rainier is but one of many outstanding peaks of the Cascade Range, a great glaciated pile of volcanic rocks that extends from northern California to northern Washington. Because it is such an extensive mountain range, the Cascades have been divided into three sections, each of which has its own distinctive mountain peaks. The *Southern Cascades Range* extends from south of Lassen Volcanic National Park (p. 233) for approximately one hundred and fifty miles northward to the vicinity of North Klamath Lake in Oregon. This section of the range is composed largely of a string of volcanic cones separated by valleys and basins. There are literally hundreds of old volcanoes sprinkled along the crest of this part of the Cascades, but the two mountains that dominate the landscape are Lassen Peak (10,457 feet) (p. 233) and Mount Shasta (14,161 feet), both of which are located in northern California.

The *Middle Cascade Range* is by far the largest of the three sections and extends northward from North Klamath Lake to a line which connects Seattle and Ellensburg, Washington. This portion of the Cascades is seen as a fairly uniform, flat-appearing platform that gradually increases in altitude to the north. In places, rising several thousand feet above this vast mountain platform, are some of the Pacific Northwest's most spectacular peaks, such as the Three Sisters (10,453 feet), 10,495-foot Mount Jefferson, graceful Mount Hood (11,245 feet), the symmetrical cones of Mount St. Helens (9,671 feet) and 12,307-foot Mount Adams, and finally majestic Mount Rainier which "tops out" at 14,410 feet. The Middle Cascade Range was also the location of now-vanished Mount Mazama, the great volcano that collapsed many thousands of years ago. Today Mount Mazama's water-filled caldera is known as Crater Lake, one of our most beautiful and popular National Parks (p. 117).

The *Northern Cascade Range* extends from the northern boundary of the Middle Cascades to a short distance north of the Canadian boundary, and this marks the end of the Cascade Range. Unlike the Middle and Southern Cascades which are comprised largely of volcanic rocks, the geologic formations of the Northern Cascades Range are mainly Paleozoic and Mesozoic sedimentary and metamorphic rocks. Moreover, most of the peaks in this part of the range have been created by stream and ice erosion rather than by volcanic eruption. However, the two most prominent mountains—ice-scoured Glacier Peak (10,436) and 10,750-foot Mount Baker —are both volcanic cones. From a distance the Northern Cascade Range appears to be a vast mountain plateau, but in many places the mountains are dissected by valleys that are as much as three thousand feet deep.

Considering the impressive array of mountains described above it is not surprising to learn that some mountaineers have called the Cascade Range the "American Alps." Even less surprising is the fact that the Cascades have also provided us with some of North America's most spectacular scenery.

Birth of the Cascades

Despite the great amount of stream and ice erosion that has been inflicted upon the Cascade Range, these are rather young mountains, geologically speaking. It is known, for example, that during Eocene time—perhaps sixty million years ago—the site now occupied by the Cascade Range in Washington was a flat lowland. Here, over a long period of time, there accumulated a great thickness of sedimentary rocks. These strata, which consist primarily of sandstones and shales interbedded with layers of coal, are believed to have been deposited in shallow basins and embayments along the western side of the present Cascade Range and in basins farther to the west. However, the oldest rocks exposed in the Park are those of the *Ohanapecosh Formation,* mainly of Late Eocene or, perhaps in part, Early Oligocene age. More than ten thousand feet thick and consisting almost wholly of volcanic debris, the Ohanapecosh also includes lenslike accumulations of lava and coarse mudflows accumulated around volcanic centers.

Following the close of the Eocene Epoch, the old flood plain was uplifted and folded to form a folded mountain range whose structure was not unlike that of the present-day Appalachians. The location of these northwest-southeast trending mountains is indicated by the presence of deformed rocks which represent the eroded roots of this ancient mountain range. As might be suspected, the Oligocene orogeny (mountain-building movement) which raised the mountains was accompanied by considerable volcanic activity and one can still see remnants of these Oligocene volcanics in the southern and central sections of the Cascade Mountains of Washington.

The next major event in the history of the Cascade Range

began during the Miocene Epoch which started about twenty-five million years ago. This chapter in the geologic story of the Pacific Northwest was a fiery one, for near the middle of Miocene time the area was the site of one of the world's greatest lava floods. The lava, which probably flowed from large fissures in the earth's crust, must have been exceedingly hot and in a very fluid condition, for it spread over a vast area in sheets ranging from a few feet to tens of feet thick. As the lava spilled out upon the surface, it accumulated first in the low places and began to fill the old valleys. However, it eventually advanced up the slopes of hills and mountains, for, in Idaho, granite hills from two thousand to twenty-five hundred feet high are covered by one thousand to fifteen hundred feet of lava from these flows. Thus, as time passed, all but the highest peaks were covered by extensive lava sheets which in places now have a maximum thickness of five thousand feet.

Today the ancient Miocene lava flows comprise a large part of the Columbia Plateau, a broad, elevated tableland that occupies the area between the Northern Rockies and the Cascade Range. This vast *lava plateau* covers a 200,000-square-mile area that encompasses southern Idaho, eastern Oregon, the southeastern quarter of Oregon, and northeastern California. In some areas the Snake, Columbia, and Spokane Rivers have slashed deep canyons in the lava plateau and the individual lava flows can easily be seen piled one upon the other in thick and regular succession. In the Mount Rainier area, evidence of Miocene volcanic activity consists largely of deposits of fragmented volcanic debris that were later covered by lava flows.

The next episode of the development of the Cascades took place during the Pliocene Epoch near the end of Tertiary time and marked the initial uplift which produced the present Cascade Range. This great crustal disturbance took place over an extended period of time during which the rocks were folded and upwarped by the cumulative effect of a long series of relatively small uplifts. It is estimated that this orogeny elevated the Cascades region from perhaps three thousand to as much as six thousand feet. The Pliocene mountain-building disturbance brought renewed volcanic activity to the Cascades region and saw the birth of many of the commanding line of volcanic cones (p. 286) that now rise several thousand feet above the surrounding Cascade mountains. Concurrent with its elevation, the great Cascade mountain block was immediately attacked by weather, wind, water, and ice. Erosion continued throughout Pleistocene time, and even today the mountains are still being worn away by the ever-present agents of erosion.

Although the Cascades were considerably eroded during the Tertiary Period, most of the sculpturing of the mountains took place during the Great "Ice Age" of the Pleistocene Epoch, when great alpine glaciers formed on the mountain peaks and extended many miles into the valleys below. During Pleistocene time the summit of the Northern Cascades is believed to have been completely covered by an icy mantle of snow and ice. It is significant that some of the nonvolcanic mountains have been more severely glaciated than the volcanic cones. This difference in the degree of glacial damage is taken as further proof that volcanic activity continued throughout much of the Pleistocene Epoch.

Mount Rainier: "The Mountain"

Having considered the geologic setting of Mount Rainier, let us now direct our attention to the mountain itself, for it, too, has a long and interesting history. "The Mountain," as Mount Rainier is often called, has long been admired—and perhaps feared—by man. To the early Indians of the region, Mount Rainier was considered to be the home of their gods, and these primitive people had many superstitions and legends about "The Mountain." However, the first evidence of its having been seen by a white man was recorded in 1792. It was in May of that year that Captain George Vancouver of the British Navy described Mount Rainier and named it in honor of his fellow naval officer Rear Admiral Peter Rainier.

Of Fire and Ice

But although the geologic history of "The Mountain" far antedates the Indians and Captain Vancouver, Mount Rainier's cone is believed to be a relatively recent addition to the landscape of the Cascades, for most of the geologic events surrounding its development occurred within the last million years. The formation of Mount Rainier began in the Quaternary Period when fire and ice—volcanoes and glaciers—joined forces to shape and reshape its face.

Yesterday's Volcano. The cone of Mount Rainier, like that of Mount Shasta, is a good example of a strato-volcano, for it is composed of many layers of lava, volcanic cinders and ash, and rock rubble from avalanches and mudflows. It is therefore neither a cinder cone like that found in Lassen Volcanic National Park or at Wizard Island in Crater Lake National Park, nor is it a shield volcano such as Kilauea in Hawaii Volcanoes National Park. Rather, the cone of Mount Rainier has the characteristics of both cinder cones *and* shield volcanoes, a feature that has prompted some volcanologists to call the cones of strato-volcanoes *composite cones.* Thus, there were two types of volcanic eruptions involved in building Mount Rainier. The earlier eruptions consisted primarily of a succession of thick intracanyon lava flows of gray *andesite,* a fine-grained, usually dark-colored volcanic rock that was first identified in the Andes of South America (hence the name andesite). Located between the andesite flows are masses of rock rubble and innumerable layers of volcanic ash, cinders, and bombs; these solid volcanic fragments were formed when particles of lava hardened in the air after having been thrown out by violent blasts of steam and gas. The arrangement of the volcanic materials in Mount Rainier's cone indicates that most of these rocks were ejected from a central vent, a fact which accounts for the symmetrical form of Mount Rainier and other typical strato-volcanoes. There is also considerable geologic evidence to suggest that there must have been numerous intermittent periods of lava outpourings and volcanic explosions. Some of

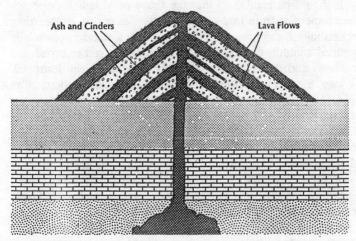

Ash and Cinders   Lava Flows

Mount Rainier is said to be a strato-volcano, or composite cone, because it is composed of alternating layers of lava and volcanic ash.

these eruptions appear to have been preceded by long intervals of quiet, for glacial sediments were deposited between certain of the ashes and lava.

When did Mount Rainier last erupt? There have been reports which suggest that "The Mountain" may have experienced as many as fourteen minor eruptions between 1820 and 1894. But because no new ash or lava was observed in any of these alleged eruptions, some geologists believe that the so-called "eruption clouds" were probably dust clouds generated by large avalanches and rockfalls. There is, however, geologic evidence that indicates that Mount Rainier's most recent major eruption occurred about two thousand years ago. The materials erupted at this time consist primarily of pumice and volcanic ash, which are widely distributed in the area east of the volcano.

Park visitors sometimes ask the Rangers, "Will Mount Rainier ever erupt again?" There is, of course, no way of telling whether the volcano is dead or merely dormant. However, steam still issues from vents in both of Mount Rainier's summit craters and these at least suggest that "The Mountain's" fires have not completely grown cold.

It should be mentioned that the shape of Mount Rainier differs somewhat from that of most composite cones. Rather than consisting of a simple cone that tapers to a slender, essentially pointed summit, the top of Mount Rainier is rather broad and rounded and the tip of the cone has obviously been destroyed. Today the summit bears three distinct peaks—Liberty Cap (14,112 feet) is on the north, Point Success, 14,150 feet, lies on the south, and on the east side is 14,410-foot Columbia Crest, the highest elevation in the Cascade Range. The truncated or "broken-off" appearance of Mount Rainier has led to the conclusion that during Pleistocene time the mountain may have been considerably higher than it is at present. This belief is further substantiated by the steeply inclined beds of volcanic materials which comprise the volcano's cone, for when the angle of these dipping beds is projected upward, they indicate that the top of the mountain may have been as much as one thousand to fifteen hundred feet higher than it is today.

Some of the major features of Mount Rainier can be seen on the above sketch. U. S. GEOLOGICAL SURVEY

We cannot be certain as to the exact manner in which the original summit was destroyed, but considerable thought has been given the problem. It was originally believed that the mountain top was literally blasted away by one or more violent explosions, but recent studies have produced no indication of the coarse volcanic debris that would have been left by such explosions. Other investigators have suggested that the summit may have collapsed to form a caldera similar to those of Brokeoff Mountain in Lassen Volcanic National Park and Crater Lake. There is also evidence to suggest that the central lava plug of Mount Rainier might have been so strongly altered that it could be more easily eroded than the more resistant lava which surrounded it. Thus, the top of the mountain might have gradually been lowered as the crater was hollowed out by glaciation and other agents of erosion. More recently, Dr. D. R. Crandell of the United States Geological Survey has suggested that the former summit of Mount Rainier was removed principally by very large avalanches that originated in rocks that had been altered and weakened by steam issuing from vents in the volcano's crater. The avalanches are believed to have been triggered by *phreatic explosion*—a volcanic explosion, usually of extreme violence, caused by the conversion of groundwater to steam. However, the landslides could have been caused by earthquakes, or perhaps slope failures due to oversteepening by glacial erosion.

Today the summit of Mount Rainier is marked by a young, virtually unglaciated cone that has been built since the last Pleistocene glaciation. Culminating in Columbia Crest, this recent cone rises some eight hundred feet above the ice-scarred, jagged rim of the old enlarged crater which has a diameter of about one and a quarter miles. Two small craters indent the top of Columbia Crest cone. The larger of the two depressions is perfectly circular, has a diameter of about thirteen hundred feet, and is tilted slightly to the east. The older western crater must have been built first, for its eastern margin is partly overlapped by the larger crater which developed to the east of it. Although both of the craters are perennially filled with snow and ice, jets of steam issuing from the floor of the eastern crater have melted the ice and snow thereby forming small caves.

Today's Glaciers. This National Park is famed for the extensive system of alpine glaciers that radiates from Mount Rainier's lofty summit. The glaciers spread their icy fingers down Mount

Rainier's deeply furrowed slopes, and together they constitute one of North America's largest single-peak glacier systems. And yet, despite the forty square miles occupied by "The Mountain's" active glaciers, the glaciers of today are but remnants of the vast rivers of ice that helped shape the present terrain of Mount Rainier National Park. Evidence of past glaciation is to be seen in virtually every sector of the Park: ridgelike moraines of glacially deposited stones and soil indicate the former position of ancient ice masses; great bowl-shaped cirques mark the heads of past and present glaciers; and the flat-bottomed U-shaped valleys that lace the mountain's flanks permit us to trace the courses of glaciers that have long since vanished from the earth. Even now glacial erosion is steadily cutting back and deepening the massive grooves in Mount Rainier's flanks; and during the great Ice Age, the mountain must surely have been more vigorously eroded than at present.

Although the marks of yesterday's glaciers are of considerable geologic interest, most visitors consider the glaciers of today to be the prime attraction of Mount Rainier National Park. There is, of course, good reason for this; for here is one national park where an active glacier can be visited with a minimum of danger and physical effort. The most accessible glacier here is **Paradise Glacier,** the best known of the Park's so-called minor glaciers. This ice mass, which can be reached via Paradise Glacier Trail, is small when compared to certain of the other glaciers; yet, it exhibits well the characteristics of a typical alpine glacier. Among the more outstanding features of this glacier are the beautiful *ice caves* that commonly open up during late summer. Melted out of the glacier's snout, the interiors of these icy grottoes are especially beautiful because refracted sunlight tints the walls a lovely shade of blue.

More spectacular—but harder to reach—is mammoth **Emmons Glacier,** the largest on Mount Rainier. About five miles long and one mile in width, Emmons Glacier flows down the northeast flank of the mountain and can best be seen from **Emmons Vista** in Yakima Park. Visitors to the south side of "The Mountain" can see the **Nisqually Glacier,** another of the Park's more accessible major glaciers. You can get a good view of this great, flowing ice mass from the Paradise area.

Persons who have visited both Mount Rainier and Glacier National Park (p. 137) frequently express surprise at Mount Rainier's many glaciers. They reason that because it is so much farther north—and because of its very name—Glacier National Park could reasonably be expected to have the most glaciers. This is not so, for the glaciers have all but vanished there while they appear to be expanding on Mount Rainier. Glacier National Park was, to be sure, the scene of great glaciation in ages past. But today the climate of these two areas is quite different, and herein lies the answer to this paradoxical situation.

In an earlier discussion of glaciers and how they are formed (p. 144), we learned that glaciers develop in areas where the rate of snowfall and the refreezing of melted snow exceeds the rate of melting. This describes well the conditions in Mount Rainier National Park, for few areas in the United States receive more snow. In fact, with the exception of the Olympic Peninsula west of Puget Sound, the coast of Alaska, and the windward Hawaiian Islands, no other part of the United States receives as much precipitation as the Mount Rainier region. Most of this precipitation, which averages more than a hundred inches per year, falls in the form of snow, and most of the snow falls below the 10,000-foot elevation. For example, at Longmire in the Nisqually Valley (2700 feet) the total winter snowfall may exceed twenty feet. Higher up the mountainside, Paradise Valley (5557 feet) is one of the world's snowiest places, for the fall of snow here averages almost fifty feet per year. In fact, one of the world's record snowfalls was recorded here when a total of eighty-five feet and four inches of snow fell during the winter of 1971–72. This figure represents the cumulative winter snowfall and the entire eighty-five feet of snow was not, of course, all on the ground at the same time. However, Paradise Valley is commonly covered by as much as twenty-five feet of snow during the winter.

Two factors—geography and climate—are responsible for Mount Rainier's heavy annual precipitation. The most basic of these is geography, for the climate of the area is the product of conditions created by the proximity of the Pacific Ocean to the Cascade Range. Thus the Pacific Ocean provides the plentiful supply of moisture carried by the prevailing winds that blow inland from the coast. And the Cascade Range acts as a natural barrier to intercept the winds, for as the moisture-laden winds ascend the western slope, the water vapor is chilled and condenses into rain and/or snow.

Because it is colder in the higher mountains, there is greater condensation and precipitation at the higher elevations in the Cascade Range. Moreover, in some areas the snow forms perennial snowfields and these are the spawning grounds of the glaciers. There is relatively little melting of this snow, and as it accumulates in an ever-thickening blanket, the snow becomes more compact and may be slowly converted to a loose aggregate of rounded granules of ice called *firn* or *névé*. This material, when covered by subsequent snows, may be gradually compressed until the bottom portion of the snowfield may be compacted into solid ice. If favorable climatic conditions persist, the firn may eventually undergo sufficient change to convert the entire mass into *glacier ice,* and in time this great ice mass may become so heavy that the lower layers begin to yield and the ice begins to move. Only then—when the ice moves— does the body of ice become a glacier.

How do we know that the glaciers are moving? The movement of this ice is not, to be sure, as obvious as the motion of a stream; but glacial motion is, nevertheless, quite real and the movement of certain of Mount Rainier's glaciers have been studied in some detail. Because the *glaciologist* (a scientist who specializes in the study of glacial ice) cannot actually observe the movement of the ice, special techniques are required to record the velocity of a glacier. One of the most effective of these methods is to implant a straight row of steel rods directly across the glacier. These stakes, which stretch from one side of the ice to the other, are then observed at specified intervals to detect signs of movement. If the ice is flowing, this will be readily apparent, for the rods will have moved some distance down valley from the site of the original line. However, not all of the stakes will have moved the same distance; rather, the rods in the center of the glacier will be located farther downslope than those near the edges. As a result of this differential movement, the row of stakes will have become bowed in the direction of ice movement and no longer forms a straight line across the ice. The new position of the stakes indicates, then, that not only do glaciers move, but that the central part of the ice moves faster than the margins of the glacier. Thus, by measuring the distance that the stakes have moved in a given period of time, the velocity of the glacier can be determined. By measuring the bending of pipes placed in vertical holes in glaciers, it has also been learned that the bottom of a glacier moves more slowly than the ice near the surface. It is believed that the marginal and basal movement is slower because of friction and drag produced when the ice comes in contact with the bedrock. Therefore, the ice of a glacier—like the water of a stream— moves more swiftly in the center and near the surface of the glacier.

In their studies of glacial ice, glaciologists have learned that glaciers may vary greatly in their respective rates of movement. For example, many glaciers move but a fraction of an inch each day, while the daily advance of others may be measured in tens of feet. There are a number of reasons why one glacier may move more rapidly than another, but the major criteria are the thickness of the ice, the atmospheric temperature, and the degree of slope down which the glacier is moving. Observations of some of Mount Rainier's glaciers suggest that their rate of advancement may vary considerably. For example, between 1953 and 1957 Emmons Glacier advanced between one hundred and two hundred feet per year, while Carbon Glacier advanced eight to twenty-six feet per year in the period of 1958 to 1961.

Interestingly enough, a glacier actually flows under its own weight, because it moves in response to the pull of gravity on its mass. However, the upper part of the ice has relatively little weight upon it, and is, therefore, almost rigid and rather brittle. When the glacier flows over an irregularity in the valley floor or over an abruptly steepened slope, the brittle portion of the ice, called the *zone of fracture,* may become severely fractured producing deep cracks called *crevasses.* These great fissures, some of which may be as much as a hundred feet deep, are not permanent features; instead, they may open or close in response to the movement of the glacier. These yawning gaps in the ice are commonly concealed beneath the snow and are among the major hazards of traveling across the surface of a glacier.

As the glaciers pursue their inexorable course down Mount Rainier's flanks, they greatly alter the surface over which they pass. In places the ice freezes to the rock and when the glacier moves, it may dislodge and pick up large fragments of the bedrock. This process, called *plucking* or *quarrying,* permits the ice at the head of the glacier to dig out a steep-walled, bowl-shaped niche in the mountainside. The development of these depressions, which are known as *cirques,* is also aided by *frost wedging,* a process whereby rocks are disintegrated by frost action. There are many well-developed cirques on Mount Rainier, and among the most impressive of these are **Willis Wall** (the cirque of **Carbon Glacier**) whose slanting headwall drops nearly three thousand feet below Liberty Cap, and **Sunset Amphitheater,** which is the cirque of **Puyallup Glacier.**

Not only do glaciers pluck rock from the valley floor and walls, the bedrock also suffers abrasion as the ice rasps and scrapes the rock over which it flows. Glacial ice is a particularly effective agent of abrasion because the *sole,* or under surface, of the glacier is studded with sharp rock fragments of many sizes. Thus the glacier literally acts as a massive flexible file with teeth formed of embedded rock debris and the larger "teeth" scour and gouge the bedrock, causing scratches and furrows called *glacial striations.* (Markings of this type are especially well displayed in **Box Canyon.**) And, in the process of abrading the bedrock, the icebound rock particles may themselves be abraded or worn flat. It is not uncommon to find rocks of this type among the glacial deposits of the Park.

But although the coarser rock fragments may scratch the bedrock, finer particles of rock may act like sandpaper to smooth and polish it. This process imparts to the rock an even, glossy surface known as *glacial polish.* It also produces *rock flour* formed of fine sand and silt produced by abrasion and the crushing and grinding action of the ice. Where streams emerge from Mount Rainier's glaciers their waters are so heavily laden with rock flour that the water may be discolored a chocolate brown.

As they flow down the mountainside, Mount Rainier's glaciers greatly modify the valleys through which they pass. Consequently, during glaciation, a somewhat winding, narrow, V-shaped, stream-cut valley may be widened, deepened, and transformed into a typical U-shaped *glacial trough.* These broad-floored, steep-sided valleys are typically more deeply eroded than are the tributary or side valleys leading into them. When the glaciers melt, the lower ends of the tributary valleys may be left suspended high above the floor of the main trough. In some parts of the Park, streams plunge from the lips of these *hanging valleys* producing beautiful waterfalls such as **Comet** and **Christine Falls.**

Some of Mount Rainier's cirques and glacial troughs have formed rather closely to one another. As each cirque or trough developed, ice sculpturing reduced the rock separating them to a series of ridges, crags, and pinnacles. Certain of these masses of volcanic rock have a rather angular, faceted appearance and are known as *cleavers;* the **Success, Puyallup,** and **Wapowety Cleavers** are typical. **Little Tahoma,** on the other hand, has been left as a freestanding rock mass called a *horn,* and it was formed in much the same manner as the celebrated Matterhorn of the Swiss Alps.

Mount Rainier's many lakes are also part of the glacial heritage of the Park. Cirque lakes are commonly developed when the glaciers disappear from these amphitheater-like depressions and **Lake George, Lake Crescent,** and **Snow, Tipsoo,** and **Mowich Lakes** have all formed in this manner. Other lakes occupy ice-eroded basins in certain of the glacial troughs and water has also accumulated in low areas behind or upon deposits of ice-deposited sediments.

But not all of a glacier's energy is expended on the process of erosion. Glaciers are also active agents of deposition, for most of the rock debris that the glacier dislodges is gathered up and carried along by the ice. Some of this material is incorporated into the sides and bottom of the glacier; however, rocks that fall from the valley walls may be transported on top of the ice.

These rock fragments range in size from finely pulverized rock flour to monstrous boulders, and they are randomly intermingled irrespective of weight, shape, or composition. Thus, when the ice begins to melt and the glacier can no longer carry its load, the heterogeneous accumulation of rock and soil is literally dumped by the receding glacier. This glacial material, called *till*, differs from sedimentary rocks in that it lacks stratification or bedding. Deposits of till can be seen at many points within the Park, in some places appearing as well-defined surface features called *moraines*. An excellent example of a moraine is visible from Emmons Vista in Yakima Park; this mound of glacial debris marks the 1700–1900 terminus of Emmons Glacier. Many of the smaller lakes and ponds in the mountain parks have formed in depressions developed on some of the moraines. Certain of the moraines also serve as effective natural dams to impound the waters of other glacial lakes. A good example is **Mystic Lake** which is dammed behind a moraine formed by a lobe of Carbon Glacier.

Near the terminus of a glacier, meltwater generally emerges from beneath the ice, and Mount Rainier's glaciers give rise to most of the Park's rivers. The water issuing from the ice generally carries a suspended load of rock flour and silt, while pebbles, cobbles, and even boulders may be rolled along the stream bed. These sediments, which have literally been washed out of the ice, are appropriately called *outwash*. And, because they owe their origin to the action of both glaciers and running water, they are also referred to as *glaciofluvial sediments*. But unlike till, outwash deposits exhibit the stratification which is typical of all stream-laid sediments. If a glaciofluvial deposit occurs as a broad plain downslope from the glacier it is called an *outwash plain*. Many times, however, a glacial trough may become partly filled with layers of outwash, in which case it is called a *valley train*.

At some points within the Park the glaciers have left behind great rocks that appear to have been transported for considerable distance. Some of the boulders are composed of rock that is quite different from the bedrock upon which they have been dropped, and these foreign boulders are called *erratics*.

Clearly, then, the scenery of Mount Rainier owes its existence to the geologic work of fire and ice: "The Mountain" is built of materials produced by the fires of volcanism—fires that may yet be rekindled in the future. And its landscape has been sculptured and modified by the glaciers of yesterday and today, a process which even now is reshaping the face of the Park.

Plants and Animals of Mount Rainier National Park

The life forms of Mount Rainier National Park are controlled by both the altitude and the time of year. But generally, flower-covered mountain meadows, luxuriant alpine forests, and a host of birds and mammals are the dominant types of life in the Park.

The trees of the forest zone grow on the valleys and slopes at elevations of no more than about five thousand feet above sea level. The lower part of this zone, the lowland forest, lies at about the 3000-foot elevation and supports dense growths of western hemlock, Douglas fir, and western red cedar. Located between the heavy lowland forests and the mountain meadows is the intermediate forest, a plant community characterized by western white pine, Pacific silver fir, Alaska cedar, noble fir, and mountain hemlock. Trees above the 5000-foot level consist primarily of scattered clumps of alpine fir and mountain hemlock, but in the Sunrise area (at an elevation of about sixty-four hundred feet) there are stands of whitebark pine and Engelmann spruce. Trees become increasingly scarce above an altitude of about seven thousand feet, and they are virtually nonexistent above the 7500-foot elevation.

However, Mount Rainier is not especially noted for its trees; rather, it is more famous for its brilliant displays of vividly colored mountain wildflowers. Luckily it is not necessary to visit the Park at any special time to enjoy these lovely alpine flowers, for there is always a display of wildflowers at some place within the Park during the summer months. However, the most spectacular wildflower displays can be seen during July and early August. The kinds of flowers that are present in the various areas depends largely upon the altitude. However, flowers are more abundant in the flat grassy alpine meadows, or *parks,* which begin at elevations of about forty-five hundred feet. Here during early July you will probably see mountain buttercup, western pasqueflower, marsh marigold, avalanche lily, and the yellow lamb's-tongue or fawn lily. The latter flower, which is also called dogtooth violet and glacier lily, is so well adapted to its mountain environment that it commonly grows right through the snow. These same meadows produce a second wildflower display about one month later. Featured during the "second show" are American bistort, lupine, valerian, cinquefoil, Indian paintbrush, and speedwell or veronica.

Wildflowers also grow in the heavily wooded mountain forests of the lower elevations. These flowers attain their top growth during July, at which time bear grass (most abundant between forty-five hundred and fifty-five hundred feet elevation), dogwood, Pacific trillium, calypso, bunchberry, and three-leaved anemone are the dominant forms.

The slopes of Mount Rainier are home to approximately one hundred and thirty species of birds and some fifty species of mammals. Included among the smaller mammals commonly seen at lower elevations in the Park are porcupine, Douglas squirrel, snowshoe hare, beaver, raccoon, and chipmunk. The more observant visitor may see the marmot (particularly among the rock slides at altitudes of more than five thousand feet), coyote, bobcat, red fox, cougar, and marten.

Among the more common larger mammals are black-tailed deer (a type of mule deer), which can often be seen near Park roads at lower elevations, and elk which inhabit the eastern side of the Park. Both elk and deer migrate with the season. In the spring these creatures follow the receding snow line as it gradually moves upslope, and they descend in the fall when the winter snows begin. During the summer months, mountain goats can commonly be seen climbing about the rocky crags near certain of the glaciers. **Skyscraper Mountain,** the **Colonnade, Emerald Ridge, Cowlitz Chimneys,** and the slopes above **Klapatche** and **Van Trump Park** are good places to look for these agile climbers. Although not too frequently observed, black bears are common in the forests and they are occasionally spotted by hikers.

The birds, like the plants and mammals, are also distributed according to altitude and they appear to prefer the mountain parks and the forest zone. Chickadees, woodpeckers, warblers, kinglets, and thrushes are typical forest-dwelling birds, while ravens, mountain bluebirds, Clark's nutcracker, and the gray jay or camp robber appear to prefer the more open mountain meadows. Living on the upper slopes of "The Mountain" are pipits, gray-crowned rosy finches, and the white-tailed ptarmigan whose mottled brown color changes to white during the winter.

Tips for Tourists

Although it is "The Mountain" that attracts the average visitor to this National Park, the enjoyment here need not be confined to looking at Mount Rainier. In addition to the superb mountain scenery, there are a variety of interpretive programs designed to help you understand and appreciate the attractions of this scenic mountain wonderland.

Mount Rainier National Park is the mecca of many mountain climbers, but only well-qualified, experienced mountaineers should consider attempting this difficult ascent. Persons contemplating this rugged two-day climb should first contact the Park Superintendent for regulations pertaining to the summit climb. The Superintendent can also provide information about professional guide service, instructions, and rental equipment which are available from the concessioner at Paradise.

The winter sports season formally opens in December, and there is good skiing until May. Winter activities center in the **Paradise area** (reached via Nisqually Entrance) where ski tows, daytime shelter, and food services are available on weekends and holidays.

Although clouds and fog may be a problem, the most photogenic subject here is "The Mountain," but don't overlook the glaciers, wildflowers, and lovely mountain lakes. The light may be brighter than you think, especially in the ice and snow, so calculate your exposures accordingly. If possible, use a telephoto lens on the distant shots of Mount Rainier, and a haze filter will also probably help sharpen up detail.

Mount Rainier National Park at a Glance

Address: Superintendent, Longmire, Washington 98397. For additional information see *A Guide to the Trails of Mount Rainier National Park,* by Robert K. Weldon and Merlin K. Potts. This book is for sale throughout the park or may be ordered from the Mount Rainier Natural History Association, Longmire, Washington 98397.

Area: 241,781 acres.

Major Attractions: Greatest single-peak glacial system in the contiguous United States radiating from the summit and slopes of Mount Rainier, an ancient volcano; dense forests, flowered meadows, wildlife.

Season: May to October depending on weather conditions.

Accommodations: Campgrounds, group campsites, and hotels. *For reservations contact:* Rainier National Park Co., Box 1136, Tacoma, Washington 98397.

Activities: Camping, fishing, guided tours, hiking, mountain climbing, nature walks, picnicking, scenic drives, and winter sports.

Services: Food service, gift shop, guide service, post office, service station, ski rental, ski rope, ski trails, telegraph, telephone, transportation, picnic tables, and rest rooms.

Interpretive Program: Campfire programs, museum, nature trails, roadside exhibits, self-guiding trails, and trailside exhibits.

Natural Features: Canyons, ice caves, erosional features, forests, geologic formations, glaciation, glaciers, lakes, mountains, rivers, rocks and minerals, volcanic features, waterfalls, wildlife, and mountain wildflowers.

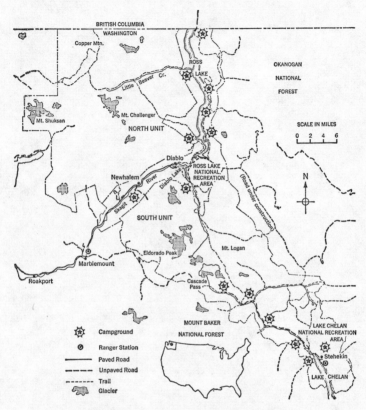

North Cascades National Park, Washington.

25. North Cascades National Park, Washington

Legacy of the Ice Age

During the Ice Age—some twenty thousands years ago—great glaciers blanketed more than 30 per cent of the land. Today ice masks only 10 per cent of the land surface. This is, nevertheless, a lot of ice. Today glaciers occupy nearly six million square miles of the earth's surface—an area almost as large as the continent of South America. Or, put another way, if we could spread the ice contained in all the world's glaciers to a uniform thickness on the surface of the land, it would form a layer of ice approximately four hundred feet thick.

We usually think of the polar regions when glaciers and ice sheets are mentioned. And rightly so, for these frigid areas are perpetually enshrouded in ice and snow. But one does not have to travel to the North or South Pole to see evidence of the Ice Age. The Cascade Range of Washington—which includes North Cascades National Park—contains approximately 756 glaciers, which cover an area of about 103 square miles. Most of these "rivers of ice" are remnants of once-mighty valley glaciers that occupied this part of the United States many thousands of years ago.

These glaciers were recently counted in a unique glacier "inventory" which revealed that the state of Washington contains about 80 per cent of the glacier-covered areas in the contiguous United States. Why bother to count glaciers in back-country wilderness areas? The North Cascade Range is the site of an intensive glacier-water-supply research program being conducted by scientists of the United States Geological Survey. A better understanding of the relation between glaciers and water supply is becoming important because of their potential as sources of fresh water. A large amount of water is locked up in glacier ice. In fact, about three fourths of all the fresh water in the world—equivalent to about sixty years' precipitation over the globe—is stored in the earth's glaciers.

But aside from their practical aspects, North Cascades National Park's glaciers have provided us with some of the world's most spectacular scenery. Indeed, the Park's ice-sculptured alpine landscape is a tribute to our planet's last great Age of Ice.

Second-Generation Mountains

As mentioned earlier, the Cascade Range is an ice-carved pile of rocks extending from northern California to northern Washington. Unlike the South and Middle Cascades, which consist mostly of volcanic rocks, the North Cascades are built of sedimentary and metamorphic rocks of Paleozoic and Mesozoic age. However, two volcanoes—Mount Baker (10,750 feet) and 10,436-foot Glacier Peak—provide evidence of volcanic activity in the northern section of the Cascade Range. Although both of these volcanic cones are outside the Park's boundaries, they have played an important part in the history of the North Cascades and have contributed greatly to the area's scenery.

The birth of these majestic mountains has already been described (pages 287–289), and geologists have studied the North Cascades in considerable detail. This area—like most of the earth's surface—has undergone many changes during geologic time. Seas invaded the land and deposited their sediments, and then the land was elevated to form the original Cascade Range. The first-generation Cascades were uplifted in Cretaceous time some eighty million years ago. This great mountain-building movement resulted in much folding and faulting of the limestones, sandstones, shales, and other sedimentary rocks that had formed on the ancient sea floor. Hot molten rock was also intruded into the heart of the newborn mountain chain. Heat and pressure from these intrusions further deformed and metamorphosed the rocks in the core of the range.

The mountains created during the Cretaceous uplift are not the North Cascades that we see today. The ancestral Cascades underwent a long period of erosion, and by the middle of the Tertiary Period—some fifteen to twenty million years ago—the craggy peaks of the original range had been eroded into relatively low, rounded hills.

The mountains we see today are not very old, geologically speaking. They were formed during the Tertiary Period about ten million years ago by a new uplift that buckled the crust upward to form a broad, north-south-trending arch. Still later, during the Quaternary Period, there was more uplift and considerable volcanic activity in the Cascade Range. However, no volcanic cones are located within the confines of North Cascades National Park. The volcanic activity of this time did have a profound effect on the Cascades. Mount Rainier, Lassen Peak, and Mount Mazama—the old volcano that collapsed to form the basin of Crater Lake—date back to these eruptions. Each of these features is spectacular enough to deserve National Park status.*

Ice Carves the Land

The high peaks of the second-generation North Cascades spawned many Pleistocene glaciers. These massive rivers of ice flowed down the flanks of the more lofty mountains and joined forces with the still larger glaciers that covered the adjacent lowlands. As time passed, the ice spread until glaciers eventually covered all but the tops of the highest mountains.

As the ice moved over the land, it drastically changed the landscape. Rounded hills were sculptured into sharp, jagged

* For more detailed information on the geology of the Cascade Range, see *Origin of Cascade Landscapes,* by J. H. Mackin and A. S. Cary (Information Circular No. 4, Washington Division of Mines and Geology, Olympia), and *Cascadia: The Geologic Evolution of the Pacific Northwest,* by Bates McKee (McGraw-Hill Book Company, New York).

peaks, and narrow, V-shaped, stream-cut valleys were scooped out to form sheer-walled, U-shaped glacial troughs such as seen at Stehekin, near the head of Lake Chelan. Indeed, the lake itself occupies an 8500-foot-deep gorge that is one of the deepest canyons in North America. Many waterfalls plunge from the hanging valleys of these steep-sided, flat-bottomed, glaciated troughs. The Park is also studded with numerous alpine lakes that occupy basins scooped out by now-vanished glaciers. This spectacular alpine scenery was not the result of a single advance of the glaciers. Great tongues of ice scoured the surface three, possibly four, times during the one and one half million years of Pleistocene time. Each major advance of the ice was followed by a warmer, interglacial period. During these warming trends the plants, animals, and climates were much as they are in the North Cascades today.

What is the status of North Cascades glaciation today? Many geologists believe that the area is now in an interglacial stage. Most of the glaciers within this section of the Cascade Range seem to be fairly stable and in equilibrium with current climatic conditions. But some of the glaciers appear to be advancing. Thus it is not impossible that ice may again cover much of the North Cascades in some yet-to-come Ice Age of the geologic future.

Plants and Animals of North Cascades National Park

The rugged terrain and varied climatic zones of the Cascade Range support a variety of life forms. Plant communities occur that range from rain forest to subalpine cone-bearing trees, lush meadows to alpine tundra, and pine forests to sunny, dry shrublands. More rain falls on the western slope of the North Cascades. Consequently, more streams and lakes are present and the vegetation is more abundant. The eastern side of the mountains receives more sunshine, and the rather warm days and cool nights support plant species more typical of a dry climate.

Because of their relative isolation, the North Cascades support extensive and varied wilderness ecosystems. Although seldom seen, moose, wolverine, cougar, and grizzly bear are present. Visitors are more likely to sight mountain goats, deer, and an occasional black bear. The white-tailed ptarmigan and many smaller birds and mammals also live in this rather inhospitable environment.

The North Cascades area abounds in streams and lakes that contain many fish. The principal game fish are trout—cutthroat, eastern brook, rainbow, and Dolly Varden. (State laws apply to fishing within the Park as well as the Ross Lake and Lake Chelan National Recreation Areas; a Washington fishing license is required.)

Tips for Tourists

Established on October 2, 1968—the same date as Redwood National Park (page 353)—North Cascades National Park is divided into two parts. The North Unit and the South Unit of the Park are separated by a narrow arm of the Ross Lake National Recreation Area (see map, page 306). The Lake Chelan National Recreation Area is south of the South Unit of the National Park. These four areas of the National Park Service encompass 174,000 acres of the more scenic part of the North Cascades mountain range near the Canadian border.

Much of the activity at North Cascades centers around the lakes, many of which provide fine fishing. Boat trips are also popular, and boats can be rented within the area. Concessioner-operated cruises are also available on certain lakes, and information about these can be obtained at a Ranger Station.

North Cascades National Park at a Glance

Address: Superintendent, North Cascades National Park, 311 State Street, Sedro Wooley, Washington 98284.

Area: 505,000 acres.

Major Attractions: Wild alpine region of jagged peaks, mountain lakes, glaciers, and wildlife.

Season: Early April to mid-October at lower elevations; mid-June to mid-September at higher elevations.

Accommodations: Tourist facilities located in larger towns and cities near the Park (contact local chambers of commerce for detailed information).

Activities: Air flights (charter), boating, boat rides, camping, fishing (Washington license required), guided tours, hiking, horseback riding, mountain climbing, nature walks, scenic drives.

Services: Boat rentals, food service, horse rentals, picnic areas, religious services.

Interpretive Program: Campfire programs, nature trails, nature walks, roadside exhibits, self-guiding trails, slide shows.

Natural Features: Canyons, geologic formations, glaciers, glacial landforms, lakes, mountains, waterfalls.

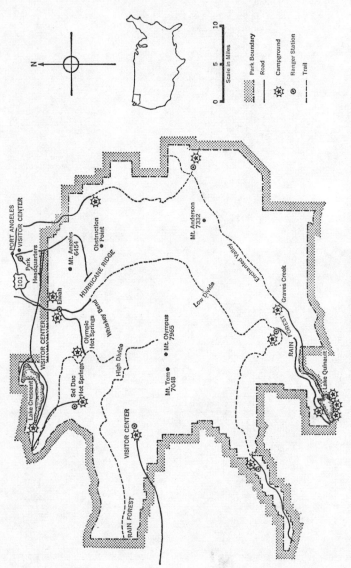

N

Scale in Miles

0   5   10

Park Boundary
Road
Campground
Ranger Station
Trail

PORT ANGELES
VISITOR CENTER
Park Headquarters
101
Elwah
VISITOR CENTER
Olympic Hot Springs
Sol Duc Hot Springs
Lake Crescent
High Divide
Whiskey Bump
HURRICANE RIDGE
Mt. Angeles 6454
Obstruction Point
Mt. Anderson 7332
Enchanted Valley
Low Divide
Graves Creek
RAIN FOREST
Lake Quinault
Mt. Olympus 7965
Mt. Tom 7048
VISITOR CENTER
RAIN FOREST

Olympic National Park, Washington.

26. Olympic National Park, Washington

Land of Contrast

Olympic National Park is a remarkable study in contrasts, for its natural beauty abounds in myriad variations. There are glistening glaciers and lush rain forests, pounding Pacific surf and placid alpine lakes, ice-scarred mountains and sparkling ocean beaches. These diverse features provide the spectacular and unspoiled setting for the westernmost National Park in the contiguous United States.

The Olympic Mountains

The Olympic Mountains are the highest and most beautiful part of the Pacific Coast Ranges, an elongated western fringe of mountains that border the Pacific Ocean from California to British Columbia. Though not marked by the volcanoes or towering granite peaks that accent the Cascades and the Sierra Nevada, the Olympics are, nevertheless, rugged mountains of unusual scenic splendor. At their highest point—Mount Olympus in the heart of the Park—the Olympics reach an elevation of 7965 feet. In addition, there are a number of other Olympic mountain peaks that exceed seven thousand feet in elevation, but most of the Olympic crest lies between five thousand and six thousand feet above sea level.

How the Mountains Were Formed

Geologically, these are young mountains, for the uplift of the present Olympics took place only about fifteen or twenty million years ago during Late Pliocene or Early Pleistocene time. This uplift coincided with the depression of the Puget Sound Basin and these two geologic events shaped much of the present topography of western Washington. But the earlier chapters in the geologic story of the Olympics were written long before the elevation of the present mountains, for the oldest rocks in the Park date back to the Cretaceous Period some 130 million years ago. These strata were formed from sediments deposited in a shallow sea that extended from California to Alaska and that covered western Washington as far north as the Cascade Range. Composed largely of rock fragments eroded from nearby landmasses and carried by streams to the sea, these sediments slowly accumulated as great thicknesses of sand and mud. Today they comprise the *Soleduck Formation,* approximately twenty thousand feet of hard gray sandstone and flinty, dark-colored slates.

The environment provided by the Mesozoic sea must not have been suitable for the development of marine life, for the rocks are virtually devoid of fossils. However, a few worms apparently became adapted to life on the muddy ocean floor and their tubes have been found as fossils. Despite their scarcity, these worm tubes provide an important clue to the age of the Soleduck Formation, for similar tubes are found in association with typical Early Cretaceous fossils in Alaska and the western Cascades. This has led many geologists to believe that the fossil worm tubes of the Olympics must also be of Early Cretaceous age. This fossil evidence—which is at best inconclusive—is about the only evidence for the age of the Olympic Mountains.

There must have been periods of submarine volcanic activity in the area during Mesozoic time, for in places there are pillow-like masses of lava interbedded with layers of impure marine limestone. "Pillow lavas" are characteristic of undersea lava flows and these flows apparently emanated from fissures which opened up on the floor of the old Mesozoic sea.

Near the end of the Cretaceous Period or the beginning of Tertiary time, mighty forces from deep within the earth elevated the sea floor and the waters were drained from this portion of western Washington. The pressure and heat which accompanied this crustal disturbance greatly deformed the rocks of the Soleduck Formation, further hardening the sedimentary strata and changing shales to slate. Concurrently, the sedimentary strata and the lava flows—which originally lay horizontal on the sea floor—were folded, fractured, and finally tilted into the position that they occupy today.

Following their uplift near the end of the Mesozoic, the newly exposed Soleduck rocks were subjected to continued and prolonged erosion. Throughout the Paleocene and Eocene Epochs, the newly raised Olympic landmass was steadily worn away until it eventually was eroded down nearly to sea level. As the land was lowered, the sea again returned to inundate the surface of western Washington. Then once again the ocean floor was rent by deep fissures from which poured great floods of molten rock. As it flowed out on the sea bottom, the lava quickly cooled and solidified, forming huge pillow-shaped rock masses. Between the lava flows in some areas there are layers of ash and other fragmental volcanic rocks—evidence that some of the eruptions were accompanied by violent explosions. Elsewhere, the submarine lava flows are interbedded with layers of sandstone, shale, and conglomerate formed from marine sediments deposited during periods of volcanic quiescence. The fossilized remains of fish, marine invertebrates, and plants have been found in some of these sedimentary layers. Volcanic activity must have been more widespread in the area that is now the eastern part of the Olympic Peninsula, for here the lava flows attain a cumulative thickness of about five thousand feet. Curiously enough, the upper flows lack the pillow structure that characterizes the lower, older lava floods. This suggests that at least some of the younger lava flows were probably extruded on dry land when the area was above sea level. This thick section of lava flows and interbedded marine sedimentary rocks has been named the *Metchosin Formation.*

The Metchosin Formation is overlain by the *Crescent Formation,* which contains fewer volcanic rocks and a preponderance of shales and sandstones. The record as indicated by these rocks suggests that volcanic activity was waning near the end of the Eocene Epoch. Moreover, as the result of becoming filled with sediment or gradual elevation of the sea floor, the Late Eocene sea gradually became much shallower.

The geologic record of the Oligocene and Miocene Epochs is not so easily deciphered. However, the Olympic region seems to have been the site of alternate advances and retreats of the sea, and deposition of marine sediments was widespread throughout the area. Eventually—near the close of Miocene time—the Olympic area was once more elevated well above sea level. This uplift was apparently greater than any that had preceded it and gave rise to a long mountain range that trended in a northwest-southeast direction from the Pacific Ocean, southeastward across western Washington to the Columbia Plateau. These mountains, which were the forerunners of the modern Olympics, were soon attacked by erosion from their summits. Finally, during Late Pliocene or Early Pleistocene time, the ancestral Olympics were subjected to one final series of mountain-building movements. It was this disturbance which produced the modern Olympic Mountains and further depressed the Puget Sound Basin, thereby creating the unusual terrain which today marks the Olympic Peninsula.

Ice the Master Sculptor

Although the major upthrust of the Olympics culminated perhaps as much as twenty million years ago, the "finishing touches" have been added to the mountains much more recently. For example, within the past one million years, there have been a number of minor, spasmodic periods of uplift in the Pacific Northwest, for earthquakes have been recorded around the margins of the Olympics for many years. Then, too, those insatiable agents of erosion—running water, wind, weather, and ice—have slowly taken their toll on the uplifted sedimentary rocks.

Glaciers of the Past. But of all the erosional agents that have left their mark on the face of the present Olympic Mountains, ice has been the most important. During the great Ice Age of the Pleistocene Epoch a massive ice sheet, or continental glacier, moved south from Canada and extended down into the United States. The southern boundary of this great glacier varied with changes in the temperature. At times the climate warmed and the ice slowly thinned and receded, only to advance again when the temperature once more began to drop. Thus, during the last one million years, this great ice sheet pushed southward at least four times, until finally about eleven thousand years ago, it completely withdrew from the Pacific Northwest.

Although the vast Canadian ice sheet had a pronounced effect on the landscape of the Olympic area, the erosional effects of the more localized mountain and piedmont glaciers are more readily discernible. During Pleistocene time, these glaciers flowed out of mountains on the mainland of British Columbia and united in the lowlands to form a huge *piedmont* (literally, "at the foot of the mountain") *glacier*. This great tongue of ice extended southward into Puget Sound and pressed against the eastern slope of the Olympic Mountains. As the ice moved south it split into two lobes, one of which flowed westward through the Strait of Juan de Fuca (see map) and pushed well up onto the flanks of the Olympics. This lobe of the piedmont glacier brought with it immense boulders that were later deposited as the ice melted. Because they consist of granite—a type of rock which does not crop out in the Olympics—it is believed that these glacial erratics were originally picked up in the mountains of western British Columbia. The presence of the granite erratics are significant in that they give some indication as to the extent and thickness of the now-vanished glacier, for some of the boulders have been found at altitudes of as much as three thousand feet. Meanwhile, the other lobe of the glacier plowed southward into the Puget Sound Basin and finally stopped about ten miles beyond what is now Olympia, Washington. There is also geologic evidence to indicate that the ice moved west along the northern border of the Olympics. As it did so it gouged and deepened an ancient valley that became filled with water as the glaciers melted. Today this glacial valley contains Lake Crescent and Lake Sutherland.

Even more spectacular, however, was the glacial sculpturing done by the Olympic mountain glaciers. Not only did these flowing tentacles of ice gouge, scour, and polish the bedrock, they also left behind thick beds of ice-deposited till, jagged, saw-toothed ridges, and many glacier-carved, lake-filled basins. Nor have the mighty rivers of ice completed their task: even now the glaciers continue to carve and reshape the mountain-sides, further altering the Olympic landscape.

Glaciers of Today. Today's glaciers are, to be sure, small when compared to the large ice masses of the Pleistocene; even so, at least sixty glaciers still exist in the Olympic Mountains today. These glaciers have a collective area of between twenty and twenty-five square miles and vary greatly in size. Six of them—Hoh, Blue, White, Hubert, Jeffers, and Hume—flow down the ice-scarred flanks of Mount Olympus. And though three of these, the Hoh, Blue, and White Glaciers, are at least two miles long, the rest of the Olympic glaciers are considerably smaller. Among the other Olympic peaks that still support glaciers are Mounts Anderson, Carrie, Christie, Ferry, Queets, and Tom. It is significant to note that most of the larger ice bodies are located in northerly valleys where they are spared the direct rays of the sun.

Why are there so many glaciers in the Olympics when they have vanished from most of the western mountains? The answer to this riddle lies in the geography of the region, for the Olympics, like the Cascade Mountains, are ideally located for the formation and preservation of glaciers and snowfields. The first, and most basic, geographic factor is the proximity of the Olympic Mountains to Puget Sound and the Pacific Ocean. These bodies of water provide an abundant supply of moisture which is picked up and carried inland by the prevailing winds. As the winds sweep across the land, they soon encounter the mountains which act as a natural barrier to intercept them. Upon coming in contact with the cooler temperatures of the highlands, the water vapor is chilled and condenses into rain and/or snow. Because it is colder at the upper elevations, snowfall is greater in the higher mountains. For example, it is estimated that during some years more than two hundred feet of snow may fall on Mount Olympus. Much of this snow at higher altitudes may never melt; instead, it may form perennial icefields which may give rise to additional glaciers. As a result of the heavy winter snowfall and generally cool summer weather, the regional snow line here is normally down to about six thousand feet. Above this "line" snow remains on the ground throughout the year. But, because the snow line may move up or down according to the rate of winter snowfall, in some years it may be as low as four thousand feet.

The Pacific Coast Area— Where Mountains Meet the Sea

Although separated by the width of the United States, Olympic and Acadia National Parks have one notable feature in common: both have rugged coasts marked by spectacular landforms produced by wave erosion. Nor is this surprising, for on the western border of Olympic National Park the land and sea are locked in a never-ending struggle that is almost as old as the earth. Here along the Pacific Coast Area, a narrow, fifty-mile ribbon of coastal land, the restless Pacific doggedly strives to wear away the land. And the land, under constant attack from the irresistible power of the waves, is slowly retreating, giving rise to a wildly beautiful, rocky coastline.

Sea versus Shore. One needs to spend but a short time observing this ageless battle to appreciate the geologic efficiency of the sea. It soon becomes apparent that the sea's most effective work is carried on with the aid of waves and wave-produced currents, for each wave that breaks on the shore erodes by a combination of processes. Most of the wave erosion is caused by the *abrasion,* or scratching, of the coastal rocks. For their abrasive tools, the waves use rock particles which are partly generated by their own action and partly brought into the sea by streams and other agents. These sediments—constantly moved back and forth by the coming and going of wave-produced currents—are gradually worn down to small size. Moreover, they also abrade the bedrock with which they come in contact.

Marine erosion is also carried on by *hydraulic action.* As great volumes of water are repeatedly hurled against the coast, the rocks are gradually dissolved or broken up, thus over long periods of time, considerable hydraulic erosion may occur. Storm waves commonly batter the sea cliffs with pile-driver-like blows that are violent enough to shatter jointed rocks along the face of the cliffs. In addition, water, dashed suddenly into rock fractures during the breaking of the wave, may compress air which will act as a wedge to loosen the rock and further enlarge the rock opening.

During storms the tempo of both abrasion and hydraulic action is greatly accelerated. High winds produce massive waves that pound the beaches and cliffs with tremendous force. The storm waves also dislodge more, and larger, rock fragments, which join the other rock particles that steadily wear away the coastal rocks.

Geologic evidence of erosion by the sea—and resistance by the land—abounds along this coastal area. At places where the waves have attacked the hilly shore, the rocks closest to sea level have been more vigorously eroded than have those above the reach of the waves. So the waves gradually undercut the cliff until the rock above does not have sufficient support to maintain itself. The undermining of the cliff face eventually permits the overhanging rock to crash into the sea, and as erosion proceeds this *wave-cut cliff* is pushed steadily back. As the cliffs retreat, a nearly level shelf or *wave-cut terrace* is developed just below the water level. Because the water of the breaking waves must cross this terrace before reaching the shore, some of their force is dissipated before reaching the cliff. Thus, the further the cliff retreats, and the wider the terrace becomes, the less effective is the erosive power of the waves. And, if the sea level remains constant, the rate of recession of the cliffs becomes steadily slower. But at best the land is afforded only temporary relief from the relentless onslaught of the sea. As the edge of the submarine terrace is gradually eroded landward, the waves will again assault the shore with renewed vigor.

How rapidly the land will be worn away will depend largely upon the resistance of the rocks and the degree of wave action that they are subjected to. If the coastline is composed of rocks of uniform resistance to erosion, the sea cliffs normally develop as a rather smooth unbroken wall. But where the rocks display varying degrees of strength, the more resistant rocks will remain as finger-like projections of land that extend out into the ocean. Such features, called *promontories,* or *headlands,* are commonly known as "points" or "heads" in the Pacific Coast Area. In areas where there are zones of weakness in the rocks, the waves will erode the shore more quickly, thereby producing indentations between the headlands. These *coves,* or *inlets,* alternate with the headlands to give the coast the rugged outline for which it is famed.

The results of differential marine erosion may also give rise to *sea caves.* These hollowed-out cavities may develop at the foot of sea cliffs in places where the rock is less resistant than elsewhere. Occasionally the cavelike hollows will cut through from both sides of a headland and may intersect to form a *sea arch.* Should the arch be enlarged until its roof finally collapses, the isolated walls become separated from the headland and remain standing as *sea stacks.*

But the sea can build as well as destroy. The material eroded from the shore must ultimately be deposited when the waves or currents suffer reduced velocity, and evidence of marine deposition is common in the Pacific Coast Area. In areas where the rocks are especially weak and sand is abundant, *beaches* have formed. Elsewhere *barrier bars* have been built across the mouth of bays and shallow inlets. These formed in places where sediments accumulated in an elongate pile when shore currents deposited their sediments upon coming into shallower, quieter waters. Thus, both marine erosion and marine deposition have joined forces to produce the picturesque terrain of the Pacific Coast Area of the Park.

Plants and Animals of Olympic National Park

Like the Park's geology, the life forms of Olympic are remarkable for their contrasting diversity. And, like the geology, the life of the Park is an outgrowth of the geographic and climatic conditions on the Olympic Peninsula. No other National Park provides refuge for such diverse creatures as seals and bears, salmon and elk, sea gulls and marmots. Plant life is equally varied, ranging from sea weed to mountain wildflowers; from jungle-like growths of mosses and ferns to sparse stands of high altitude alpine fir.

Because of the wet winter climate that prevails on the western side of the Olympic Peninsula, this region supports an incredible array of plant life. This is primarily due to the high rate of rainfall, which in places may exceed one hundred and forty inches per year. This does not include the heavy snowfall which blankets Mount Olympus and the high country each winter. So the total annual precipitation is much greater, thus producing the wettest winter climate in the conterminous United States. Yet only fifty miles away, on the northeast side of the Peninsula, lies one of the West Coast's driest areas. Indeed, some of the land is so arid that farmers must irrigate their crops.

To explain the paradoxical situation of a rain forest on one side of the mountains and a "rain shadow," or arid area, on the opposite side, we must again refer to the geography of the area. As noted earlier, the moisture-laden prevailing winds are deflected upward as they approach the Olympic Mountains. Here the cooler temperatures cause the water vapor to chill and condense into the rain or snow responsible for the high rate of winter precipitation on the western slope. But by the time the winds have passed over the Olympic peaks and reached the northeast side of the peninsula, they have left behind most of their moisture and are nearly dry.

It is not surprising, then, that an unusually large concentration of plant life is found in the western valleys of the Olympic Mountains. This is the Olympic rain forest: a complex, moss-draped plant community of trees, shrubs, ferns, and fungi. The western hemlock and Sitka spruce are the most characteristic trees of the rain forest. The latter, a species which grows along the coast from Alaska to California, occasionally attains a height of three hundred feet and a diameter of as much as thirteen feet. The western hemlock is not much smaller—some of them may reach a height of one hundred and seventy-five feet and a record-size specimen with a trunk diameter of nine feet has been found in the Park. Two other trees, the western red cedar and Douglas fir, are also common in the rain forests and both have attained record size in Olympic National Park. Other trees that are common in the Park include the red alder, big-leaf maple, and black cottonwood; these seem to prefer the banks of streams. Ferns, mosses, vine maple, and numerous smaller plants grow in tangled profusion beneath the umbrella of trees, and the forest floor is carpeted with moisture-loving fungi, mosses, and liverworts. The rain forest is a silent, mysterious, beautiful plant "underworld" that normally is pervaded by a yellow-green glow. Here, thanks to the National Park Service, you can lose yourself in a green-lit fairyland and enjoy the solitude of one of our nation's truly remarkable plant preserves.

At higher elevations within the Park, the climate is different and so is the vegetation. Above the lowland rain forests, at an elevation of about two thousand feet, there are stands of western hemlock, Douglas fir, western white pine, and Pacific silver fir. The undergrowth of these forests consists largely of small shrubs, herbs, and flowering plants. The mountain slopes between thirty-five hundred and five thousand feet above sea level are characterized by yet another flora. The dominant trees here are alpine and Pacific silver fir, Alaska cedar, mountain hemlock, and Douglas fir. These trees are typical of Olympic's high country, and they occur in forests that are occasionally separated by flower-filled alpine meadows. Although mountain wildflowers grow at virtually all elevations within the Park, they are most varied and abundant in these high-country mountain meadows. Cinquefoil, aster, arnica, false hellebore, shooting star, avalanche and lamb's-tongue or fawn lilies, blue lupine, larkspur, columbia lily, and bluebell—these are but a few of the colorful wildflowers that grow at different places in the highlands. In the northeast part of the Park, especially in the vicinity of **Hurricane Ridge** and **Mount Angeles,** there are unusually fine displays of mountain wildflowers.

Among the many animals that inhabit this Park, the most noteworthy is the Olympic or Roosevelt elk. These huge beasts—a full-grown bull may weigh as much as a thousand pounds—can sometimes be seen in the western valleys or in the high-country meadows. Columbian black-tailed deer are more easy to observe. Look for them during early morning or late afternoon in the lowlands and meadows, especially at **Deer Park** and **Hurricane Ridge.** Although not as common as deer, black bear can sometimes be seen feeding in the mountain meadows. A popular, rather easy-to-see animal is the Olympic marmot which lives in the rock slides of the high country. They are especially numerous at Deer Park, but the observant visitor is apt to see them among boulder piles at many places in the higher parts of the Park.

More than one hundred and forty kinds of birds have been known to summer here, and they, like the plants, seem to vary with the elevation. Near sea level along the Pacific Coast Area look for ravens, herons, gulls, cormorants, and flocks of crows. If you happen to be hiking along roadless stretches of the beach, keep your eye peeled for a glimpse of the bald eagle, for they nest in trees near the shore. Other birds, for example the water ouzel or dipper, the belted kingfisher, and the colorful western harlequin duck, are more likely to be seen along the banks of streams. Among the birds that live in and near the forests are wrens, flickers, thrushes, hummingbirds, ruffed grouse, Steller's jays, and a variety of woodpeckers. At higher elevations in the mountain meadows near timber line, you may see hawks, ravens, Oregon jays or camp robbers, horned larks, sooty grouse, mountain bluebirds, and the Oregon junco. The snowy mountain peaks are not hospitable environments for birds, but at least one species—the sparrow-like rosy finch—is sometimes seen at these elevations.

In addition to the larger mammals and birds, a number of smaller animals make their home in Olympic National Park. Among these are toads, salamanders, and a few snakes, none of which are poisonous. Moreover, a host of marine animals thrive in the salty waters bordering the Pacific Coast Area. Some of these animals live along the rocky shore or in intertidal pools; others prefer the sandy beaches. The more adventurous visitor will take time to prowl along the shore, for here is a veritable seaside museum of fascinating marine life. While hiking along the beach, you will also probably see a number of sea birds, and—if you are lucky—you might get a glimpse of a seal sunning on the offshore rocks. From time to time deer and bear can be seen in the woods near the shore, and raccoon and skunk are fairly common.

Tips for Tourists

A unique experience awaits the visitor who explores this unspoiled wilderness from the wild Pacific Coast Area, through the humid rain forest and flower-decked mountain meadows, to the sparkling glaciers of the high country. But whether you stay in Olympic for two days or two weeks, the National Park Service has provided interpretive facilities which will greatly enhance your understanding and enjoyment of this remarkable scenic area.

The ideal place to get the "feel" of Olympic is at one of the three strategically located Visitor Centers. The museum exhibits here have been prepared to introduce you to the general features of the Park. Here, too, you can purchase maps of the trails, books that will help you to identify Olympic's plants and animals, and Ranger-Naturalists are on duty to provide additional information and help you plan your stay.

The many faces of Olympic National Park offer limitless

photographic opportunities. Mountain panoramas, seascapes, wildlife, and wildflowers are but a few of the many subjects that are available. On long-distance shots of the ocean or mountains, a haze filter will be helpful. And, because the light here may be brighter than you think, check your exposure readings carefully. Especially unusual photographs may be taken in the rain forest. However, long exposures are often required so it is advisable to take a tripod along. Close-up or telephoto lenses will permit you to take much better pictures of small plants and wildflowers. Moreover, the latter will be of great help in photographing the wild animals.

Olympic National Park at a Glance

Address: Superintendent, 600 East Park Avenue, Port Angeles, Washington 98362.

Area: 896,599 acres.

Major Attractions: Mountain wilderness containing finest remnant of Pacific Northwest rain forest; active glaciers; wilderness beach; and rare Roosevelt elk.

Season: Year-round.

Accommodations: Cabins, campgrounds, hotels, motels, and trailer sites. For latest information on accommodations in the Park write the Park Superintendent for the latest copy of the "Facilities, Accommodations, and Services" sheet. Accommodations are also available in the town of Port Angeles; for a complete list of accommodations write for a "Directory of Visitors' Accommodations and Service," obtainable from Washington State Ferries, Seattle Ferry Terminal, Seattle, Washington 98104.

Activities: Boating, camping, fishing, guided tours (in summer), hiking, horseback riding, mountain climbing, nature walks, picnicking, scenic drives, swimming, water sports, and winter sports.

Services: Boating facilities, boat rentals, food service, gift shop, post office, religious services, service stations, ski rental, ski tow, telephone, transportation, picnic tables, rest rooms, and general store.

Interpretive Program: Campfire programs, guided hikes, museum, nature walks, roadside exhibits, self-guiding nature trails, and trailside exhibits.

Natural Features: Canyons, forests, geologic formations, glaciers, lakes, marine life, mountains, rivers, rocks and minerals, seashore, unusual plants, waterfalls, wilderness area, wildlife, rain forest, and submarine lava flows.

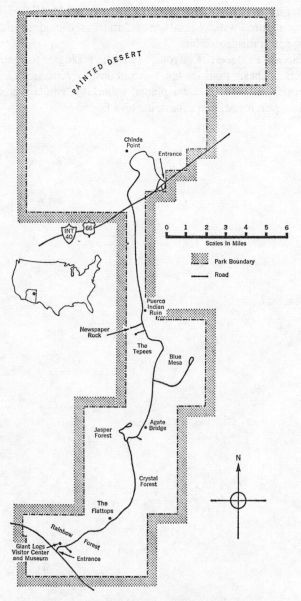

PAINTED DESERT

Chinde
Point

Entrance

INT
40   66

0 1 2 3 4 5 6
Scales in Miles

Park Boundary

Road

Puerco
Indian
Ruin

Newspaper
Rock

The
Tepees

Blue
Mesa

Agate
Bridge

Jasper
Forest

Crystal
Forest

The
Flattops

Rainbow

Forest

Giant Logs
Visitor Center
and Museum

Entrance

N

Petrified Forest National Park, Arizona.

Rainbows in the Rocks

In the semidesert country of northeastern Arizona lies one of
the world's famous "forests." Here is Petrified Forest National
Park, the most colorful and extensive petrified wood display
known to man. Here, too, is the Painted Desert—a bizarre,
multicolored wonderland of steep-walled gullies, sharp pinnacles, and rounded knobs.

The Great Stone Trees

Although petrified wood is found in almost every state in the
Union and in many countries throughout the world, there are
no "stone trees" that are quite so well known or as easily
accessible as those of Petrified Forest National Park. Within
the Park's 94,189 acres there are six major "forests" of fossil
trees which were felled by nature and now lie scattered upon
the Painted Desert. Each of these forests is appropriately
named and each has its own special charm and appeal. There
is, for example, beautiful **Jasper Forest,** so-named because of
the many trees that have been converted to jasper, a colorful,
opaque subvariety of the mineral quartz. And, as might be
expected, the fossil trees in the **Long Logs** area are noted for
their unusually great length—logs more than a hundred feet
long are not uncommon here. **Crystal Forest** is also well named,
for some of the fossil logs contain openings in which have

been deposited perfectly formed crystals of clear quartz and, more rarely, crystals of *amethyst,* a purple- or violet-colored variety of quartz. Interestingly enough, these relatively valuable crystals were instrumental in prompting President Theodore Roosevelt to create Petrified Forest National Monument in 1906. Prior to the geological study which preceded this decree, Arizona's fossil trees were systematically being destroyed or removed by souvenir hunters, commercial jewelers, and amateur gem collectors. In fact, there were even plans to erect a crushing mill to pound these timeless relics into fragments to be used in making abrasives for sandpaper. It was the latter development that prompted a conservation-minded group of nature lovers to press for federal protection of this wondrous exhibit of petrified wood. Fifty-six years later, on December 8, 1962, this area became known officially as Petrified Forest National Park, further assuring the permanent preservation of the vast stone forest and its colorful landscape.

The above forests, along with **Blue Mesa** and **Rainbow Forests,** are located on a 28-mile scenic road that extends from the Painted Desert in the north to Rainbow Forest Visitor Center on the south. **Black Forest,** an area noted for its interesting array of dark-colored fossil wood, is well off the "beaten path" in a roadless and desolate part of the Painted Desert.

As you drive through the colorful desert and visit the fossil forests, many questions will no doubt cross your mind. You will almost certainly wonder how and why the trees have turned to stone. You may also marvel at the broad spectrum of colors that beautify the fossil wood. And perhaps you will also speculate as to why the trees are not standing upright and how does it happen that most of the logs are cracked and broken into sections? The best place to find the answers to these questions is, of course, in the fossil forests, along the trail, and in the Rainbow Forest Visitor Center Museum. The Park has a well-planned and comprehensive interpretive program and there are museum displays, wayside exhibits, and other educational devices to help the visitor to understand better the story of the fossil logs. And, as in all of the

National Parks, Park Ranger-Naturalists are ever ready to answer your questions. But in the meantime, let us turn back the pages of time and attempt to answer some of the questions that we have raised.

Kinds of Trees in the Fossil Forests

The fossil record clearly indicates that near the close of the Triassic Period, some 175 million years ago, this region was probably a rather low flood plain which was periodically deluged by flooding streams and its climate was probably of a tropical to subtropical nature. There were no cactus or desert-dwelling plants growing here then; rather, there appears to have been an abundance of moisture-loving plants such as ferns, mosses, and rushes. You might not have recognized these plants, however, for some of the rushes, or horsetails, grew to be as much as thirty feet tall and their stems were almost a foot in diameter.

Here and there on scattered hummocks and ridges above the swampy lagoon grew small, thick-set clumps of the trees whose fossilized remains we find today. The most common of these was a cone-bearing tree kindred to the modern araucarias, a group of pine-related trees that today are native only to Argentina, Brazil, Chile, eastern Australia, and a few other scattered locations in the southern hemisphere. *Paleobotanists* (scientists who specialize in the study of fossil plants) have named these trees *Araucarioxylon arizonicum* and they believe that these huge trees must have dominated the swamps in both size and number, for specimens up to one hundred and twenty feet long and seven feet in diameter have been discovered in the Park. However, most of the fossilized tree trunks are considerably smaller, averaging some three to four feet in diameter. In addition to *Araucarioxylon,* two other species, *Woodworthia arizonica,* another extinct conifer, and the peculiar *Schilderia adamanica* inhabited the Triassic swampland of northern Arizona. A rather small tree whose trunk probably did not exceed four feet in diameter, *Schilderia* is distinguished by a series of peculiar radiating rays. Curiously enough, paleobotanists have been unable to show a close affinity between this fossil and any living plant species. But though the three trees listed above are most commonly found as fossils, many more plants undoubtedly flourished here, for more than forty-two fossil species have been recognized within the Petrified Forest. These include the remains of ferns, flowering plants, cycads (a group of plants with palmlike leaves), and a variety of coniferous plants.

The type of environment described above would most assuredly have attracted many water-loving animals and this, too, is borne out by the fossil record. Had you visited this boggy lowland you would surely have noticed the *phytosaurs,* for some of these great crocodile-like reptiles grew to be twenty-five feet long and weighed almost a ton. A rather formidable beast, this animal had a long, slender snout armed with many sharp teeth and a heavy body protected by a covering of dense, bony plates. Despite their superficial resemblance to the crocodiles, the phytosaurs are a separate and distinct group of reptiles that lived only during the Triassic Period. Another odd—and long-extinct—denizen of the Arizona flood plain was the *labyrinthodont,* or *stegocephalian,* an early thick-skulled amphibian that is related to the modern salamanders. Clumsy, short-legged, alligator-like creatures, these animals were about five feet in length and weighed several hundred pounds. There is also evidence that a variety of other amphibians and reptiles lived in and among the luxuriant swamp vegetation, for their fossilized remains have been found in conjunction with the petrified wood. This, then, was the environment of northern Arizona some 175 million years ago when the now-petrified trees were flourishing in the warm, humid swamplands.

Log Jams and Sandbars

Even the earliest visitors to the Great Stone Forests noticed one obvious characteristic of the fossil logs: virtually all the trees have been stripped of their roots, bark, and branches, and they lie randomly scattered about the colorful badlands. This has led geologists to conclude that most of the logs represent fallen trees that were transported into the area by flooding Triassic streams which originated in an area of low hills that rose to the west and southwest of what is now Petrified Forest National Park. Most of these trees, like those in present-day forests, were probably killed by natural means such as disease or fires caused by lightning and volcanic activity. The occurrence of petrified charred tree remains and fossil wood fragments which contain undisputed evidence of attack by insects, are considered as proof that the trees did die from such causes. What happened to the trees that originally grew in the flood plain? They probably died of similar causes, but decayed rather quickly after falling on the damp swamp floor and being exposed to the warm, humid atmosphere and the destructive action of insects and fungus. But, as noted earlier, standing stumps with root systems in place have been discovered in the Park so it is known that similar trees *did* grow in the area. However, the general lack of fossilized cones, bark, roots, and foliage, plus the worn appearance of the logs is considered ample evidence that most of the logs were rafted in by flooding streams.

As the streams floated the logs into the shallow lagoons and swamps, many accumulated in certain areas in much the same manner that a log jam might develop in a stream of today. The trees in these dense log packs eventually became water-logged and sank to the bottom of the marshy lagoons where they were buried rather rapidly by the sediments which were being brought in by the streams. Other logs became stranded on sandbars or along the shore; these, too, were ultimately covered by the sands and muds of the fast-flowing streams.

Geologic evidence indicates that the above events must have occurred repeatedly over untold numbers of years, for more than four hundred feet of sandy muds eventually accumulated in this ancient flood plain and tree remains were probably deposited throughout the sediments. As time passed the sediments were slowly converted into shales and sandstones which are now known a the *Chinle Formation*. It is these rocks that comprise the surface geology of the Park today.

Meanwhile, the atmosphere was periodically clouded by dense clouds of ashes and dust ejected by volcanoes which erupted in the distance. These volcanic materials were blown into the area, where they settled in the swamps and lagoons. Additional volcanic ash was transported into the area by streams, and today we see these thin layers of water-deposited volcanic ash interbedded with the stream-laid shales, sandstone, and conglomerates of the Chinle. The deposition of this volcanic material was an important event in the history of the Petrified Forest, for it provided much of the mineral material that was later deposited in the buried trees. There were also other signs of crustal unrest in the western United States; for as the Triassic Period drew to a close and Jurassic time began, this part of the continent was slowly subsiding and the area was gradually inundated by a shallow sea. Thus, during the more than 100 million years that the Mesozoic seas were in this region, more than three thousand feet of marine sediments were deposited on top of the Chinle Formation and its buried trees. It was probably during this long interval of time that petrification of the logs took place.

From Wood to Stone

As mentioned earlier, petrified wood is found the world around and is without doubt the best known and most commonly encountered type of plant fossil. It should be noted, however, that trees and their foliage may be preserved in a variety of ways other than petrification. Moreover, as we shall learn later, petrification is not the only type of fossilization that has preserved the plants of our ancient Triassic forest. Yet, petrification—a word that literally means "to make into rock"—is by far the most common type of preservation that has taken place in the Park.

Before attempting to explain how the logs were turned to stone, it should be noted that petrification is a rather complex chemical process and there is much that is not known about it. We do know, however, that the wood in the Petrified Forest is *silicified,* that is the plant remains have been preserved by *sil-*

ica (SiO_2), an oxide of silicon. Silicon occurs in nature as five distinct minerals, and it is the second most abundant (after oxygen) element in the earth's crust.

Practically all of the fossil wood in the Park has been preserved by quartz, a mineral that is transparent to translucent when pure, but may be violet, rose, black, and other colors when impurities are present. There are two basically different types of quartz and both occur in the Park's fossil trees. The most common quartz variety found in the petrified wood is *chalcedony* (kal-sed'-nee), a flintlike variety of quartz that occurs in layers rather than crystals. There are several varieties of chalcedony, and they vary considerably in their color and color patterns. For example, the trees in the Jasper Forest are *jasperized,* for they consist primarily of *jasper,* a dark, opaque form of chalcedony that may be red, brown, yellow, green, or blue. Other trees are composed largely of *agate,* a colorful variety of chalcedony showing parallel bands of several hues. Agatized wood is especially beautiful, because the agate commonly appears as wavy, curved, or circular rainbows of color.

The second variety of quartz, *crystalline quartz,* is not as commonly found as chalcedony. However, in some trees such as those of the Crystal Forest, beautiful crystals of *clear quartz* (sometimes called rock crystal) and *amethyst* can be seen.

Park visitors frequently ask why the fossil wood is so brilliantly colored. Unless it is stained or contains some mineral impurity, quartz is normally colorless or white. But in many instances, minor amounts of other minerals were held in solution with the silica at the time it was deposited. Hence, if manganese oxide or carbon was present, a black, purple, or some dark-colored quartz was formed, whereas silica solutions containing iron oxides tended to produce fossil wood that is yellow, brown, and red. Not all the color was added at the time of petrification; some of the coloring agents were introduced as secondary fillings in cracks and holes in the logs after petrification had occurred.

The quartz that was deposited in the petrified logs is believed to have been derived from silica dissolved from volcanic ash deposited in this area as a result of the earlier-described explosions of the ancient Triassic volcanoes. As water filtered downward through the ash-laden sediment, it dissolved silica from the volcanic ash and carried it along in solution. The buried trees soon became saturated with the mineral-bearing waters, and as the silica solution became more concentrated, quartz was slowly deposited. At first quartz formed only in openings of the empty cells, but as silicification proceeded, the cell walls themselves were sometimes penetrated. Thus, the walls of the original plant cells served as a framework to enclose the quartz that was being deposited in their cavities. Because the cell walls and much of the other plant tissue are literally embedded in quartz, some of the wood has been so faithfully preserved that the cellular structure and annual rings can be studied in great detail.

Though the vast majority of the Petrified Forest's plant remains have been preserved by petrification, other types of plant fossils have also been found here. For example, fossil leaves of nineteen plant species have been found in the form of *impressions* and *compressions*. The former generally show only the shape or external structural features of the leaf and little, if any, of the plant tissue is preserved. An impression may be produced when a leaf is buried in sediment which is subsequently hardened into rock. However, after the leaf has decayed, the imprint of its form will be retained in the stone. It is not uncommon to see an impression of a recent leaf in a sidewalk or driveway. These formed when the leaf settled on the surface of wet cement before it hardened, a process that is somewhat analogous to that described above.

Compressions are plant fossils formed when organic material is preserved as a thin carbonaceous film. Known also as *carbonaceous residues,* compressions are formed by *carbonization* or *distillation,* the same process by which coal is formed. "Carbon copies" of this type are produced as organic tissue slowly decays after burial. With the passage of time, the organic matter gradually loses its liquids and gases, leaving behind a thin carbonaceous film which duplicates closely the flattened plant leaf or similar organic object that it represents. Certain compression fossils are so well preserved that the outer layer of

Nisqually Glacier, a valley glacier that flows down the side of Mount Rainier, is typical of the mighty "rivers" of ice that have shaped the face of many of the National Parks. NATIONAL PARK SERVICE

A variety of unusual experiences await visitors to the National Parks. Here a group of visitors take a lunch break in Mammoth Cave's famous underground Snowball Dining Room. KENTUCKY TRAVEL DIVISION

At an elevation of 5,820 feet above sea level, Waterrock Knob Parking Overlook is one of the highest points along the more than four-hundred miles of Blue Ridge Parkway that stretches between Shenandoah and Great Smoky Mountains National Parks. NORTH CAROLINA TRAVEL INFORMATION DIVISION PHOTO BY CHARLES A. CLARK

Evidence of past glaciation—such as these glacial striations—can be seen at a number of places along the Bumpass Hell Self-guiding Trail, in Lassen Volcanic National Park. JENNIE A. MATTHEWS

The end of this log in Petrified Forest National Park is exposed where the surrounding rocks have been removed by erosion. JENNIE A. MATTHEWS

Our multisplendored National Parks are many things to many people. Some, such as these hikers in the high country of Olympic National Park, take to the trails to enjoy the quiet and solitude of the wilderness areas. NATIONAL PARK SERVICE

Bumpass Hell, a barren, sulfur-stained, craterlike depression, is one of the major attractions in Lassen Volcanic National Park. This interesting area lies at the end of the Bumpass Hell Self-guiding Nature Trail. JENNIE A. MATTHEWS

Located in the extreme southern part of Yosemite National Park, Mariposa Grove is famous for its many giant sequoia trees. Note the man (arrow) standing at the base of the tree in the middleground.
NATIONAL PARK SERVICE

Located on a bald, granite dome in Yosemite National Park, this battered old pine tree has withstood the ravages of time, weather, and the attacks of vandals. JENNIE A. MATTHEWS

Avalanche chutes (arrow)—slick, steep-sided chutes that represent the cracks of avalanches—are common in Kings Canyon National Park, in the High Sierra. CALIFORNIA DIVISION OF MINES AND GEOLOGY PHOTO COURTESY OF DR. GORDON B. OAKESHOTT

The special type of weathering known as exfoliation is well displayed at California's Yosemite National Park. Above are the Royal Arches and Washington Column (arrow) viewed from the valley floor. Massive North Dome towers above them. U. S. GEOLOGICAL SURVEY PHOTO BY F. E. MATTHES

Checkerboard Mesa—a well-known landmark in Zion National Park —shows typical cross-bedding intersected by vertical joints. NATIONAL PARK SERVICE PHOTO BY GEORGE A. GRANT

In places, igneous rocks have been intruded into sedimentary rocks to form sills such as seen here in the ice-carved Garden Wall of Glacier National Park (see arrow). Grinnel Glacier lies to the right of and below the arrow, and Josephine Lake is in the foreground. MONTANA HIGHWAY COMMISSION

cells may be studied in great detail, thus providing much valuable data about the fossil and the rocks in which it was found. In some instances, compressions have been removed intact from the rock in which they were embedded. This process, known as *maceration,* is usually accomplished by soaking the specimen in hydrofluoric acid, thereby dissolving the rock and releasing the organic material. By using this technique on plants from other areas, paleobotanists have recovered well-preserved leaves more than 150 million years old. Some of these have even been glued on sheets of paper in much the same way that recent plants are mounted for scientific study.

The Past Revealed

Following their burial in Mesozoic time, the Triassic trees lay entombed in the rocks for millions of years. Then forces deep within the earth gradually began to elevate the western United States high above sea level. There were several such episodes of mountain building, and as the Sierra Nevada and Rocky Mountains were uplifted the land in between the mountain ranges was also elevated. Thus the area that is now Arizona was slowly raised thousands of feet above its original location near sea level. This great change in altitude was accompanied by changes in climate and the Colorado Plateau, upon which the Petrified Forest is located, became quite dry and desertlike. No sooner had the area been uplifted than the forces of erosion began their inevitable attack; and in due time, rain, wind, and frost had reduced the surface rocks to loose sand and gravel. These rock fragments were steadily washed away by rain and running water until the strata that had once covered the petrified logs were finally stripped away.

As time proceeded, the relatively soft rocks of the log-bearing Chinle Formation were also subjected to erosion, and the landscape was soon cut into a series of gullies and deep ravines. It was then that the long-lost treasure of ancient stone trees began to weather out of the surrounding rocks. Today these same agents of erosion continue to turn back the pages of earth history as they uncover still more silicified Triassic logs. It is interesting to note that despite the great abundance of petrified wood that is found in the Park, the rocks of the Chinle have given up but a fraction of their fossil wood. It has been estimated that only one-fourth of the available fossil wood has been exposed. The remainder lies concealed beneath the surface in the some three hundred feet of Chinle rocks that are still present in the area.

Why the Logs Are Broken

Curiously enough, most of the fossil logs are riddled by numerous transverse fractures and they have been broken into segments of rather uniform length. In past years there has been considerable speculation—and some disagreement—as to what caused the fracture of the logs. At the present time most geologists believe that the cracks were caused by rhythmic vibrations generated by earthquakes that accompanied the intermittent subsidence and elevation of the Park region.

As the shock waves passed through the rather brittle silicified logs, many fine cracks developed at rather regularly spaced intervals and perpendicular to the long axis, or length, of the fossil logs. The cracks were further widened after the logs had been exposed on the surface of the desert and the surrounding rocks were eroded from under and around them. The washing away of the supporting rocks caused the logs to sag and the cracks continued to spread farther apart until the logs were finally broken into many different pieces.

The Painted Desert—Colorful Badlands

As you drive along the scenic road that passes through this part of the Painted Desert, you will see one of the world's most spectacular erosional landscapes. This is *badlands* topography —a terrain that is developed in arid regions nearly devoid of vegetation.

Rainfall in this part of Arizona often occurs as torrential cloudbursts accompanied by heavy runoff. As flash floods surge down the normally dry arroyos, their waters carry a large load of abrasive rock fragments that slice the land into an ever-expanding and intricate maze of narrow ravines and sharp crests and pinnacles. Such rapid erosion prevents most plants from establishing a foothold, so there are few roots to bind the soil together and impede erosion. Moreover, the volcanic ash from which this soil was derived is virtually devoid of organic matter, hence it is unable to sustain vegetation. Thus the brief heavy rains alternating with long dry periods in between are essential factors in the formation of the Painted Desert's landscape.

Yet, despite the significance of the climate in the development of the Painted Desert badlands, the composition of the rocks plays an even more important role. Like those of the Petrified Forest, the rocks of the Painted Desert belong to the Chinle Formation of Late Triassic age and are composed of water-laid beds of volcanic ash interbedded with thin layers of river gravel, sandstone, and shale. You will recall that it was the decomposition of volcanic ash that yielded the silica which was deposited as quartz in the buried trees. This same process also converted the ash into *bentonite*, a type of soft porous clay that is abundant in the strata of the Painted Desert. When wet, bentonite absorbs great quantities of water, and when sufficient moisture has been absorbed the clay will disintegrate into a fine flowing mud. Upon drying, however, the bentonite again becomes hard and strong. Consequently, the bentonitic rocks and loose soil are rapidly sculptured during the short-lived heavy cloudbursts, but they dry quickly and harden between rains to preserve the sharp ridges and deep gullies.

During the winter another agent of erosion helps sculpture the landscape of the Painted Desert. Water from rain or melting snow seeps into cracks and crevices in the rock and fills them. When the water freezes, it expands and exerts pressure that further widens the cracks and weakens the rock. Eventually the repeated freezing and thawing process will pry loose part of the rock, and it falls to the bottom of the gully, later to be washed away. Wind has also had a hand in molding the badlands' terrain, for during the dry, hot summer, blustery winds sweep over the Painted Desert and fine sand and dust may be transported for considerable distances.

At some localities in the Park, steep-sided mesas and buttes rise abruptly above the surrounding flat land. These landforms have developed because of the presence of a resistant cap rock of sandstone or lava that was harder than the underlying rock, which eroded away much faster. This process accounts for the presence of Blue Mesa and the Flattops, two interesting areas along the Park's scenic drive. You may also see pieces of petrified wood balanced on top of small spires, columns, and pyramids. In this type of erosional remnant, the petrified wood provides the resistant "cap rock" which protects the softer pedestal upon which the fossil is perched.

But it is the color of the badlands that makes the Painted Desert truly unique. The colors here, like those in the fossil wood, are the result of mineral impurities which were contained in the volcanic ash. Most of the color is due to the presence of iron oxides which have stained the normally white bentonite varying shades of red, brown, blue, and yellow. However, certain other chemical compounds have also played a part in the "painting" of the clays.

To see the Painted Desert to best advantage, visit during early morning or late afternoon if at all possible. And don't complain if it showers, for the colors are most vivid when they are wet.

Plants and Animals of Petrified Forest National Park

The fauna and flora of the northern Arizona desert is markedly different from those which inhabited the ancient coastal plain of Triassic time. The moisture-loving plants and animals of that period have long since been replaced by plants that have adapted themselves to the semiarid environment that prevails today. Resident mammals of the Park include the white-tailed antelope, ground squirrel, bobcat, jackrabbit, cottontail, coyote, skunk, prairie dog, kit fox, and the pronghorn ("antelope"). A number of reptiles, among them the large, but harmless collared lizard and the not-so-harmless prairie rattlesnake, make their homes in Petrified Forest National Park. The rock wren, horned lark, phoebe, house finch, and several species of sparrows are the dominant songbirds of the area. Also present are

Swainson's hawk, marsh hawk, and golden eagle, especially in the area from the U. S. Highway 66 overpass to Blue Mesa.

Plant life is sparse and inconspicuous, but some typically produce rather fragile, beautiful flowers. In the spring look for displays of cactus, yucca, and mariposa lily; however, sunflowers, painted cup, rabbit brush, and asters show their blooms throughout much of the summer.

Tips for Tourists

The interpretive program at Petrified Forest National Park is designed to help you understand the story of the Great Stone Trees, how they originated, and how they were preserved.

If you enter the Park from U. S. Highway 180, plan to make the Rainbow Forest Visitor Center your first stop. Ranger-Naturalists are on duty here and they will be glad to help you plan your visit and point out areas of unusual interest. In addition, the photographs, paintings, dioramas, and actual specimens will add greatly to your enjoyment and further your understanding of the processes of nature and the geologic events that occurred here.

Visitors entering the Park from U. S. Highway 66 should stop at the **Orientation Building** for information and a look at the educational exhibits pertaining to the area. From here the road leads through the colorful badlands of the Painted Desert and there are a number of overlooks from which there are excellent views. **Kachina Point** is a good place for general orientation and a strategic point from which to photograph the badlands.

The Painted Desert and the fossil forests abound in picture-taking opportunities. Color film will, of course, help you capture the true flavor of the area—especially after a rain—but don't overlook the possibilities of some good black and white photographs. By composing your photographs carefully and making use of back-lighting and cloud patterns, some spectacular black and white photos can be secured. A yellow (cloud) filter can be used to dramatize the cloud shots, and a haze filter will be useful with either black and white or color film. Midday lighting is usually "flat" and provides little depth; early morning and late afternoon are the best times to photograph this area.

Here, perhaps more than in any other National Park, Park Ranger-Naturalists must display the patience and vigilance for which they are so well known. Because of the apparent abundance of the fossil wood, some visitors find it difficult to refrain from taking a "sample" home as a souvenir, a practice which if carried out by each visitor would soon strip the Park of its major attractions.

Petrified Forest National Park at a Glance

Address: Superintendent, Holbrook, Arizona 86025.

Area: 94,189 acres.

Major Attractions: Extensive natural exhibits of petrified wood; Indian ruins and petroglyphs: portion of colorful Painted Desert badlands.

Season: Year-round.

Accommodations: None available in Park.

Activities: Hiking, nature walks, picnicking, and scenic drives.

Services: Food service, gift shop, service station, telephone, picnic tables, and rest rooms.

Interpretive Program: Museum, nature trails, roadside exhibits, self-guiding trail, and trailside exhibits.

Natural Features: Deserts, erosional features, geologic formations, petrified wood, rocks and minerals, volcanic features, and Indian ruins.

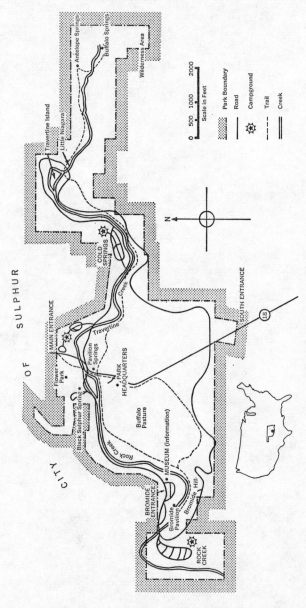

Platt National Park, Oklahoma.

28. Platt National Park, Oklahoma

Prairie Oasis

Our nation's smallest National Park lies nestled in the tree-shaded, gently rolling foothills of the Arbuckle Mountains of southern Oklahoma. Platt National Park commemorates a landmark in natural resource conservation, for it is the first and only area (one and a half square miles) ceded to the United States by an Indian tribe for conservation purposes.

Here, in a unique geological setting, where the eastern hardwood forest and the western prairie meet, there is an abundance and variety of living things seldom found elsewhere.

Why the Springs Are Here

The rocks from which Platt's springs derive their water were deposited in a widespread sea that covered this part of the country during Early Pennsylvanian time about 300 million years ago. Here, over a long period of time, more than two miles of sediments were deposited on the floor of this ancient ocean. Later, near the end of the Pennsylvanian Period and after the sediments had hardened into rocks, southern Oklahoma was subjected to severe earth movements. This crustal disturbance was accompanied by great uplift during which the sedimentary rock strata were severely folded and fractured. The more obvious evidence of this mountain-building event has long since been removed by erosion; now one sees only the Arbuckle Mountains—an elevated tableland or plateau which ranges from seven hundred and fifty feet to thirteen hundred feet above sea level.

Today the crumpled, broken, tilted rocks are mute evidence of the violent geologic history of this area. These rocks also make possible the famous Platt springs for certain of the rock layers beneath the Park have been tilted upward where they are exposed at the surface a short distance to the north and east of the Park. Here, rain falling on the ground soaks into the exposed rocks. Some of the rainwater that falls in the intake area passes through the tilted permeable strata and part of it returns to the surface by means of fractures and fissures in the rocks.

Some of this spring water is fresh and some of it is *mineral water;* that is, it contains numerous dissolved minerals. In general, the type of water that flows from each spring is determined by the composition of the rock through which it has flown. Thus, the so-called bromide or sulfur water in the Park contains varying amounts of bromide or sulfur which has been dissolved out of the rock through which it passed. On the other hand, water flowing from the freshwater springs (such as Antelope and Buffalo Springs) flows through rocks which contain relatively small amounts of these substances.

An unusual geologic phenomenon associated with certain of the springs is the deposition of *calcareous tufa* or *travertine.* This spongy porous limestone deposit is precipitated by the evaporation of water containing appreciable amounts of calcium carbonate. In some places these thick accumulations of travertine cause small waterfalls and rapids as at **Travertine Falls;** or it may be deposited in great masses as around **Travertine Island.**

It is interesting to note that the surface deposits of travertine in Platt are composed of the same type of material as the flowstone deposits of Carlsbad Caverns and Mammoth Cave. The latter have been deposited underground whereas the travertine deposits in Platt National Park are a surface manifestation of the geologic work of groundwater.

Plants and Animals of Platt National Park

The primary animal attraction at Platt is a small herd of bison. These shaggy creatures, which once roamed Oklahoma in great numbers, are representative of the once vast "Southern Herd" that was almost exterminated by 1900. A good trail leads to points from which the bison may be seen. In addition to the herd of bison, more than forty species of smaller mammals make their homes in the Park. Among the more common of these are the opossum, raccoon, armadillo, eastern fox squirrel, southern flying squirrel, rabbit, and gray fox.

Large numbers of birds inhabit Platt National Park and almost one hundred and fifty species have been sighted within the limits of the Park. Among the more common of these are the blue jay, eastern bluebird, cardinal, nighthawk, southern yellow-shafted flicker, eastern meadowlark, orchard oriole, brown-headed cowbird, common goldfinch, and several species of woodpeckers and sparrows.

An interesting variety of plants grow in the Park's wooded valleys and provide the ideal setting for sparkling streams and cold-water springs: cottonwood, redbud, sycamore, dogwood, willow, hickory, plum, and several species of oak. Present also are a host of wildflowers including asters, goldenrod, larkspur, and prairie lily.

Tips for Tourists

The Platt National Park Museum in the **Bromide Springs** area has interesting exhibits explaining the springs, rocks, plants, animals, and history of the Park. These are all designed to help you understand better this area and its features.

Each morning (weather permitting) there is a nature walk from the Museum to the top of **Bromide Hill** where there is a fine panoramic view of the countryside. While on this interesting 1½-hour hike, the Park Naturalist will identify some of the birds and plants and point out certain other interesting features of the area.

The **Buffalo Springs-Antelope Springs Nature Trail** is a short self-guiding trail between two of the more interesting places in the Park. The features indicated by markers along the trail are identified and described in the interpretive pamphlet that has been prepared by Park Naturalists.

In addition to pointing out some of the more interesting plants of the Park, this trail leads you to **Buffalo Springs,** named for the thousands of buffalo who once watered here, and **Antelope Springs,** once a favorite watering place of antelope. These freshwater springs which feed **Travertine Creek** normally have a combined flow of more than five million gallons of water per day.

Platt National Park at a Glance

Address: Superintendent, Box 201, Sulphur, Oklahoma 73086.

Area: 911.97 acres.

Major Attractions: Numerous cold mineral springs with distinctive properties, including several bromide springs.

Season: Year-round.

Accommodations: Campgrounds, group campsites, and picnic areas in Park. Overnight accommodations available in nearby city of Sulphur, Oklahoma.

Activities: Camping, fishing, guided tours, hiking, nature walks, picnicking, scenic drives.

Services: Picnic tables and rest rooms. (Food service, laundry, post office, rental camping equipment, etc., is available in Sulphur.)

Interpretive Program: Campfire programs, museum, nature trails, nature walks, roadside and trailside exhibits, and Saturday morning children's program.

Natural Features: Forests, geologic formations, prairies, unusual birds and mammals, unusual plants, waterfalls, streams, cold artesian mineral water springs, cold freshwater springs, and travertine formations.

Redwood National Park, California.

29. Redwood National Park, California

A Living Link in Geologic History

Scientists have described the redwood (*Sequoia sempervirens*) and giant sequoia (*Sequoia gigantea*) trees as "links" between recent and geologic time. And rightly so, for these are the sole remaining species of a group of trees that flourished in many parts of the Northern Hemisphere in the geologic past. During the Oligocene Epoch of the Tertiary Period the redwoods were evolving and thriving in vast forests that blanketed the coast of California. But today—more than thirty million years later—these magnificent trees are fighting to stave off extinction as man encroaches ever deeper into their natural environment.

In mid-Tertiary time the original redwood forest must have covered many millions of acres. Extending in a belt along the California coast, the redwoods reached northward to British Columbia, and scattered forests must have also occurred in North and South Dakota, Montana, and as far east as Yellowstone National Park in Wyoming. The petrified tree trunks of prehistoric redwoods found in Yellowstone are silent testimony to the past distribution of the sequoia family.

The sequoia trees—both giant and redwood—have two major distinctions: they are the largest and oldest living objects on earth. Interestingly enough, the scientific names of the two surviving species support these claims. One does not need to be a Latin scholar or a botanist to deduce that *Sequoia gigantea* translates as "giant sequoia." The redwood—*Sequoia semper-*

virens—has an equally fitting name, for these Latin words mean "evergreen sequoia." The sequoias, like all conifers, are truly "evergreen" trees; they seem to stay green almost forever!

The word *Sequoia* is derived from the name of Chief Sequoyah, a Cherokee Indian who devised a phonetic alphabet system that enabled members of his tribe to read and write. In recognition for Sequoyah's contribution to his tribe, the Austrian botanist who first described the "Big Trees" named them in the Chief's honor. (The change in spelling of his name was made to conform with botanic nomenclature, which requires that all proper names be Latinized.) It is a fitting tribute that our nation's noblest trees, with their proud, erect, massive red trunks, be named in honor of one of our more distinguished original Americans.

The heroic proportions of the sequoias is well demonstrated among the coast redwoods growing in Redwood National Park. These include the three tallest trees on earth. The tallest of these is the 367.6-foot Howard Libbey Tree, followed by the Harry W. Cole Tree (365.4 feet) and the National Geographic Tree, which tops out at 364.5 feet. The world's largest tree, a giant sequoia named the General Sherman Tree is located in Sequoia National Park, California. Standing 272.4 feet tall, this giant has a circumference of eighty-three feet eleven inches and is estimated to be more than 3,500 years old.

The sequoias are also thought to be the oldest organisms on earth. Some of the trees in Redwood National Park probably first sprouted as much as two thousand years ago. Studies made on giant sequoias in other National Parks suggest that many of the larger trees are more than thirty centuries old. Not only are the individual sequoias extremely old, but their entire family tree dates far back into the geologic past. In fact, their ancestry can be traced back to the Mesozoic Era, the "Age of Dinosaurs." During their long history on earth the sequoias were widespread, probably as widespread and numerous as pine trees are today.

Botanists estimate that at least a dozen species of sequoias were living during earlier geologic time. Why has such a successful group of plants been reduced to two surviving species?

No one knows for sure, of course, but scientists have made speculations on their decline. There is evidence, for example, that many of the giant trees were killed by widespread volcanic eruptions which plagued the Pacific Northwest throughout much of the Tertiary Period. Later in geologic history this area also underwent drastic climatic changes and deserts formed in regions where the great trees had once flourished. But these are only "scientific guesses." Unfortunately, the more recent history of the remaining species is not known, for few fossil trees of this period have been discovered.

Ecology: Key to Survival?

It is only within recent years that we have heard much about ecology and its increasingly important role in man's future welfare. For most organisms—including the redwoods—ecological balance has been the difference between survival and extinction. For the sequoias, climate has perhaps been the key ecological factor. The redwood country is characterized by moderate temperatures, heavy rainfall during the winter months, and dry summers with considerable fog. It has been suggested that there may be some relationship between the distribution of the redwoods and the so-called coastal "fog belt." The range of the redwoods and the "fog belts" do coincide geographically. However, scientists are not sure what, if anything, this has to do with the range of the redwoods.

Geologically, the rocks and soils of the redwoods region are quite varied. The high slopes and geologic changes in the area produce rather unstable conditions, and the earth's surface is subject to considerable erosion. In general, it seems that the climate, soil, and plant and animal life of the redwood-forest ecosystem play interacting roles in the survival of the redwoods.

Plants and Animals of Redwood National Park

As the name implies, Redwood National Park's reason for being is its magnificent stand of redwood trees. But though *Sequoia sempervirens* is the "star" of Redwood's plant communities, its role is bolstered by a large "supporting cast" of other interesting plants and animals. Indeed, life is so varied in this National Park that four major ecosystems are clearly defined. These four different plant and animal communities are present because of climatic differences, variations in altitude (from sea level to 1500 feet), and other factors that affect the physical environment.

As the name implies, the *redwood-forest* ecosystem consists of the redwood forest, its streams, the coastal forest border, and the forests that form the inland border of the redwoods. *Sequoia sempervirens* makes up about 60 per cent of the trees in this ecosystem, and there is an under covering of smaller trees and shrubs.

Animal life in the redwood forest is not as abundant as might be suspected. The Townsend chipmunk, western Douglas squirrel, beaver, raccoon, and bobcat are common in some groves. Among the larger game animals are Columbian blacktailed deer, mountain lion, and black bear. Roosevelt elk are found in the northern part of the Park. Sport fish include rainbow trout, steelhead, and salmon. As in most forest environments, the redwood forest contains many species of birds. However, their presence is not always apparent, for many spend their lifetimes in the crowns of trees that loom more than two hundred feet above the forest floor.

The *cutover-forest* ecosystem is in stark contrast to the redwood-forest community. The trees in this area have been cleared or nearly cleared, creating a totally different environment. Hopefully, these clearcut areas will go through a more or less predictable series of ecological adjustments that will eventually result in the establishment of a new redwood-forest ecosystem. It is estimated that it takes at least a century for this cycle to be completed.

Where the redwood forest thins, as it nears the rocky Pacific coastline, the *north-coastal-shrub* ecosystem takes over. This is a harsh environment characterized by rocky soils with poor drainage and almost constant salty winds. Only low-growing trees, herbaceous plants, and woody shrubs seem to be able to adjust to this rather inhospitable habitat.

The *marine-and-shore* ecosystem is an interesting mixture

of land, ocean, and life. Here, where the Pacific Ocean meets the California coast, are sand dunes, beaches, pools of fresh and of salt water, marshes, and rocky cliffs. Also included is the submerged land up to one fourth mile out from mean high tide level. Seals, sea lions, and sea birds live on rock masses located offshore. Other water-dwelling animals inhabit the tidal pools and salt- and fresh-water marshes.

This is an interesting life zone and one that is well worth visiting and exploring. Visitors are urged to watch the tide in this area, and one should move to higher ground as the tide starts to come in. The presence of rough surf, heavy undertow, and rocky, rapidly descending beaches make this part of the Park totally unsafe for swimming and surfing.

Tips for Tourists

Redwood is one of our newest National Parks, and this great natural preserve is undergoing many developments. Three California State Parks—Jedediah Smith, Del Norte, and Prairie Creek—are located in the forty-six-mile coastal strip that reaches a width of seven miles. Many acres of private land are also located within the boundaries of the Park. Consequently, there are still some places within the Park that are "off limits" to visitors. Meanwhile, there is much to see and do in this area, for the State Parks, which are adjacent to the National Park, provide many recreational opportunities for the visitor. Consequently, check the Park Headquarters in order to learn more about them.

Redwood National Park at a Glance

Address: Superintendent, Redwood National Park, Drawer N, Crescent City, California 95531.

Area: 57,094 acres.

Major Attractions: Coast redwood forests containing virgin groves of ancient trees; world's tallest tree (367.6 feet); and forty miles of scenic Pacific coastline.

Season: Year-round.

Accommodations: Cabins, campgrounds, motels, trailer parks, and other facilities are available on nearby private lands or State Parks.

Activities: Boating, boat rides, camping, fishing (California license required), hiking, horseback riding, nature walks, scenic drives, swimming (river and pools only; swimming and surfing in the ocean are *not* recommended).

Services (mostly on nearby private lands): Auto repairs, boat rentals, food service, gas stations, general store, gift shop, health service, horse rental, ice, laundromat, museums, picnic areas, post office, propane service, public showers, religious services, sewage-dump station, telephone.

Interpretive Program: Campfire program, nature trails, roadside exhibits, self-guiding trails, slide shows.

Natural Features: Redwood forests, ocean beaches, rocky coastline, marine life, forest life.

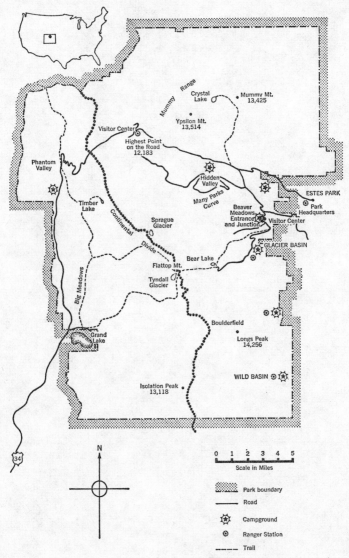

Rocky Mountain National Park, Colorado.

30. Rocky Mountain National Park, Colorado

Where the Mountains Are

Located high in the magnificent Front Range of the Rockies is one of the nation's most popular National Parks: Rocky Mountain National Park—four hundred and ten square miles of sparkling lakes, ice-carved valleys, alpine meadows, and snow-capped mountains.

The terrain of the Park is unusually rugged and presents a varied scenic picture. Ice-sculptured mountains rise thousands of feet above deeply incised lake-filled valleys, and the mountainsides—extending as they do through several life zones—support a host of unusual plants and animals. Here, as in most National Parks, the main attraction is scenery. And here, as in other Parks, the magnificent mountain panoramas have been produced by geologic processes that have been in operation for eons.

Creation of a Landscape

Rocky Mountain National Park is appropriately named, for the face of the Park is studded with stony peaks, eighty-four of which exceed eleven thousand feet elevation. These mountains, like those of Grand Teton and Glacier National Parks to the north were formed by crustal uplift near the end of the Cretaceous Period some sixty million years ago. But the rocks forming the core of the Rockies are much older: they date back to Precambrian time, more than a billion years ago.

We have already become familiar with the formation of volcanic, faultblock, and erosional mountains; let us now consider the development of *geosynclinal* or *folded mountains*. Most of the world's major mountain systems were formed in this way, and geologists consider the Rockies to be a good example of this type of landform.

The "Ancestral Rockies"

Geologically the present-day Rocky Mountains are relatively young, but their "ancestors" date far back into geologic history. The earliest of these ancestral mountain ranges developed along the same zone of crustal weakness occupied by the Rocky Mountains today and were elevated hundreds of millions of years ago during Precambrian time. The uplift that created the first mountain range must have been accompanied by considerable igneous activity, for great quantities of molten rock were injected into the Precambrian formations.

Concurrent with the uplift of the "first generation" Rockies was their gradual destruction by the agents of erosion. Running water, freezing and thawing, wind and rain, landslides and avalanches, gradually reduced the mountainous landscape to an expansive rolling plain.

Much later, during the Paleozoic Era, the Rocky Mountain belt underwent another period of uplift, and the old erosional plain was elevated to form a "second generation" of mountains. Geologic field studies indicate that these mountains were of considerable height, but the true nature of the mountains can only be inferred; by the middle of the Mesozoic Era erosion had completely destroyed the mountains and the area was covered by the Cretaceous sea.

Birth of the Rocky Mountains

Following the erosion of the "second generation" Rockies, there was a broad subsidence of the area that they had once occupied; this marked the beginning of the *Rocky Mountain Geosyncline*. As time passed, this great elongated downfold in the earth's crust gradually sank below sea level, permitting the trough to be flooded by a shallow inland sea. Extending from the Arctic Ocean to the Gulf of Mexico, this great seaway covered thousands of square miles and existed for many millions of years. During that time, rivers from the surrounding area emptied into the sea, dumping their loads of silt and sand. Ocean currents distributed the sediments over the ocean floor and to these were added limy deposits precipitated from the sea itself. As the sediments continued to accumulate, the center of the geosyncline slowly began to sink; this gradual deepening of the basin made it possible for still-thicker deposits of sediments to accumulate there.

Much later in geologic time, forces within the earth caused the floor of the geosyncline slowly to rise, draining the water back into the ocean basins. This gradual uplift finally reached its climax about sixty million years ago in the great crustal movement called the *Laramide Revolution*. This disturbance did not occur suddenly; instead, it was the culmination of tremendous internal forces which buckled the earth's crust forming immense "wrinkles" of rock which bulged upward. In places, the rocks were so strongly compressed that they were faulted and great segments of the crust slid up and over younger rocks as at Chief Mountain in Glacier National Park. Thus, the geosynclinal sediments which had been piling up for hundreds of millions of years are now found on the tops of mountains.

As you enjoy the magnificent mountain vistas of Rocky Mountain National Park, you may wonder what forces could have been great enough to form such massive "wrinkles" in the earth's crust. Geologists have long asked this same question, but they are still unable to answer it. They have, however, advanced several theories which cast some light on the formation of geosynclines and their subsequent deformation into folded mountain ranges. A few of the more prominent theories are briefly reviewed below.

One of the oldest of these, the *Contraction Theory,* assumes that the earth is cooling from a former molten condition and that contraction produced by cooling has caused the crust to shrink and wrinkle like the skin of a drying apple. Another, the theory of *Continental Drift,* was first proposed about 1910

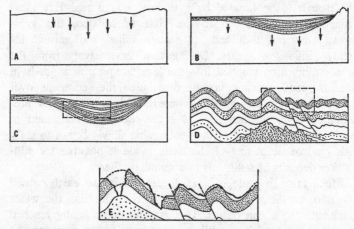

Development of geosynclinal mountains. The spectacular peaks and valleys of the Rocky Mountains as we know them today are made of rocks that record a story that began more than 600 million years ago. At that time, part of western North America began to sag downward to form an elongated trough as in A.

Rivers poured sand, silt, and gravel into the lowland area. Downwarping continued until the trough was filled with a shallow sea, into which poured a steady flow of sedimentary materials, as in B.

Downsinking continued, but it seems to have been at a rate that corresponded closely to the rate of filling, so that sedimentation was always into shallow marine waters. The mass of sedimentary materials slowly changed to sedimentary rock, as the load on top increased, until it had a form like that in C.

For reasons we do not yet understand, the trough area was then severely compressed so that the rocks in it were folded and broken. At about this time in the history of such mountains, great masses of molten materials commonly appear in the cores of the folded and broken rock, eventually solidifying into granite. (D is what an enlarged section of C would look like.)

Uplift accompanied the folding and faulting, and as soon as the rocks emerged from the sea they were subjected to erosion. Rivers and glaciers carved the valleys and formed the peaks as shown in E, an enlarged part of D. This is the stage of development of our Rocky Mountains now.

Reproduced from *Jasper National Park: Behind the Mountains and Glaciers,* by Donald M. Baird, by permission of the Geological Survey of Canada.

but has gained renewed attention within recent years. This theory suggests that at some time in the geologic past an ancient continent broke into several segments which drifted apart. As each continental block moved forward, its front would drag against the subcrustal material, causing the continental margins to crumple, thus forming the folded coastal mountain ranges of Europe and North and South America. Although there are several objections to this proposal, if you look at a globe you will see how the idea originated. Notice that the coastlines along both sides of the Atlantic Ocean match surprisingly well; note also that some of the older mountain belts in the Americas appear to be continuations of similar mountain belts in the eastern continents.

One of the more recent proposals is the *Convection Theory* which argues that convection currents beneath the crust may cause the rocks to expand and push upward. Such currents produce massive flows of earth material which might be compared to movements produced in a pot of boiling oatmeal. It is thought that the heat to produce these currents may be derived from radioactive elements such as uranium. Within recent years many geologists have come to favor a combination contraction-convection theory which utilizes the best ideas of each. According to this hypothesis, the crust is expanding as the earth cracks along fracture lines through which molten rock material would well up. As the crustal segments moved apart, tremendous pressures would be exerted upon them, thereby causing the rocks to buckle upward.

It should be noted that although numerous objections have been advanced against each of the above theories, each has strong points which must be taken into consideration. Moreover, new theories are constantly being proposed in the light of recent research and the answer to this baffling problem may be found in the not too distant future.

Uplift and Erosion

Since the close of the Laramide Revolution, the Rocky Mountains have been subjected to intermittent periods of uplift followed by extensive intervals of erosion. Park visitors are often surprised at the relatively flat summits of mountains such as **Deer** and **Flattop Mountains.** These are evidence of past periods of erosion during which the Park's surface area was worn down to a fairly flat upland; the most recent such period of erosion occurred only a few million years ago.

Periods of uplift were apparently accompanied by volcanic activity in the western part of the Park. **Specimen Mountain,** which can be seen from Trail Ridge Road, is an extinct volcano; it is responsible for the thick lava flow which is exposed in the cliffs above **Iceberg Lake.** In addition, there are extensive accumulations of volcanic rocks in the rugged mountains of the **Never Summer Range.**

Why do we see only remnants of this ancient plain and its associated volcanic materials? Within fairly recent geologic time—perhaps the Pliocene Epoch—this area underwent renewed uplift. Streams were rejuvenated, thereby increasing their power to deepen their canyons and further alter the landscape. Accelerated stream erosion was followed by glaciation, and these processes united to all but obliterate this once-extensive plain.

Remnants of the Ice Age

Although the scenery of Rocky Mountain National Park is the product of a variety of geologic agents, it was the work of ice that added the crowning touch to the landscape. But do not expect to see the mighty valley glaciers that created this striking glacial scenery, for they have all but vanished from the Park. Yet, in a few sheltered places at higher elevations, there are small glaciers that may be remnants of once-powerful rivers of ice. Two of these, **Andrews** and **Tyndall Glaciers,** can be visited by trail (after a strenuous hike), and several can be seen from Park roads.

But this was not the case twenty-five thousand years ago—during Pleistocene time glaciers were common here. However, the glaciers appear to have been local in extent and underwent maximum development at altitudes of between eight thousand and ten thousand feet and they did not reach the lower plains. Originating at the heads of mountain valleys, the glaciers relentlessly advanced down V-shaped, stream-incised gorges, converting them into typical flat-bottomed glacial troughs such as **Spruce Canyon.** *Cirques,* amphitheater-like depressions which indicate the point of origin of the glaciers, can be seen at many places in the Park. One of these is occupied by **Chasm Lake** at the base of the East Face of Longs Peak; a smaller, but more accessible, cirque contains **Iceberg Lake** on Trail Ridge Road.

In places, the glacial valleys were more deeply eroded than were the tributary valleys which led into them. When the glaciers melted, the ends of these side valleys were left suspended in midair, thus forming features called *hanging valleys*. Water dropping from hanging valleys into the main valley below is responsible for such picturesque waterfalls as **Horseshoe Falls** of Roaring River.

As the glaciers ground their way downslope, rock fragments embedded in the ice eroded the bedrock over which they moved, scratching it in some places and polishing it in others. In addition, immense boulders of granite were brought in from distant areas and randomly deposited on ice-smoothed exposures of schist and gneiss. Such boulders are called *glacial erratics* and one can be seen resting on ice-polished bedrock along the **Bear Lake Nature Trail.**

Upon reaching lower elevations, the glaciers began to melt, thereby depositing the rock debris that had been picked up along the way. This material, a heterogeneous mixture of boulders, gravel, and clay, was transported within and on top of the ice. Accumulations of ice-transported debris deposited in a ridge along the glacier's front are called *terminal moraines* and glacial sediments occurring in long ridges along the flanks of the glacier are called *lateral moraines*. Many classic examples of moraines can be seen in Rocky Mountain National Park, but perhaps the most conspicuous of these is **South Lateral Moraine** in Moraine Park. Moraines also formed at various points as a glacier receded or became temporarily stable. In places, these landforms, called *recessional moraines,* formed natural dams across valleys, thereby creating lakes such as **Dream Lake,** whose waters are impounded by a moraine.

Most of the other lakes in the Park are also of glacial

origin. Some have formed in depressions on the surface of moraines; others occupy ice-gouged basins or cirques.

What does the future hold for Rocky Mountain National Park? Will the glaciers some day return to reshape the landscape? Or will the present-day Rocky Mountains be reduced to a plain like the ancestral mountains before them? We can only guess, of course. But if geologic processes continue—as surely they must—the familiar peaks, lakes, and tree-covered moraines will eventually vanish from the face of the Park. However, it is comforting to know that such changes take place so very slowly as to be imperceptible in an individual's lifetime; thus, only man's carelessness and indifference will prevent future generations of Americans from enjoying this beautiful area.

Plants and Animals of Rocky Mountain National Park

The majestic setting of Rocky Mountain National Park is the home of many interesting plants and animals. They are found in three life zones which range from flower-filled mountain meadows, through forested mountain slopes, to the rolling tundra above timber line. Because of marked variations in climate and duration of growing seasons, each life zone has its own characteristic plant and wildlife population. One of the best ways to become familiar with the life zones in the Park is to browse through the interpretive exhibits at the Alpine Visitor Center. In addition, the differences in vegetation of the three life belts are dramatically displayed along Trail Ridge Road.

The plants of Rocky Mountain National Park vary with changes in elevation, climate, and soil conditions. At elevations below about ninety-five hundred feet, the climate is relatively warm and dry; plants typical of this belt, called the *Montane Zone,* can be seen in the Moraine Park area. In this zone the predominant trees are ponderosa pine (common in the lower part of the zone) and the Douglas fir, which is widespread on the cool north-fronting slopes. In damp areas and along streams there are stands of blue spruce, willow, cottonwood, alder, and birch trees. Cool groves of tall quaking aspen trees thrive in moist, sheltered areas, and spreading ground-hugging Rocky Mountain junipers (erroneously called red cedars) grow in dry rocky areas and on canyon walls. Sagebrush, antelope brush, boulder raspberry, currants, and black chokecherry are some of the more abundant shrubs of this belt. Open areas of dry grassland also are present below ninety-five hundred feet; among the more common grasses are bluegrass, grama, needlegrass, and downy cheatgrass.

Wildflowers bloom at this altitude throughout the summer. Look for wild rose, shrubby cinquefoil, pasqueflower, Rocky Mountain iris, black-eyed Susan, tall pentstemon, and firewood.

Higher in the Park you enter the *Subalpine* or *Canadian Zone* which extends from about ninety-five hundred feet to timber line. As you ascend Trail Ridge Road you will note the increase of coniferous trees. Look especially for the stately Engelmann spruce, gnarled ragged limber pines, and subalpine fir. Trees in the upper part of the Subalpine Zone (from 10,500 to 11,500 feet elevation) grimly reflect the rigors of harsh climate. Because of high winds, low temperatures, and lack of moisture, trees at the upper edge of this zone are commonly stunted, grotesquely deformed, and often grow close to the ground. You will see many trees of this kind between Rainbow Curve and Forest Canyon Overlook on Trail Ridge Road. In addition to conifers, willow, bog birch, and a number of shrubs and wildflowers inhabit the Subalpine Zone.

Beginning at timber line is the *Alpine* or *Arctic-Alpine Zone,* an expanse of treeless rolling terrain where climatic conditions approximate those of Siberia, Alaska, and other Arctic regions. This is the alpine tundra: a windswept, frozen wilderness in winter, a flower-carpeted meadowland in summer. The climate here is unusually severe; temperatures remain below freezing all winter and blizzards are common. In addition, tundra life is continually subjected to high winds (from fifty to seventy miles per hour at times) and periods of drought may occur during winter or summer. Because of these conditions, the growing season in the tundra is short, ranging from less than six to as much as twelve weeks duration.

Tundra plants are remarkably small and many of the dwarf plants so typical of this area have taken several hundred years to attain their present size. Indeed, botanists believe that some of the tiny plants that you see here may be several hundred years old. On your drive over Trail Ridge Road plan to walk **Tundra Trail,** a short, self-guiding nature trail starting at Rock Cut, on Trail Ridge Road. Here you will be introduced to some of the fragile plants which have successfully adapted to one of nature's more hostile environments.

Climate and elevation also affect the animals, but not in the same way that they affect the trees and flowers. Some of the animals move with the seasons, occupying the high country in the summer and the more protected mountain slopes in winter. Thus, mule deer and the American elk (or wapiti) migrate to the subalpine meadows and tundra region for summer grazing but return to the lower elevations when winter storms begin. The careful observer may spot elk (especially during early evening) on the edge of forests below Fall River Pass or Rock Cut on Trail Ridge Road. Mule deer are more numerous and are commonly seen along Park trails or grazing near forests or on the tundra. Although Rocky Mountain National Park is famous for its bighorn or mountain sheep, they are not seen by most visitors. However, the patient observer may occasionally see bighorn at Sheep Lake in Horseshoe Park (in the early morning), on Specimen Mountain, and near Milner Pass on Trail Ridge Road. Among the other animals that are present, but seldom seen, are black bear, marten, cougar, bobcat, and coyote.

The larger mammals are interesting, but as you hike or drive through the Park you are much more likely to see some of the smaller mammals that live here. Chipmunks, ground squirrels, and yellow-bellied marmot can often be seen at turnouts along Trail Ridge Road. You will also probably notice evidence of beaver for their dams are common in Glacier Basin, Moraine and Horseshoe Parks, and along streams at lower elevations in the Park. But you are not likely to see the industrious creatures that built the dams: they do their work at night and rest during the daylight hours.

Many birds can be seen in the Park each summer, but most of them are migratory. Among those that live here throughout the year are Clark's nutcracker (commonly seen at Bear Lake and turnouts at lower elevations along Trail Ridge Road), Steller's jay, nuthatches, chickadees, owls, woodpeckers, grouse, and a limited number of golden eagles. Some species, for example, the ptarmigan, rosy finch, pipit, and horned lark, are more likely to be seen above timber line.

The more common fishes in the Park are rainbow, brown, cutthroat, and brook trout. The latter can usually be seen swimming near shores of Bear Lake.

Tips for Tourists

It is the matchless scenery that attracts most visitors to this Park, but the enjoyment here is not alone in the seeing. There are mountains to be climbed, streams to be fished, trails to hike, and isolated wilderness areas for rest and solitude. Most of the visitor activity is concentrated on the east side of the Park near Estes Park village. However, the Shadow Mountain National Recreation Area near Grand Lake (see map) is becoming increasingly popular and the National Park Service also provides naturalist services and interpretive programs there.

Every National Park has its "must" and in Rocky Mountain it is **Trail Ridge Road.** Truly one of the great "highroads" of the world, this modern, hard-surfaced highway is the highest continuous automobile road in the United States: more than four miles of the road is at a height of twelve thousand feet and eleven miles are above timber line (approximately 11,500 feet). Following an old Indian trail which crossed the mountains over Trail Ridge, this fifty-mile scenic drive offers an unbroken but ever-changing panorama of mountains, valleys, and lakes.

From Deer Ridge Junction, the beginning of Trail Ridge Road, the highway drops into Hidden Valley and then steadily climbs upward past a variety of carefully selected observation turnouts. The first of these, **Many Parks Curve,** is at an elevation of 9620 feet; from here you can see a number of parks, as these mountain-enclosed meadows are called, and there is an excellent view of the snow-covered Mummy Range to the north. This stop is also popular with the youngsters because of the presence of many ground squirrels, chipmunks, and birds.

Next comes **Rainbow Curve** (altitude 10,829 feet), which permits a sweeping view of the Great Plains to the east and the meadows of Horsehoe Park almost a half mile below. After passing through a rugged area of stark, weathered skeletons of trees burned in a forest fire many years ago, the road climbs above tree line and enters the alpine tundra. Soon you are at **Forest Canyon Overlook** (11,716 feet elevation), an excellent vantage point from which to see the effects of glaciation and to get a close look at tundra vegetation.

At **Rock Cut** you are 12,110 feet above sea level, an elevation that affords one of the Park's most spectacular views. Stop here long enough to walk the half-mile **Tundra Trail**, a self-guiding nature trail that will acquaint you with the fascinating and unusual plants of the tundra. Two miles farther is **Iceberg Lake**, a cirque which contains a mass of snow resembling ice. The reddish cliffs surrounding the lake are composed of relatively recent lava and are evidence of past volcanic activity in the Park. To the west is a sweeping view of the Never Summer Range which is sixty miles away.

Upon reaching **Fall River Pass** (11,796 feet), stop to visit the new **Alpine Visitor Center** (p. 360), which has interpretive exhibits describing the features of the alpine tundra. Just beyond Fall River Pass is **Specimen Mountain,** the remains of an extinct volcano believed to be the source of the lava that forms the cliffs at Iceberg Lake.

You cross the Continental Divide at **Milner Pass,** an elevation of 10,758 feet, and from this point on the road continues downhill to Grand Lake some eighteen miles away.

Numbered markers along the road call attention to features of special interest and these correspond with numbers in the "Self-guiding Auto Tour of Trail Ridge Road" section of the Park brochure which you receive upon entering the Park. If you follow this log and make the stops suggested, your drive over Trail Ridge will be much more interesting.

One of the most unique features of this Park is **Tundra Trail,** located near the Rock Cut parking area on Trail Ridge Road. The world's highest nature trail, this half-mile path is the ideal place to learn more about the dwarfed grasses, sedges, herbs, and other unusual plants which make up the alpine tundra. The scenery here is breath-taking and so is the altitude; you are more than two miles above sea level and a slow leisurely pace is recommended. Here, as along all National Park trails, you are asked to stay on the path. But this precaution is especially meaningful on this trail because tundra vegetation is extremely fragile and some of the tiny plants living here have taken several hundred years to reach their present size.

Hidden Valley, ten miles west of Estes Park, is the hub of winter recreation in the Park. From mid-December to mid-April you can enjoy ice skating, skiing, snowshoeing, and platter sliding. A cafeteria, skating rink, ski tow, and rental-equipment shop are in operation. Bus service is available.

Shadow Mountain Recreation Area is a convenient lake and mountain recreational area which borders the southwest corner of Rocky Mountain National Park. About twenty-nine square miles in extent, the area is part of the Bureau of Reclamation's Colorado-Big Thompson Project and includes Lake Granby and Shadow Mountain Lake, man-made reservoirs which are linked by channel to Grand Lake, Colorado's largest natural body of water. Camping, fishing, hunting, picnicking, horseback riding, and water sports are popular recreational activities here. In addition, naturalist services, including nature walks and campfire programs, are offered during the summer.

In Rocky Mountain almost any time of day is good for picture taking although it may cloud up in the afternoon. Trail Ridge Road offers many photographic possibilities, as does the vicinity of Bear Lake, but light is intense at these high altitudes so gauge your exposures accordingly. Interesting shots of chipmunks, squirrels, and birds are available at some of the turnouts and the slanting, twisted trees at timber line also making interesting subjects.

Rocky Mountain National Park at a Glance

Address: Superintendent, Rocky Mountain National Park Service Group, Estes Park, Colorado 80517.

Area: 262,324 acres.

Major Attractions: One of the most magnificent and diversified sections of the Rocky Mountains, with eighty-four named peaks in excess of eleven thousand feet elevation. Shadow Mountain National Recreation Area joins southwest corner of Park.

Season: Year-round.

Accommodations: Campgrounds and group campsites. There are no overnight accommodations within the Park under government supervision. A large number of hotels, motels, lodges, and camps are located near the Park; for information about these write to Chamber of Commerce at Estes Park or Grand Lake, Colorado.

Activities: Camping, fishing (license required), guided tours, hiking, horseback riding, mountain climbing, nature walks, picnicking, scenic drives, and winter sports.

Services: Guide service, mountaineering school, religious services, ice skating rink, ski rental, ski tow, ski trails, picnic tables, and rest rooms.

Interpretive Program: Campfire programs, museums, nature walks, roadside exhibits, self-guiding trails, and trailside exhibits.

Natural Features: Canyons, forests, glaciation, glaciers, lakes, mountains, rivers, tundra plants, volcanic features, wilderness area, and wildlife.

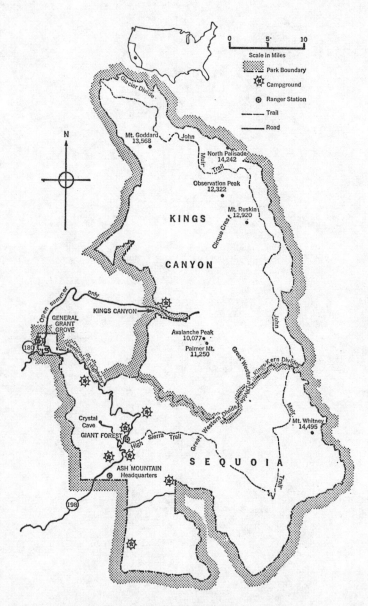

Scale in Miles
0   5   10

Park Boundary
Campground
Ranger Station
Trail
Road

N

Glacier Divide

Mt. Goddard
13,568

John

Muir

Trail

North Palisade
14,242

Observation Peak
12,322

Mt. Ruskin
12,920

KINGS

Cirque Crest

CANYON

Open summer only

GENERAL GRANT GROVE

KINGS CANYON

Avalanche Peak
10,077

Palmer Mt.
11,250

John

180

General Highway

Great Western Divide

Kings-Kern Divide

Muir

Crystal Cave

GIANT FOREST

High Sierra Trail

Great Western Divide

Mt. Whitney
14,495

ASH MOUNTAIN
Headquarters

SEQUOIA

Trail

198

Sequoia-Kings Canyon National Parks, California.

31. Sequoia-Kings Canyon
National Parks, California

Home of the Big Trees

Like Yosemite National Park, their neighbor to the north, Sequoia and Kings Canyon National Parks are located high on the western slope of the Sierra Nevada. Because of their proximity, both of these Parks bear certain similarities to Yosemite. All three of these National Parks are characterized by giant forests of aged sequoia trees, all have valleys similar in form and origin, and all of these areas are marked by the typical High Sierra granite cliffs and domes.

Sequoia is the oldest of the two Parks, having been established in 1890, thus making it the second oldest National Park, after Yellowstone. The former General Grant National Park (now a part of Kings Canyon National Park) was also created in 1890. However, the present Kings Canyon National Park was not established until 1940 and is one of the newer National Parks. Although the two areas have many similar features and are administered as a single unit, the observant visitor will soon recognize that each Park has its own distinctive character.

Snowy, Saw-toothed Mountains

Truly one of the world's great mountain complexes, the Sierra Nevada dominates the eastern part of California and is the longest and highest mountain range in the conterminous United States. When first seen by a party of Spanish missionaries in

1775, these mountains were described as *sierra nevada*. Sierra (which literally means "saw") was a term used by the Spaniards to describe mountains which had a serrated, or saw-toothed, profile; the word *nevada* means snowy. Thus, Sierra Nevada, literally "snowy, saw-toothed mountain range," came into usage to describe these mountains and the name was eventually officially adopted.

The Shifting Crust

As noted earlier, the earth's crust is far from stable; rather, it is undergoing continual adjustment and change. There is evidence for this in any mountain range, for mountains have been formed during every stage of earth history. And mountains are still being created today, especially in areas of crustal unrest, where volcanoes and earthquakes are common. Some mountain ranges, such as the Rocky and Appalachian Mountains, are formed from great folds in the earth's crust; others, like the Cascade Range, are essentially volcanic in origin. But the Sierra Nevada was not created by folding or volcanism; it is a *fault-block mountain range* like the Teton Range of Wyoming. Mountains of this type consist of great segments of the crust which have been tilted and elevated with respect to the surrounding land. Certain of these fault blocks, such as those of the Tetons and Sierra Nevada, have produced some of the world's most spectacular scenery.

The block producing the Sierra Nevada is one of the largest known. Ranging from forty to eighty miles in width and more than four hundred miles long, this single, tilted block forms an unbroken mountain chain almost as extensive as the combined Swiss, French, and Italian Alps. The steep eastern front of the Sierra Block juts more than eleven thousand feet above the floor of Owens Valley, which lies east of the Parks. Its broad western flank slopes more gently westward to the Great Valley of California and thence under the Pacific Ocean. Thus, the abrupt eastern edge, where the faulting occurred, rises more than two miles above sea level and its western margin is nearly five miles below the ocean's surface.

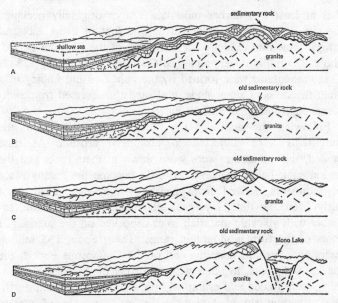

Sequoia, Kings Canyon, and Yosemite National Parks are located in the Sierra Nevada. These cross-sections show the positions of the Sierra block from the first uplift to the present day. (A) shows the Sierra Nevada after the granite intruded the metamorphic and sedimentary rocks of the first and second ancestral mountain systems; (B), (C), and (D) are the same section of the Sierra Nevada in the vicinity of Yosemite National Park. In (B), after the first uplift, most of the sedimentary and metamorphic rock has been eroded and the rock debris has washed down into San Joaquin Valley. In (C), after the second uplift, much more of the metamorphic rock is worn away and added to the existing sediments in San Joaquin Valley. (D) shows the changes after the third uplift and the downfaulting of the land to the east in the vicinity of Mono Lake. All land heights are exaggerated. Reproduced by permission from *The Yosemite Story*, by Harriet E. Huntington, Doubleday & Company, Garden City, N.Y. (1966).

Although the final uplift of the Sierra Block occurred during geologically recent times, the earlier history of the Sierra Nevada reaches far back into the geologic past. It is now known that at least two earlier mountain ranges originally occupied the site of the present-day Sierra. The first of these ancestral ranges came into existence near the end of the Permian Period, about 225 million years ago (see Geologic Time Scale, p. 34). These mountains were formed by the upheaval and folding of a great thickness of slate, shale, and sandstone formed from sediments deposited in an ancient Paleozoic sea.

Following the uplift of the "first generation" Sierra, the mountains were subjected to prolonged erosion. As time passed, the mountains were worn down to their roots and the region sank below sea level to again become the resting place of sediments derived from nearby landmasses. Throughout early Mesozoic time, layers of silt, sand, and limy ooze, together with volcanic material, were deposited on the submerged roots of the former mountain range. Then, about 135 million years ago, perhaps near the end of the Cretaceous Period, the marine sediments, together with remnants of the earlier mountains, were folded and crumpled into a series of parallel, northwest trending ridges. Concurrently, great masses of molten granite invaded the folded strata from below and slowly crystallized into solid rock. Thus was produced the "second generation" Sierra, an impressive system of mountains that occupied most of what is now eastern California.

During the Cretaceous Period (which followed the Jurassic), the second mountain system underwent a long period of stream erosion. The folded sedimentary strata were stripped away from large areas, exposing the granite which had been intruded during Jurassic time. Finally, near the end of the Cretaceous Period, the Sierra region had been reduced to a lowland bearing northwest-trending rows of hills—the remains of the "second generation" Sierra.

Thus, as the Tertiary Period began, the Sierra region was a vast lowland extending far inland from the Pacific Coast and the stage was set for the uplift of the great Sierra Block. The elevation of the block was not accomplished in one great upward movement along a single rift in the crust. Instead, there were four major uplifts and numerous intervening minor mountain-building movements, each of which elevated the range several thousands of feet and further steepened the westward slope. These disturbances, of course, took place over many millions of years.

Following each major elevation there were lengthy intervals of stability during which time the lands were eroded. The landscapes produced during these periods of erosion were characterized by fairly deep, broad valleys which were cut in the western flank of the range. Geologic evidence indicates that several such erosion surfaces were developed during the Cenozoic Era and their remnants are recognizable in the Parks today.

The major dislocation of the Sierra Block took place along a series of vertical fractures which were produced by internal disturbances within the crust. This series of faults parallels the eastern front of the range; and in places along the rift zone, the elevation of the Sierra Block was accompanied by depression of the adjacent fault block. For example, this is known to have taken place in the Owens Valley area where Mount Whitney rises almost eleven thousand feet above the floor of Owens Valley, which is situated on a depressed block.

The Work of Ice

Here, as in Yosemite National Park, it was the Ice Age glaciers that added the finishing touches to the Sequoia-Kings Canyon region. This area apparently underwent three distinctive periods of glaciation, and the effects of glaciation are obvious at many places within both Parks. Polished and grooved granite surfaces, erratic boulders, glacial quarrying (p. 147), and ridgelike moraines of glacially transported rock debris provide clues as to the size and extent of the glaciers.

Features of the Landscape

The spectacular landscape of Sequoia and Kings Canyon National Parks will probably remind you of nearby Yosemite, as their more characteristic landforms bear striking similarities. For example, the presence of U-shaped glacial troughs, hang-

ing valleys, cirques, and glacial lakes typify all three of the
Sierra Parks. Even more conspicuous are the helmet-shaped
granite domes which are so typical of this part of the Sierra
Nevada. Such outstanding scenic attractions as **Moro Rock, Fin
Dome, Alta Peak, Beetle Rock,** and **Tehipite Dome** are typical
examples of features developed by exfoliation, a process which
has shaped other well-known Sierra landforms such as Yosemite's famed Half Dome. (The role of exfoliation in dome formation is discussed in the chapter dealing with Yosemite National Park.)

Among the more unique geological features of the Sequoia-
Kings Canyon region are an unusually large number of well-
developed *avalanche chutes*. Although not commonly recognized by most visitors, these slick, steep-sided chutes represent
the tracks of avalanches where heavy winter snows have broken
loose the outer layers of rock and sent tons of snow, trees, and
rocks crashing into the canyons below. Avalanches move tremendous quantities of earth materials, and they further erode
and polish their chutes each time rock debris is dragged over
them. Because they occur in the more isolated parts of the High
Sierra, avalanches are not a hazard to most Park visitors. But
in certain parts of the world avalanches are a great peril; some
of these great slides have been clocked at speeds of nearly one
hundred miles per hour and have been known to destroy
whatever towns, houses, or persons that lay in their paths.

Avalanche chutes develop most readily in massive, unfractured granite, but they can also be formed in granite that is
rather uniformly fractured. The rocks exposed along parts of
the Great Western Divide, particularly near **Bearpaw Meadow,
Kern Canyon,** and **Mount Whitney,** contain many such chutes.
In some parts of the Sierra, chutes are so numerous that the
walls of certain cirques and canyons appear to be fluted.

Mountain Caverns

Because Sequoia National Park is considered to be one of the
"mountain" Parks, most visitors are amazed to learn that there
are several interesting caverns within its boundaries. These include **Clough Cave, Palmer Cave, Paradise Cave,** and **Crystal**

Cave, but only the latter is easily accessible and open to the public.

Even geologists are surprised to learn of the caves in Sequoia, for caves are usually associated with sedimentary rocks such as limestone and dolomite, and the High Sierra is noted for its extensive exposures of granite-like rocks. But the caves do not occur in these igneous rocks; instead, they have formed in marble which was produced by the metamorphism of limestone. Originally deposited as limy ooze when prehistoric seas covered the area millions of years ago, the limestone was later recrystallized and converted into marble. Unfortunately, any fossils that might have been present in the limestone were destroyed by metamorphism, hence its geologic age has not been determined.

Marble, like limestone and dolomite, is composed of calcite, one of the more soluble minerals. The caverns in Sequoia's marble rocks, were created and decorated by the same cave-forming processes that are discussed in conjunction with Carls-bad Caverns and Mammoth Cave. Briefly, the underground chambers and passageways have been dissolved from the marble by water which made its way beneath the surface by means of fractures, bedding planes, and other openings in the rocks. Most of the marble was dissolved below the *water table,* that is, within the zone in the earth's crust wherein all openings in the rock are completely filled with water.

The cave formations such as stalagmites, stalactites, and draperies were formed later after the water had been drained from the cavern. Even so, the cave decorations also owe their origin to the geologic work of groundwater, for they are composed of minerals deposited by water dripping from the ceilings of the caves. **Crystal Cave** contains many interesting cave formations and should be visited if at all possible.

Plants and Animals of the Parks

Sequoia and Kings Canyon National Parks, like all National Parks, are sanctuaries for the plants and wildlife that inhabit them. Here can be found an assemblage of typical High Sierra plants and animals living in their natural habitat, relatively undisturbed by outside agencies.

Most visitors to Sequoia and Kings Canyon National Parks are so spellbound and overwhelmed by the giant sequoias that they fail to notice the many animals that make up a vital part of the Sierra forest community. Yet, living in and among the big trees is an important wildlife society that plays a significant part in maintaining the biological balance necessary to the long life of the sequoias.

Among the larger mammals that might be seen are the abundant California mule deer, which live in many parts of both Parks, and the black bear which is common in forested areas. Present, but rarely seen, is the Sierra bighorn sheep. Typical smaller mammals include the golden-mantled ground squirrel, the Sierra chickaree, chipmunk, marmot, raccoon, ringtail, and mountain beaver.

The giant forests are especially attractive to the birds; Steller's jay, western bluebird, red-shafted flicker, white-headed woodpecker, western robin, Stephen's fox sparrow, and grouse are among the many species present. If you are lucky you may see a golden eagle in the higher, more remote areas of the Parks, and the gray dipper, or water ouzel, can often be seen darting among the waters of certain fast-flowing streams and waterfalls.

The Giant Forests

Each National Park has its hallmark: in Yellowstone it is Old Faithful, in Yosemite it is Half Dome, and in Sequoia and Kings Canyon it is the giant sequoia tree. In fact, Sequoia National Park was established for the express purpose of preserving these magnificent trees which in 1890 were rapidly being destroyed by lumbermen.

The giant sequoias, known scientifically as *Sequoia gigantea,* are the largest of all living things. And, on the basis of present verified scientific evidence, they are possibly the oldest living things on earth. Unfortunately, many unconfirmed and contradicting claims have been made as to the age of the sequoias, but as yet there is no accurate way to determine the age of these gigantic trees while they are alive and standing. The age of a tree can be determined accurately only by counting the annual *growth rings* as seen on a cross section of the stump or butt log after the tree has been cut down. Tree rings are useful for this purpose because most trees add a narrow layer of new wood just under the bark each year. Moreover, the wood produced in the earlier part of the year is different from the wood added later in the summer; thus, it is possible to distinguish between one year's growth and the next. When seen in cross section, this series of annual rings can be counted and each of these rings represents approximately one year's growth. Although estimates of the age of standing trees can be obtained by counting the growth rings of fallen or cut trees of comparable size and which grew under similar conditions, experience indicates that trees of the same approximate size may vary as much as six hundred years in age.

There can be no doubt, however, about the great antiquity of the Big Trees. Actual ring counts on many sequoias show that the age of these trees may exceed three thousand years and one is known to have been at least 3210 years old. On the basis of this evidence, many botanists believe that some of the larger trees are in excess of thirty-five hundred years old.

In addition to the great age of individual trees, the giant sequoias have a long family history. They were apparently abundant throughout the Mesozoic Era, "The Age of Dinosaurs," for their fossils have been found in Mesozoic rocks in many parts of what is now the United States and in central and western Europe. Moreover, fossilized sequoia stumps are present in the petrified forests of Yellowstone National Park in Wyoming and among the stony trees of Petrified Forest National Park, Arizona.

The giant sequoias are not only the largest and perhaps the oldest trees known, they are also among the most limited in range and number of individuals of any major tree species. They are native only to central California on the western slopes of the Sierra Nevada, and occupy a total area of probably no more than fifteen thousand acres. Restricted to elevations ranging from about four thousand to eight thousand feet above sea level, the sequoia groves occur in a narrow, discontinuous belt which extends north and south for a distance of about two hundred miles.

Although the giant sequoias have many unusual and distinctive characteristics, it is their size that is outstanding, for fully grown specimens may average 275 feet in height and 25 feet in diameter. Among the larger and more famous "Big Trees" are the General Sherman Tree in Sequoia's **Giant Forest** and General Grant Tree in **Grant's Grove** at Kings Canyon National Park. The General Sherman Tree stands 272.4 feet high and measures 101 feet around the base. It's age, based on growth ring studies of trees of similar size, is estimated at thirty-five hundred years. As an indication of its enormity, this huge tree has one limb that is 150 feet long and 6.8 feet in diameter —a single branch that is larger than the largest specimens of many of the more common tree species! Although only 267.4 feet tall, the General Grant Tree has a circumference of 107.6 feet, six feet more than that of the General Sherman Tree.

To become more familiar with these unusual trees and to appreciate more fully their phenomenal size, be sure to visit the 2387-acre **Giant Forest** in Sequoia National Park. This is the location of the famous General Sherman, Lincoln, and President Trees. Here, too, is the **Congress Trail,** an inspiring self-guiding nature trail through the heart of the world's largest and finest sequoia grove. In King's Canyon National Park you should visit **General Grant Grove,** the home of the General Grant, General Lee, and California Trees, and the **Big Stump Self-guiding Trail.** Another feature of King's Canyon is **Redwood Mountain Grove,** a 2500-acre forest which contains thousands of giant sequoias including the Hart Tree, the fourth largest known sequoia.

Tips for Tourists

There is much to see and do in the 1300-square-mile area that comprises these two interesting Parks. Some of the more unusual attractions may be seen along the main roads, but the real charm of Sequoia and Kings Canyon National Parks lies in the back country. These Parks are classic examples of wilderness areas—unspoiled natural preserves where one can still see and enjoy a precious remnant of the once vast wilderness that was America's heritage.

One of the more unusual conducted trips in Sequoia National Park is the tour of **Crystal Cave** (below), which is located about nine miles northeast of Giant Forest (see map). From the Crystal Cave parking area a self-guiding nature trail leads to the cavern entrance, which is guarded by a unique iron gate that resembles a giant spider web. The trail drops 320 feet along its half-mile length, but it has a hard surface and is not difficult if you take your time.

Cavern tours led by Ranger-Naturalists are scheduled twice daily, and a safe, well-kept, 1600-foot trail provides access to the more scenic parts of the cavern. A well-concealed system of indirect light is used to illuminate the cave and show the cave formations to best advantage. You may wish to take along a light sweater or jacket: the temperature remains at about 50° F at all times. All persons entering Crystal Cave must be accompanied by uniformed personnel of the National Park Service. There is a nominal fee for the guide service provided by the government, but children under twelve years of age are admitted free of charge.

Crystal Cave is a relatively recent attraction in Sequoia National Park. Despite the fact that it was discovered in 1918, the cavern was not made open to the public until 1949—almost sixty years after the Park was established.

As noted earlier, Crystal Cave has been dissolved from marble of unknown geologic age. As you walk down the trail which approaches the cave you will have an opportunity to examine this marble which was derived from an ancient limestone which formed from limy ooze deposited on the bottom of a prehistoric sea. Your National Park Service guide will tell you more about the rocks and the formation of the cave; he will also be glad to answer your questions. Even more information is available in *Crystal Cave in Sequoia National Park,* a publication of the Sequoia Natural History Society, which may be purchased in the Park. Written by Frank R. Oberhansley, former Sequoia National Park Naturalist, this easy-to-read, authoritative pamphlet is well illustrated and treats all facets of this interesting cavern.

Visitors to Sequoia will certainly want to take the **Congress Trail,** a carefully planned self-guiding trail that loops through the heart of the **Giant Forest.** Covering a distance of approximately two miles and requiring about two and a half hours to complete, this trail is an ideal place to learn more about the giant sequoias and other trees typical of the Sierra Nevada. The trail begins at the entrance to the upper parking area near the General Sherman Tree (see map).

The Congress Trail derives its name from the interesting grouping of some of its trees. Dominating what might be called the "Government Group" is the **President Tree** which represents the executive branch of the government. This 230-foot-tall behemoth has a diameter of more than thirty feet and, appropriately enough, stands apart from and somewhat above the **Senate** and **House Groups** which are located a short distance down the trail. **Circle Meadow** lies near the Senate and House Groups, providing proper separation of these two groves of trees which symbolize the legislative branches of our government.

Other highlights of the Congress Trail include the **Chief Sequoyah Tree,** which was named for the Cherokee Indian who devised a phonetic alphabet system that enabled members of his tribe to read and write. This alphabet contributed greatly to the advancement of the Cherokees, and in recognition of this achievement the Austrian botanist who first described the Big Trees named them in Sequoyah's honor. The change in spelling of his name was made to conform to botanic nomenclature, which requires that all proper names be Latinized.

Also of interest are the **Room Tree,** so named because its interior has been gutted by fire, leaving a "room," and the **McKinley Tree,** whose 291-foot height makes it the tallest tree in the Giant Forest.

Growing in close association with the Big Trees are sugar pine, white fir, chinquapin, and a host of smaller but interesting shrubs and bushes. Visitors wishing to take a conducted tour of Congress Trail may do so during the summer season; check bulletin boards for details.

Although not on the Congress Trail, the **General Sherman Tree** is located near the trail in this part of the Park.

The **Big Stump Trail** in Kings Canyon National Park relates to a different and more sobering chapter of the Big Trees story. Here, in Big Stump Basin in the southwest corner of General Grant Grove, is a dramatic but silent reminder of man's depredation on nature. During the late 1800s, Big Stump Basin was the site of extensive logging operations which slashed their way through the virgin sequoia groves and which, if they had not been checked, might well have led to the extinction of these magnificent trees. Few places in our National Parks portray more graphically the importance of conservation or better exemplify the importance of a system of National Parks, Monuments, and Forests.

Features of the Big Stump Trail include the **Burnt Monarch,** the fire-gutted stump of a giant sequoia which is ninety-seven feet in circumference and estimated to be 2350 years of age, and the **Mark Twain Stump,** which is twenty-four feet across. The latter is the remains of a tree which was felled in 1891; a cross section of its trunk may be seen today in the American Museum of Natural History in New York City. Also of interest are relics of the "boom" days of the turn of the century when Big Stump Basin was the site of a bustling lumber camp. The location of the old saw mill, sawdust piles, and partially cut trees are marked by numbered stakes along the trail. In addition, historical notes about these features are presented in the informative trail guide booklet that is available at the head of the trail.

The scenic trails in these two Parks are ideal for the horseman. Saddle horses may be rented at concessioner-operated corrals in Giant Forest, General Grant Grove, and in Cedar Grove.

In addition to riding horses, pack stock may be rented in Cedar Grove. Stock travel is a popular means of travel for those visitors who do not wish to carry their own equipment or for families with small children. Trips of this type vary greatly, but they depend, in general, upon the amount of time, energy, and money that one cares to expend. For example, there are those who prefer to travel on foot but rent a burro or mule to carry their camping gear. Others prefer to hire a packer, horses, and pack animals to pack them into a base camp, making arrangements for the packer to return for them at a later date. This method permits one to establish a "home base" and to explore many trails from a central point. It has the added advantage of eliminating the responsibility of feeding and caring for the saddle horses and pack animals. Finally, there are the so-called touring trips. Under this type of arrangement the packer and stock remain with the party throughout the entire trip. Although this type of wilderness travel is more expensive, it does require considerably less work on the part of the visitor. Moreover, it enables him to cover a maximum amount of country in a minimum period of time.

If you are planning a stock trip, you should make reservations with an authorized packer at the earliest possible date. To obtain a list of reliable packers who have permits to pack and rent walking stock in these Parks, write: Superintendent, Sequoia and Kings Canyon National Parks, Three Rivers, California 93271.

These two Parks offer the photographer many fine viewpoints from which to photograph the spectacular scenery. Most of the turnouts and overlooks along the main roads are worthy of consideration, and opportunities are especially numerous along the wilderness trails.

The most popular subjects are, of course, the Big Trees, for they hold a special attraction for the photographer. But their most noteworthy feature—their great size—often proves to be a photographic liability, for unless you use a wide-angle lens it is difficult to photograph the entire tree. You can, and should, move some distance away, but even so, other trees are likely to obstruct your view. In some places there are clearings from which you may shoot, but you may have to photograph the larger trees in sections. This problem is not so serious for the person making moving pictures: he can "pan" up and down the tree and record it in its entirety. Incidentally, many of the trees have signs which give their statistics; after you have photographed the tree take a picture of the sign and you will have available some interesting facts and figures about your subject.

Good light is often at a premium in the sequoia groves and exposures are best determined by an exposure meter. In general, the diffused sunlight of a hazy day is better than a bright day. Bright sunlight will accentuate the forest shadows, thus blocking out detail in your pictures. In most areas, midday (from about ten-thirty to twelve-thirty) is the best time for pictures, but the creative cameraman will find suitable subjects regardless of time of day or season of the year.

Regardless of when or from where you photograph the sequoias, be sure to include people in your pictures. They will give scale and perspective to the picture and will give a better indication of the size of the trees themselves.

Sequoia-Kings Canyon National Parks at a Glance

Address: Superintendent, Three Rivers, California 93271.

Area: 460,331 acres.

Major Attractions: Great groves of giant sequoias, world's largest and possibly oldest living things; magnificent High Sierra scenery, including Mount Whitney, highest mountain (14,495 feet) in conterminous United States.

Season: Year-round.

Accommodations: Cabins, campgrounds, trailer park. *For reservations contact:* Sequoia and Kings Canyon National Parks Company, Visalia, California 93277.

Activities: Camping, fishing (license required), guided tours, hiking, horseback riding, mountain climbing, nature walks, picnicking, scenic drives, and winter sports.

Services: Food service, gift shop, post office, religious services, service station, ski rental, ski tow, telegraph, telephone, picnic tables, rest rooms, and general store.

Interpretive Program: Campfire programs, visitor center museum, nature trails, roadside exhibits, self-guiding trails, and trailside exhibits.

Natural Features: Avalanche chutes, canyons, caverns, forests, geologic formations, lakes, mountains, rocks and minerals, wilderness area, and wildlife.

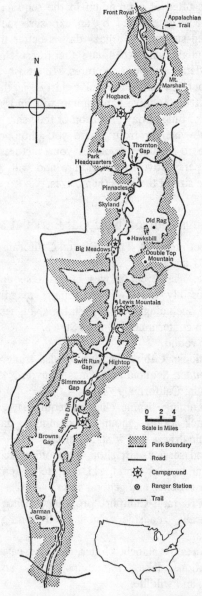

Shenandoah National Park, Virginia.

32. Shenandoah National Park, Virginia

Daughter of the Stars

Shenandoah National Park—one of our most accessible and popular Parks—extends approximately eighty-five miles astride the crest of the fabled Blue Ridge Mountains of Virginia. This area, which derives its name from an Indian word meaning "daughter of the stars," is in many ways similar to Great Smoky Mountains National Park, which lies to the southwest. They have, for example, the same geological heritage, for both the Great Smokies and the Blue Ridge are part of the great Appalachian Mountain Range. There are also distinct similarities in the plants and animals of the two Parks, and both areas have been the home of many generations of hardy mountain people. And, because of their strategic location and many natural and historical attractions, Great Smoky Mountains and Shenandoah are two of the most commonly visited National Parks.

But the strongest tie between the two Parks is the Blue Ridge Parkway, a beautiful highroad that serves as a connecting link between these two scenic areas.

Geologic Story of the Blue Ridge

Like the Great Smokies of the southern Appalachians, the Blue Ridge Mountains bear the distinctive characteristics of mountains that are very, very old. The gently rounded summits, folded and fractured sedimentary rocks, and the unmistakable signs of metamorphism and igneous activity are evidence that the Blue Ridge has not always been what it is today. But unlike the Smokies, which were carved out of sedimentary rocks, the Blue Ridge is composed of older and more resistant igneous and metamorphic rocks.

The core of the Blue Ridge Mountains consists of igneous rock derived from molten rock material that hardened deep within the earth. This molten rock, called *magma,* was injected into the earth's crust where it gradually cooled and crystallized into the granitic rocks that typify the Blue Ridge today. These rocks are of two different types. One kind, a fine-grained granite, is exposed largely along the crest of the Blue Ridge and on the western slope of the mountains. You can get a good look at this rock where it crops out at Marys Rock Tunnel. The other type, a much coarser grained granite, is well exposed in the area east of the crest of the Blue Ridge and on Old Rag Mountain, the locality for which the granite was named. Both the Old Rag granite and the granodiorite were subjected to considerable metamorphism, and in places they present a layered or laminated appearance. This type of rock structure, called *gneissic texture,* is evidence of the changes brought about by the metamorphism of the ancient Precambrian granites.

With the passage of time, the ancient rocks slowly solidified many miles beneath the surface, and the Blue Ridge area underwent considerable erosion. One can only guess at the amount of time required to strip the great thickness of overlying rocks from the now-exposed granite and granodiorite, but it must have taken many millions of years. Finally, the deeply buried rocks were laid bare and were themselves subjected to the forces of erosion. It was then that the surface of the ancient basement rocks was carved into a rugged landscape of mountains, hills, and canyons. As time passed and erosion proceeded, the granitic mountains were steadily lowered and the valleys gradually became filled with sand and gravel. During this chapter in earth history, the face of the eastern United States differed greatly from what we see today. Life as we know it barely existed then and primitive one-celled plants were the only life forms that could be found in the barren Precambrian seas—a far cry from the luxuriant Blue Ridge flora of today.

The next chapter in the Blue Ridge story is a fiery one and it began more than 600 million years ago. This episode is recorded in the rocks of the *Catoctin Formation,* an extensive series of lava flows. But unlike the lava flows found in certain parts of the world, these flows did not emanate from the crater of some ancient volcano. Instead, the magma reached the surface by means of great cracks, or *fissures,* in the earth's crust. From these fissures came great outpourings of lava called *fissure flows,* a type of eruption similar to those which have formed great *lava plateaus* such as the Columbia River Plateau of the Pacific Northwest.

These flows, which averaged about one hundred and fifty feet in thickness and were at least six miles long, continued to be extruded onto the earth's surface over a long period of time. As the layers became successively thicker, the flows, which had at first filled the valleys, eventually swallowed even the highest mountain peaks. We can get some ideas as to the extent of the lava flows in the Big Meadows-Stony Man area where there are at least twelve separate flows with an accumulated thickness of eighteen hundred feet. But, as impressive as this is, it by no means represents the full thickness of the ancient lava flows, for the upper sequence of flows has been removed by erosion.

Some of the lava flows exhibit a distinctive characteristic called *columnar jointing,* a type of structure developed as the lava cooled and contracted, thus producing a set of fractures with a more or less clearly defined hexagonal pattern. These cracks, or *joints,* extended downward as lava in the lower part of the flow cooled, thereby producing a set of prism-shaped, many-sided columns. Good examples of columnar jointing can be seen at Crescent Rocks and Franklin Cliffs.

Later geologic events produced drastic changes in temperature and pressure which greatly modified the basaltic lava and transformed the original igneous rock to greenstone schist —a metamorphic rock.

Finally, after untold thousands of years, the eruptions ceased and the lava flows were subjected to a period of prolonged erosion. During this period of geologic time the Blue Ridge area probably resembled the flat lava plateaus of Idaho and Oregon. But here and there the monotonous plain may have been broken by one of the ancient granite peaks that had been uncovered by erosion. And, although millions of years had passed, the barren landscape was apparently still devoid of life—except perhaps for scattered colonies of algae.

The later chapters in the geologic history of the Blue Ridge are not clear, for the rocks in which the record is to be found have been eroded from much of the Blue Ridge proper. But the geologic history of the Appalachian Mountain Range, of which the Blue Ridge is a part, is known with some certainty and from this point in earth history the development of the Blue Ridge roughly parallels that of the Appalachians. Thus, by correlating the geologic evidence from one part of the Appalachians to another, we can reconstruct, with varying degrees of accuracy, the sequence of missing events.

And so it seems that about six hundred million years ago, the Blue Ridge area—like the Great Smokies—was the site of an ancient shallow sea. This sea extended for thousands of miles from Alabama to Newfoundland, and occupied an elongate, troughlike depression called the *Appalachian Geosyncline*. Here, over a period of about three hundred million years, there accumulated some thirty thousand feet of marine sediments. In the beginning the old Appalachian sea was devoid of all but the lowest type of plant life. But as time went on, these lower life forms gave rise to more complex organisms until the sea was swarming with life. Today we can find the fossilized remains of these ancient sea creatures preserved in certain of the geologic formations which are exposed in the Appalachians.

As the sediments accumulated on the sea bottom, the floor of the trough was gradually warped downward until more than six miles of rock debris had been deposited in the slowly subsiding geosyncline.

But from where, you may ask, did this great volume of sediments come? Although they no longer exist, geologic evidence suggests that throughout much of Paleozoic time there must have been landmasses to the southeast of the old Appalachian seaway. And, as the floor of the geosyncline sank, there was an accompanying vertical uplift of the nearby southeastern highlands. Meanwhile the countless layers of loose sediment which had settled in the geosyncline were gradually cemented and compacted to form layer upon layer of sedimentary rocks. Other sedimentary rocks were derived from the remains of countless marine organisms.

The next clearly decipherable geologic event in the Appalachian story took place near the end of the Paleozoic Era, about 200 million years ago. This was a period of extensive mountain-building in the eastern United States and the rocks of the geosynclinal area were crumpled into a series of giant folds. As the strata yielded under the force of relentless horizontal compression, the crust of the earth was perceptibly shortened. Indeed, it has been estimated that if the folded strata in Pennsylvania were returned to their original horizontal position, they would fill an area about 25 per cent wider than they occupy at present.

These lateral pressures came from the southeast, and produced a long series of northeast trending folds. Moreover, as the rocks were deformed, they were subjected to great extremes of heat and pressure; thereby converting certain of the igneous and sedimentary rocks to metamorphics such as quartzite (from sandstone), slates (from shale), and greenstone (from the basaltic Catoctin lavas).

Not all of the rocks folded. In places the pressure was so great that the strata were not only bent but were broken and shoved over and above younger rocks, thereby producing great overthrust faults like that in Great Smoky Mountains National Park to the southwest.

This great *orogeny,* or mountain-building movement, has been named the *Appalachian Revolution,* and gave rise to the "first generation" Appalachians, a range believed to have been much higher than the present-day Appalachians. How much higher were these late Paleozoic mountains than those of today? There is, of course, no way of knowing for sure, but if we reconstruct certain of the folds of central Pennsylvania, it appears that the early Appalachians may have risen at least twenty thousand feet above sea level. This is as high as the towering Alaskan ranges, the highest in North America today.

The Appalachian Revolution not only elevated the ancestral Appalachians, but to the west of the mountains, as far south as the Great Plains region, the ocean floor was raised and the waters drained away. This uplift brought into existence all the lands of the Central United States, much of which has not been beneath the sea since.

But like all mountains, the majestic ancestral Appalachians were destined for destruction. They had no sooner raised their lofty heads than they were attacked by the forces of weathering and erosion. And so, during the more than 100 million years of the Mesozoic Era, the Appalachians were almost completely worn away and only their erosion-scarred roots remained as evidence of the former mountain range. In some regions, including the Blue Ridge and Great Smoky Mountains areas, the broad, featureless late Mesozoic landscape contained many *monadnocks,* isolated hills where the ancient, more resistant, crystalline rocks had staved off final destruction by erosion.

Near the beginning of the Cretaceous Period—after the entire Appalachian region was reduced to a nearly level plain—a series of vertical uplifts elevated the site of the old Appalachian Geosyncline so that its central portion rose about two thousand feet above the surrounding territory. The increase in elevation rejuvenated the streams and hastened the development of the valleys which were carved out of the softer rocks. Thus, the Shenandoah Valley is an extensive lowland that was excavated by the large streams that flow across exposures of relatively soft shale and limestone.

But the more resistant rocks, such as greenstone (a type of green schist) and gneissic granite, were left standing as ridges. In general, the Blue Ridge is high where there are broad exposures of greenstone, and its elevation decreases to the north as the band of greenstone begins to narrow. Another resistant formation, the quartzite of the *Chilhowee Group* (the same group which crops out in Great Smoky Mountains National Park) caps the western foothills of the Blue Ridge. And so, like the Great Smokies, the present landscape and drainage of the Blue Ridge are the result of fairly recent geologic events.

Plants and Animals of Shenandoah National Park

A profusion of plants inhabit this mountain National Park and they range from tiny, delicate summer wildflowers to ancient hemlocks more than four hundred years old and three feet in diameter.

Virtually all of the Park is wooded, and much of this forest is composed of red, scarlet, and chestnut oaks. Also present are hickory, maple, birch, yellow poplar, and white, Virginia, and pitch pine. In higher parts of the Park there are red spruce and balsam fir.

Spring is a beautiful season in the Blue Ridge. Flowering redbud, azaleas, dogwood, and mountain laurel add their color to the myriad violets, marigolds, trilliums, hepaticas, and jack-in-the-pulpits that bloom in the Park. Summer has its own special flowers including columbine, primroses, asters, daisies, Turk's-cap lilies, purple thistles, and black-eyed Susans. But the fall of the year is the favorite of many of Shenandoah's visitors. Here, when the leaves turn from green to the golden hues of autumn, the mountainsides are ablaze with a dazzling array of color.

Birds dominate the wildlife of the Park and as many as two hundred species may enter and leave Shenandoah each year. The larger birds include ravens, buzzards, vultures, and wild turkeys. Among the smaller forms that you are likely to see are woodpeckers, mourning doves, jays, and towhees. If you know your songbirds perhaps you will recognize the catbirds, chickadees, red-eyed vireos, black-throated blue warblers, and brown thrashers.

The mammals of Shenandoah National Park are relatively small in size and in number. You may see white-tailed deer in the early morning or evening near the Whiteoak parking area at Skyland, at Big Meadows, and near the edge of the forests throughout the Park. The population of bears, foxes, and bobcats is steadily increasing, and chances are that you may see some of these larger mammals, as well as raccoons, skunks, chipmunks, woodchucks, and a variety of squirrels.

Tips for Tourists

The visitor to Shenandoah can select his activities from a broad and diversified program. Much of the attraction here revolves around Skyline Drive (p. 400–402), which extends the length of the Park from Front Royal to Rockfish Gap. While it is true that "motor mountaineers"—those visitors who confine their visit to their automobiles—will be rewarded with superb mountain panoramas at almost every turn of the Drive, those who would savor the true beauty of Shenandoah will stop long enough to leave their car and their cares behind them. These visitors will take time to hike the mountain trails, join a Ranger-Naturalist on a nature walk, or take a short "self-study course" along one of the Park's self-guiding nature trails. And in so doing they will come to know the true charm of Shenandoah National Park.

When you reach the Park you may wish to obtain a copy of the *Park Guide,* a very comprehensive publication that describes the area's many attractions and tells where and how they can be seen. These may be purchased throughout the Park or ordered by mail from the Shenandoah Natural History Association, Shenandoah National Park, Luray, Virginia 22835.

The principal drive through Shenandoah National Park is Skyline Drive, a 105-mile highway that winds safely and gently along the crest of the Blue Ridge Mountains. Skyline Drive begins at Front Royal and ends at Rockfish Gap where it intersects U.S. 250. This also marks the beginning of the famous Blue Ridge Parkway.

As you drive along **Skyline Drive** you will notice milepost signs that are keyed to the Park map. These have been erected in order that you might more easily locate points of interest along the road. There are several areas on Skyline Drive where lodging, food service, gasoline, and other facilities are available. These include **Skyland, Big Meadows,** and **Lewis Moun-**

tain. Service stations, camper's stores, gift shops, and coffee shops are located at other conveniently spaced points along Skyline Drive. These, too, are indicated on the Park maps. In addition, the starting points of most of the nature walks and hiking trails will be found along Skyline Drive. At more scenic points along the way, you will find parking overlooks which will enable you to park your car and enjoy the spectacular scenery. Moreover, a number of pleasant picnic areas, complete with tables and running water, have been set aside for the convenience of Park visitors.

Some of the "must" stops along Skyline Drive include **Shenandoah Valley Overlook** (Mile 2.8), **Dickey Ridge Visitor Center** (Mile 4.6), **Range View Overlook** (Mile 17.1), **Hogback Overlook** (Mile 21.0), **Panorama** (Mile 31.5), **Marys Rock Tunnel Overlook** (Mile 32.4), **Little Stony Man Parking Area** (Mile 39.2), **Crescent Rock Overlook** (Mile 44.5), **Upper Hawksbill Gap Parking Area** (Mile 46.6), **Byrd Visitor Center** of Big Meadows (Mile 51.1), and **Rockytop Overlook** (Mile 78.2). These are points from which there are particularly fine vistas of the Blue Ridge and Shenandoah Valley, but any of the overlooks are worthwhile places to stop and each offers a different perspective of the surrounding valleys and mountains.

As mentioned above, Skyline Drive has its terminus at Rockfish Gap, which also marks the starting point of the **Blue Ridge Parkway.** This scenic highway, which is about three-quarters completed, will eventually extend from Shenandoah National Park in Virginia to the North Carolina entrance of Great Smoky Mountains National Park, a distance of almost five hundred miles. The road traverses one of the most beautiful parts of the eastern United States, and the excellent grades, gradual curves, and absence of commercial vehicles and billboards further enhance this lovely highroad. There are no commercial enterprises along the Parkway, but intersecting state roads lead to nearby towns where food, lodging, gasoline, and other services are readily obtainable.

At convenient intervals along the Parkway you will find picnic areas, campgrounds, and hiking trails which have been established for the unhurried visitor. The country along the Parkway is not only noted for its natural history—plants, animals, and geologic features—it is also rich in the mountain folklore of one of America's most historic regions. As you drive the Blue Ridge Parkway take time to stop at the overlooks to enjoy the scenery and to read the interpretive signs which indicate and explain points of special interest. You should also plan to take advantage of the Parkway's interpretive program, which includes six visitor centers (which explain the human and natural history of the Appalachian), eleven self-guiding trails, and, during June, July, and August, conducted nature walks and campfire programs. The latter are scheduled at **Peaks of Otter, Doughton Park, Price Park,** and **Crabtree Meadows** (schedules are available at points of visitor concentration along the Parkway). The National Park Service has prepared a detailed map, road log, and descriptive brochure which describes the features of the Parkway. These may be obtained at Visitor Centers and ranger stations. Remember that while you are on the Parkway you are in an area administered by the National Park Service and subject to the same regulations as the National Parks.

There is no limit to the photographic opportunities that the alert photographer will discover in this Park. Unsurpassed mountain panoramas appear at most of the scenic lookouts along Skyline Drive, and the cameraman who takes to the trail will be even more rewarded. Because of the characteristic blue haze from which the Blue Ridge derives its name, you should use a haze filter for most long distance or panoramic views. In general, your pictures will be sharper and have more perspective if they are taken in the morning or late afternoon. During the middle part of the day the light is very flat and your photographs will have very little definition or depth.

Shenandoah National Park at a Glance

Address: Superintendent, Box 387, Luray, Virginia 22835.
Area: 195,538 acres.
Major Attractions: Outstanding portion of Blue Ridge Mountains, with Skyline Drive traversing crest; magnificent vista of

historic Shenandoah Valley and Piedmont region of Virginia; hardwood forests; wealth of wildflowers.

Season: Year-round.

Accommodations: Cabins, campgrounds, group campsites, hotels, motels, trailer sites, but no utility hookup for trailers. *For reservations contact:* Virginia Sky-line Company, Inc., P. O. Box 191, Luray, Virginia 22835.

Activities: Camping, fishing (Virginia fishing license required), hiking, horseback riding, mountain climbing, nature walks, picnicking, scenic drives.

Services: Food service, gift shops, laundry, public showers, religious services, service stations, telephone, picnic tables, rest rooms, and general store.

Interpretive Program: Campfire programs, museum, nature trails, roadside exhibits, self-guiding trails, trailside exhibits.

Natural Features: Forests, geologic formations, mountains, rivers, unusual birds, unusual plants, waterfalls, wildlife.

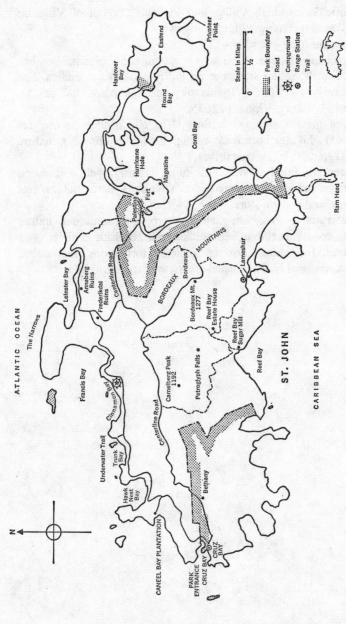

Virgin Islands National Park, St. John.

33. Virgin Islands National Park, St. John

Caribbean Paradise

Snow-white, palm-fringed beaches, unusual plants, and exotically beautiful coral reefs—these are but a few of the attractions of this island National Park. A relatively recent addition to the National Park System—it was added in 1956—Virgin Islands National Park is located on the island of St. John hundreds of miles southeast of the continental United States. The quiet beauty of this heavenly isle is greatly enhanced by its tranquil setting, for the tree-covered mountains jut sharply out of the surrounding blue-green waters of the Caribbean Sea. There are three major islands in the American Virgin Islands: St. Thomas, the most populated and the capital of the Territory; St. Croix, the largest of the United States group; and St. John, the most beautiful and least disturbed of the three. There are, in addition, some fifty smaller islands and islets, some of which are just large enough to land on in a small boat.

St. John Island, the site of the Park, is about five miles long, nine miles wide, and has an area of approximately nineteen square miles. Its highest point—1277-foot Bordeaux Mountain —rises abruptly from the sea and from its summit can be seen magnificent views including St. Thomas Island to the west, St. Croix to the south, and several of the British Virgin Islands to the north.

St. John is famous for its climate and relatively low humidity. Rainfall, mostly in brief showers at night, ranges from very

light on the dry eastern end of the major islands to approximately
fifty inches per year near the center of the larger islands. And,
thanks to the tempering effect of the northeast trade winds,
days are pleasant and the nights delightfully cool.

The Geologic Story

Despite the fact that certain of the Virgin Islands have been
studied in considerable detail, the geologic history of the islands is still imperfectly known. Moreover, that part of their
evolution that has been deciphered is most complex and the
findings of the few geologists who have studied the area do
not always agree. In general, however, it appears that the
Virgins are underlain by mildly deformed volcanic rocks and
limestone of Cretaceous—or possibly older—age. During Late
Cretaceous or Early Tertiary time, the volcanics and limestones
were intruded by great masses of molten rock which later
solidified beneath the surface at relatively shallow depths.
As the hot magma came in contact with the cooler rocks
into which it was injected, the surrounding rocks were metamorphosed, thereby converting them to new rock types.

The earliest geologic episode recorded in the geologic history of the area was the extrusion of almost two miles of
volcanic rocks on what was perhaps the original ocean floor.
The rocks formed from these great submarine lava flows are
known today as the *Water Island Formation,* which consists
primarily of lava flows and minor amounts of tuff (a rock
formed of compacted volcanic fragments). Yet, despite the
fact that approximately fifteen thousand feet of the Water Island Formation is exposed in the Virgin Islands, the age of
the formation has only recently been determined. On the
basis of data gained from radioactive dating, it is believed
that the top of the formation is at least 108 million years old
and is probably Late Cretaceous in age. Water Island volcanics
can be seen in road cuts near the intersection of **Centerline
Road** and **Bordeaux Mountain Road,** on the points near
Chocolate Hole, and along the trail from **Centerline** to **Reef
Bay** (see map).

Near the beginning of Late Cretaceous time, this area underwent relatively rapid uplift and the ocean floor was raised above sea level. This portion of the geologic record is represented by the *Louisenhoj Formation,* which consists largely of volcanic debris that was deposited on land. In places, layers of stream-laid conglomerate or coarse beach deposits are interbedded with the volcanic debris; these, plus the weathered Water Island rocks, provide further proof that the area was not submerged during this part of earth history. There are good outcrops of the Louisenhoj Formation on the western part of St. John, especially near **Cruz Bay** and at **Rendezvous Bay** on the southwestern shore.

Following the uplift and erosion that occurred during Louisenhoj time, there was a period of volcanic quiescence and the land again slipped beneath the sea. It was then that the marine sediments which now comprise the *Outer Brass Limestone* were deposited. This formation is composed primarily of thin-bedded, dark, silica-bearing limestones which contain the fossilized remains of one-celled microscopic animals called *radiolarians.* The limestone, which was probably deposited offshore in relatively deep water, is in places metamorphosed by the Late Cretaceous and/or Early Tertiary igneous intrusions that were mentioned earlier. Locally there are thin layers of tuff interbedded with the marine limestones, but these account for only about 10 per cent of the total formation. Exposures of metamorphosed Outer Brass Limestone can be seen at **Annaberg Point** and in road cuts and along trails from **Cruz Bay** to **Caneel** and around to **Cinnamon Bay.**

The next chapter in the geologic story of the Virgin Islands is marked by another period of uplift during which the land was sharply tilted to the north. With the sea floor once again exposed to the atmosphere, the rocks of the Louisenhoj and Outer Brass Formations were rather rapidly eroded. Thus, the *Tutu Formation* which overlies the Outer Brass Limestone consists mostly of weathered volcanic debris from the Louisenhoj Formation mixed with fragments of limestone. There are good exposures of the Tutu Formation on St. John Island from **Maho Bay** to **Leinster Point** and on the north end on Francis Bay accessible by the Francis Bay Trail.

After the deposition of the Tutu Formation the area appears to have been subjected to great crustal deformation, which resulted in folding and faulting. In addition, this structural deformation seems to have been accompanied by considerable volcanic activity. And so near the end of the Cretaceous Period the Virgin Islands—which at that time may have been a continuous landmass with Puerto Rico—were elevated and subjected to prolonged erosion. As time passed and erosion continued, this ancient Virgin landmass was reduced to a rolling plain, but no sooner had the mountains been worn away than the land was raised once more, only to again be eroded down to near sea level. Indeed, geologic evidence indicates that there were at least three such periods of uplift and erosion which are believed to range from Late Cretaceous to Early Tertiary in age.

The latest episode in the history of the Virgin Islands probably began near the end of Pleistocene time. It was then that the waters became warm enough to permit the growth of the corals which formed the beautiful reefs that fringe the island. This, too, was the time that marked the start of the development of the predominant features that characterize the island landscape of today.

Man Comes to the Virgins

The human history of the Virgins is almost as colorful and fascinating as the islands themselves. They were discovered and named by Christopher Columbus, who sighted the islands in 1493 while on his second voyage to the New World. But though Columbus "officially" discovered the Virgin Islands, St. John had been inhabited by the Arawak Indians centuries before his arrival.

Following Columbus's "discovery" in 1493, the islands were visited by ships of many lands, but it was not until early in the eighteenth century that the Danes established a permanent settlement on St. John. From a humble beginning in 1717, these early inhabitants spread out over the island establishing plantations which were worked by imported slaves. By the year 1726 all of St. John had been set aside as large estates which were the source of profitable crops of sugar cane and lesser amounts of cotton and tobacco. All was well until 1733, at which time the slaves revolted against their Danish masters. The rebellion was finally put down but only after extensive loss of life and property, and finally—in 1848—Denmark abolished slavery. Deprived of the slave labor that was the backbone of the island economy, the plantation owners soon abandoned St. John and the land once again returned to its original wilderness condition as it reverted to tropical forest. However, the ruins of Danish sugar mills, such as that at Annaberg, still remain as points of interest in the islands today.

The Virgin Islands became United states territory in 1917, at which time they were purchased from Denmark for $25 million. However, it was not until 1956 that most of the island of St. John was dedicated as a National Park. The establishment of the Park was made possible when Jackson Hole Preserve, Inc., (of which Laurance S. Rockefeller, grandson of John D. Rockefeller, was president) presented more than five thousand acres of Park land to the citizens of the United States.

Plants and Animals of Virgin Islands National Park

More than three quarters of Virgin Islands National Park is blanketed with vines, shrubs, flowers, and trees. The trees, however, are not virgin timber, for most of the island's trees were cleared away by the slaves of the early Danish plantation owners. Among the interesting tropical trees seen on the island are palms, breadfruit, mango, soursop, mahogany, kapok, and mangrove. Color is added by flowering trees and shrubs such as frangipani, hibiscus, bougainvillea, and flamboyant. In the interior highlands, there are dense forests of evergreen hardwoods; the drier slopes are forested by various broad-leaved trees. In addition, several species of cactus thrive on the island, and in certain parts of the islands these dominate the flora.

Birds are the predominant form of animal life on St. John and almost one hundred species have been observed here. Shore-dwelling birds such as pelicans, gulls, herons, egrets, terns, frigate birds, and ducks are fairly numerous, but land birds are more common. Among those that you are likely to see are forest doves and pigeons, thrashers, hummingbirds, mockingbirds, warblers, smooth-billed ani, banana quits, and perhaps a hawk.

Bats are the only native mammals that inhabit the island, but a number of introduced forms such as the mongoose are common. Present also are several species of frogs, lizards, sea turtles, and a few nonpoisonous snakes.

Offshore waters abound in marine life of all types. Among the showy reef fish that are likely to be seen by snorkelers are jacks, grunts, banded butterfly fish, blue tang, trumpet fish, four-eyed butterfly fish, angelfish, parrot fish, and squirrelfish. In addition, the shells of clams and snails are commonly found on the beaches, and the coral reefs are literally alive with an amazing array of colorful and unusual animals such as sponges, shellfish, and corals.

Tips for Tourists

There is a fine interpretive program in operation here, and the snorkel trips, nature walks, self-guiding trails, campfire programs, and museum exhibits enable the visitor to become better acquainted with the natural and human history of St. John Island. A schedule of interpretive programs is posted weekly at the Cruz Bay Visitor Center and at Cinnamon Bay Campground Commissary, or you may ask any Ranger-Naturalist for information about the current activities.

The roads on St. John are unimproved, but jeep taxi service or jeep tours are available. **Centerline Road** traverses the island from west to east and leads to major points of interest within the Park. As you follow the road along the northern shore of the island, you will pass **Hawksnest Bay, Trunk Bay** (site of the submarine self-guiding trail), **Maho Bay, Annaberg Ruins,** and **Reef Bay Valley.** A side road leads to 1277-foot **Bordeaux Mountain** and **Coral Bay Overlook** from where there are magnificent vistas of the surrounding territory.

The clear, warm waters and the beautiful sand beaches are a

natural invitation to the swimmer, and the self-guiding underwater trail is especially designed for the snorkeler.

In addition to the unusual underwater self-guiding trail, snorkel trips are another feature unique to Virgin Islands National Park. Conducted snorkel trips along the coral reefs are regularly scheduled from the east end of **Cinnamon Bay Beach** and **Hawksnest Beach.** These trips, which last about forty-five minutes, provide a wonderful opportunity to learn more about the marine life that is so abundant in the shallow waters off the island's coast. Another good locality for snorkeling is **Trunk Bay.**

A host of marine fishes live in the offshore waters of St. John and fishing is rapidly becoming a popular pastime. Visitors interested in deep-sea fishing should inquire at Caneel Bay Plantation or the Visitor Center at Cruz Bay.

There is an infinite variety of interesting photographic subjects here. Beautiful seascapes, colorful sunsets, and shots of tropical plants are the favorites of most photographers. However, those persons who have the proper equipment will find the coral reef to be a veritable paradise for underwater photography. National Park Service personnel will be glad to answer your questions about photography and offer suggestions for suitable subjects.

Virgin Islands National Park at a Glance

Address: Superintendent, Box 1707, Charlotte Amalie, St. Thomas, Virgin Islands 00801.

Area: 15,150 acres.

Major Attractions: An island of lush green hills and white sandy beaches; rich in tropical plant and animal life; prehistoric Indian petroglyphs; remains of colonial sugar plantations; coral reefs and marine life.

Season: Year-round.

Accommodations: Campgrounds, group campsites, tents and beach-type cottages. For reservations for camping equipment contact Cinnamon Bay Camp, U. S. Virgin Islands 00830.

Activities: Camping, fishing, guided tours, hiking, nature walks, picnicking, scenic drives, swimming, and snorkeling.

Services: Food service, picnic tables, and rest rooms.

Interpretive Program: Campfire programs, nature trails, and self-guiding trails.

Natural Features: Forests, marine life, seashore, unusual plants, coral reefs, and sugar mill ruins.

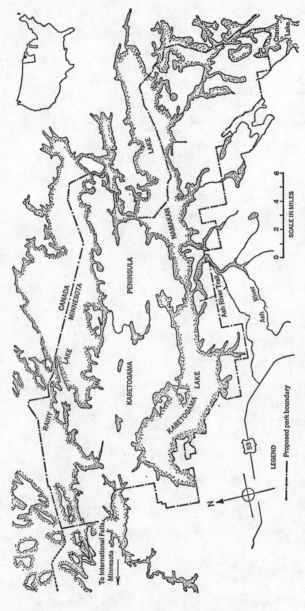

Voyageurs National Park, Minnesota.

34. Voyageurs National Park, Minnesota

Land of Lakes

If we could turn back the geological calendar ten thousand years, many surprising changes would greet us. The northern part of the United States would probably remind us of the great white continent of Antarctica, for this section of our nation was blanketed by a massive sheet of glacier ice. Although that great continental glacier has long since vanished, it left many tracks on the land over which it passed. The most noticeable of these are the many lakes that have developed in depressions that were scooped out by the advancing ice. These include the Great Lakes, the Finger Lakes of New York State, and the eighty thousand acres of lakes in Voyageurs National Park.

This part of Minnesota's north country was scraped bare by ice sheets that flowed out of Canada on four different occasions. When the last tongue of ice retreated some ten thousand years ago it left behind glaciated exposures of ancient bedrock and a maze of streams and lakes. This glaciated landscape is mute testimony to the efficiency of continental glaciers as shapers of the landscape.

Les Voyageurs

Geologic history is not the only kind of history recorded in this rock-ribbed, forested National Park. This lake-studded country was the scene of a most interesting chapter in the history of

North America. During the eighteenth and nineteenth centuries, French-Canadian trappers and fur traders—*les voyageurs*—plied the lakes and streams of Minnesota's Kabetogama Peninsula. Using the many interconnecting lakes as their highway, the colorful *voyageurs* paddled huge canoes full of trade goods and furs between Montreal and the far Northwest.

Today the French-Canadian canoemen are gone, but their wooded lake country looks much as it did during those frontier days. The surface of the ice-carved land is laced with a seemingly endless system of waterways. Rocky promontories jut out from the shores, and countless islands dot the waters of Voyageurs' fifty or more lakes. Three of these—Kabetogama, Rainy, and Namakan lakes—dominate the Park's waterways. Kabetogama Peninsula, which encompasses some seventy-five thousand acres, is the main body of land with Park boundaries. It will take time for this land of lakes to be developed to its full potential. But the National Park enthusiast of the future will doubtless find Voyageurs National Park to be one of our country's more interesting natural areas.

Plants and Animals of Voyageurs National Park

Logging began in this area, in the 1880s, and much of the original forest has been removed. Even so, there is good woodland cover today, and aspen, birch, balsam, white spruce, and maple grow on the uplands. In boggy areas left by Ice Age glaciers are dense stands of black spruce, white cedar, ash, and tamarack trees.

The second-growth forest has protected the area, keeping streams and lakes clear and providing food and shelter for a wide variety of animal life. Moose are seen occasionally, and white-tailed deer and black bear are common. The area also shelters some of the few remaining timber wolves in the entire nation. Of the original animal population, only the woodland caribou, wolverine, and pine marten have completely vanished. Smaller animals in the Park include beaver, mink, otter, bobcats, and rabbits, and many other species are found in sizable numbers.

Many of Voyageurs' lakes are accessible only by foot trails. There are many sport fish in these Park waters, and the major ones are the walleye, the northern pike, and the smallmouth bass. Some lakes also contain muskellunge and lake trout. Rainbow trout are found in some of the smaller lakes and streams. A rare and endangered species, the lake sturgeon, lives in some of the larger lakes in the Park.

The ruffed and the spruce grouse are principally upland birds, with mallard, black duck, and merganser the main migratory species nesting in the Park. The area also provides good nesting habitat for the common loon, Minnesota's state bird. Among rare and endangered bird species are the northern bald eagle and the American osprey. These, along with the sturgeon and the timber wolf, will benefit from the protection that will be afforded them by the National Park Service.

Tips for Tourists

Located just east of International Falls, Minnesota, along the international boundary with Canada, our thirty-sixth National Park covers an area roughly forty miles long and from three to fifteen miles wide. Its more than 219,000 acres have great promise, but full development of Voyageurs will not be completed for many years. Although many of the to-be-expected facilities will eventually be provided, waterways will always be the primary means of travel within the Park. Thus, by canoe or boat the Park visitor can quickly leave civilization behind for a relaxed camping and/or fishing expedition. Perhaps more important, these watery highways will afford much-needed protection from the burgeoning numbers of automobiles that are polluting and plaguing most of our National Parks.

Voyageurs National Park at a Glance

Address: Superintendent, Voyageurs National Park, P. O. Drawer 50, International Falls, Minnesota 56649.
Area: 219,431 acres.

Major Attractions: Beautiful northern lakes and forests; interesting landscape produced by continental glaciation during the Ice Age.

Season: To be announced.

Accommodations: Available in nearby communities.

Activities: Boating, camping, canoeing, fishing, hiking.

Services: Services and activities currently being provided by other government agencies and by private enterprise in and near future Park.

Natural Features: Glaciated landscape, glacial landforms, lakes, wildlife.

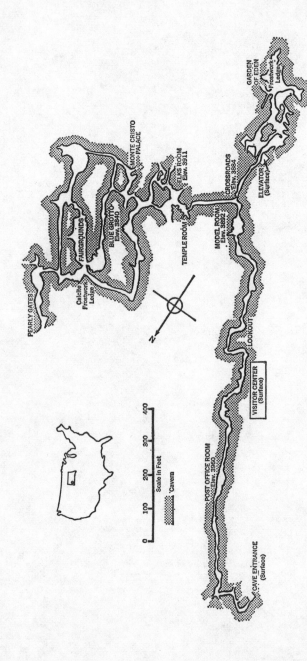

Wind Cave National Park, South Dakota.

35. Wind Cave National Park, South Dakota

Sacred Cave of the Winds

Located in the famous Black Hills of South Dakota, Wind Cave National Park has been referred to as a natural "underground barometer." The cave derives its name from the strong air currents that blow alternately in and out of the mouth of the cave. These wind currents, believed to be controlled by changes in the atmospheric pressure outside the cave, blow outward when the barometer is falling, but when the barometer is rising strong air currents rush into the cave. Thus, the direction of the air flow may roughly indicate changes in the weather.

Although it was not officially discovered until 1881, early Indian legends suggest that this is the sacred "Cave of the Winds" so highly venerated by the Sioux. It is believed that the desire to protect this sacred cave had much to do with the strong resistance of the Sioux to the invasion of this area by white men.

Origin of the Cave

The area in which the Park is established is located on the southeast flank of the Black Hills, an elongate dome-shaped geologic structure which rises abruptly above the South Dakota plains. The central core of this great dome is composed of complex Precambrian igneous rocks which may be as much as one billion years old. The granite core of the dome is flanked by beds of steeply dipping sedimentary rocks of Paleozoic and Mesozoic age; these form sharp ridges around the outer rim of the uplift.

The geologic formation from which the cave has been dissolved is known as the *Pahasapa Limestone;* it was deposited during Early Mississippian time about 325 million years ago. In addition to Mississippian deposits, great thicknesses of Pennsylvanian, Permian, and Mesozoic rocks are also present in the area.

The uplift of the Black Hills, which probably began in Early Tertiary or Late Cretaceous time (some sixty to seventy million years ago), subjected the Pahasapa Limestone and surrounding rocks to great stress and strain. These strong forces severely fractured the rocks of this area, thereby creating many channels by which groundwater could invade the subsurface formations. In the meantime the surface of the Black Hills was being subjected to continued erosion, which eventually laid bare the granite core and carved the concentric ridges and valleys which encircle the uplift.

As groundwater circulated through the complicated fracture system developed during the elevation of the Black Hills, the limestone adjoining the fracture was dissolved. This caveforming process, which has previously been described, greatly enlarged the crevices and developed solution cavities and channels which eventually created the cave.

Deposition in the Cave

The most predominant decorative formation of Wind Cave is quite unlike any of the features that have been discussed in conjunction with Carlsbad Caverns and Mammoth Cave. This cave is famous for its intricate *boxwork*—delicate veins of calcite which have been deposited in the countless fractures which crisscross the Pahasapa limestone. Groundwater dissolved the more soluble limestone, leaving the finlike intersecting veins of calcite projecting from the wall, and this differential erosion has resulted in the intersecting honeycomb pattern of boxes and compartments for which the cave is renowned. The calcite partitions which form the sides of the boxes exhibit a variety of colors ranging from yellow, pink, and brown to blue; they are particularly striking when an electric light shines through them.

Another characteristic deposit of Wind Cave is known as *frostwork,* a formation composed of myriad small, icy-looking, calcareous crystals. Some of this material is seen in groups of minute needles closely resembling frost. But frostwork also occurs in more massive clusters which resemble snowballs. In some places, frostwork literally covers the walls and ceilings of the cave giving the impression that the chamber has been carved from ice and snow.

Stalactites, stalagmites, and other dripstone formations are noticeably absent in Wind Cave; however, deposits of flowstone and sparkling crystal-lined solution cavities are present in certain parts of the cave.

The Guided Tour

Although about seven miles of Wind Cave's passages have been explored, only about one and a quarter miles are open to the public. The cave trails are hard-surfaced and not too strenuous, but the wise visitor will wear low-heeled shoes, preferably with rubber soles. You will enjoy the tour more if you wear a light jacket or sweater for the temperature in the cave is a constant 47 degrees.

The underground trip takes about one to one and a half hours to complete and is electrically lighted all the way; a uniformed tour guide is in charge of each trip. During the tourist season (April 1 through October 31) tours are scheduled daily; public tours are not conducted during the winter. Trip tickets and tour schedules can be obtained at the Visitor Center.

Features of the Trip. Guided tours either walk in via the **Cave Entrance** or descend by elevator in the **Elevator Building.** Underground chambers visited along the route vary greatly in size; they range from small rooms tens of feet in height and width, to a vast hall the size of an auditorium. At its deepest point the trail is about 324 feet underground.

You will see excellent exposures of boxwork in many parts

of the cave, but they are especially noteworthy in the **Post Office Room** and the **Elks Room.** Look for frostwork in the **Garden of Eden** and in the **Fairgrounds.** At the latter location you will have an opportunity to see a frostwork ledge illuminated by ultraviolet, or "black light." This permits the visitor to view a beautiful display of mineral fluorescence as the frostwork glows varying shades of green, blue, red, and purple. In the **Garden of Eden** there are good examples of cave "popcorn," unusual limy deposits resembling "petrified popcorn balls." More delicate frostwork, this time in the form of hairlike needles of calcite, occurs near the **Pearly Gates.**

Here, as in all caves under the jurisdiction of the National Park Service, you are asked not to touch or handle the cave formations.

Other Tours

In addition to the traditional cave tour described above, a number of other interesting underground activities are available here.

"Spelunking" tours are special cave-crawling tours into a primitive section of the cave. Tours are limited to ten persons, who must register in advance. Several hundred feet of passageway will be explored using basic spelunking techniques and equipment.

Unique **candlelight tours** are held daily in June, July, and August. If you would like to experience a cave tour reminiscent of the turn of the century, this activity is a must. Limited to twenty-five people, on a first-come, first-served basis, the hour-long tour features a leisurely walk through the Garden of Eden section of the cave.

Special cave tours may be arranged in advance for the physically handicapped, educational groups, photographic tours, etc. Please write to the Superintendent outlining your request.

Plants and Animals of Wind Cave National Park

A variety of plants and animals inhabit the forty-four square miles within the boundaries of the Park. The main animal attractions are the prairie dogs, which live in their own well-established little "towns," and the large herd of bison. Unfortunately these great shaggy creatures almost became extinct as a result of man's senseless slaughter, but they are now protected by the Government. It should be remembered that Wind Cave bison are *not* tame animals; it is not safe to approach them on foot, and it is dangerous (and unlawful) to feed, tease, or molest these animals in any way.

Other animals present include the pronghorn (a graceful antelope-like creature), deer, elk, badgers, coyotes, and raccoons. Some of the birds that you are likely to see are woodpeckers, meadow larks, magpies, doves, sharp-tailed grouse, kingbirds, and bluebirds.

The vegetation of the Park ranges from wide-open rolling grassland to stands of ponderosa pine and Rocky Mountain juniper. Other trees in the area include American elm, ash, cottonwood, oak, and quaking aspen. Also present are cactus, yucca, wildflowers, and a variety of grasses.

Tips for Tourists

On the surface of the Park there are a number of activities of general interest to the tourist. There is an exhibit room in the Visitor Center with appropriate exhibits relative to the geology, flora, fauna, and human history of the Park. Free illustrated lectures are presented in the amphitheater at **Elk Mountain Campground** each night from late June through Labor Day.

The self-guided nature trail to **Rankin Ridge** is about one and a quarter miles long; it will introduce you to many of the natural features of the Park. The numbered stakes along the trail correspond to similarly numbered paragraphs in *Rankin Ridge Trail Guide Booklet.* Along the trail you will see objects of geological and biological interest as well as some fine scenic views of the Black Hills.

Special photographic tours are not conducted through Wind Cave, but the alert camera enthusiast will be able to get some interesting flash exposures in the cave. Aboveground, the bison and prairie dogs are both likely subjects; a telephoto lens will be useful here.

Wind Cave National Park at a Glance

Address: Superintendent, Hot Springs, South Dakota 57747.

Area: 28,059 acres.

Major Attractions: Limestone caverns exhibiting unusual boxwork formations of crystalline calcite; large bison herds and prairie dog towns.

Season: Year-round.

Accommodations: Campground; motels and trailer parks in towns of Hot Springs (ten miles) and Custer (twenty miles).

Activities: Camping, guided tours, hiking, nature walks, picnicking, and scenic drives.

Services: Food service, gift shop, guide service, telephone, picnic tables, and rest rooms.

Interpretive Program: Campfire programs, cavern tours, museum, nature trails, roadside exhibits, self-guiding trails, trailside exhibits, and wildlife caravans.

Natural Features: Caverns, erosional features, forests, geologic formations, mountains, prairies, rocks and minerals, wilderness area, and wildlife.

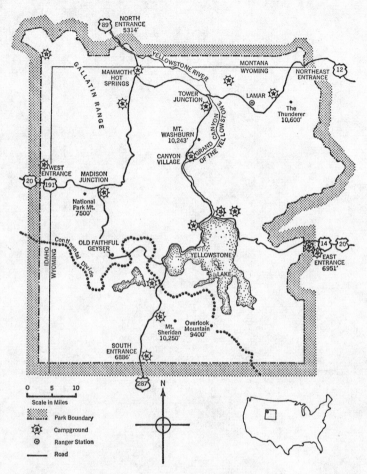

Yellowstone National Park, Wyoming.

36. Yellowstone National Park, Wyoming

Queen of the National Parks

The name Yellowstone, perhaps more than any other, is synonymous with National Park. And there is good reason for this: not only is Yellowstone our largest and oldest National Park, it is one of the nation's most popular vacationlands. Moreover, it was around a campfire in what is now Yellowstone National Park that a group of farsighted men had discussions that later led to the establishment of our National Parks.

We have already learned that pioneer mountain men like John Colter and Jim Bridger were awed by the unusual sights in the fabulous "Yellow Rock" country. What they had seen was so unusual that no one would believe them. Who had ever seen such things as a cliff of black glass, spouting hot springs, and hills that belched sulfurous steam? But today Yellowstone's attractions are world famous; indeed Old Faithful (which might have been one of Bridger's "spouting hot springs") is virtually a household word.

The reasons for Yellowstone's great popularity are myriad, for there are hundreds of unusual features crowded into this 3472-square-mile wonderland. Spouting geysers, boiling hot springs, sputtering mud volcanoes, delicate limestone terraces, sparkling waterfalls, and colorful canyons are only a few points of geologic interest. Wildlife also abounds: moose, deer, American elk, the rare trumpeter swan, and—last but not least—the famous (but dangerous) Yellowstone bears. Combine these outstanding natural attractions with Yellowstone's well-developed camping areas, highways, trails, interpretive program, and accommodation centers and it is easy to understand why millions of Americans have made Yellowstone their favorite National Park.

The Record in the Rocks

The geologic history of the Yellowstone region dates back to Precambrian time, and although rocks of this age are not conspicuous in the Park, they have provided a firm foundation for the events of later geologic periods.* Strata of Paleozoic age are also found here; these sedimentary rocks indicate that a sea covered this area during the Paleozoic Era. These formations, which are several thousand feet thick, suggest that the Paleozoic history of Yellowstone is similar to that of the rest of the Rocky Mountain region.

During Mesozoic time, much of the Rocky Mountain area was again covered by a shallow sea but in places there were swampy lowlands. Dense forests grew in the warm, moist, low-lying areas and coal and fossilized plant remains now mark the site of these ancient woodlands. Animal fossils have also been found in the Park; marine invertebrate remains indicate the position of ancient seas and dinosaur remains have been found in continental deposits.

Near the end of the Mesozoic Era, the Rocky Mountain region was elevated well above sea level and the Cretaceous seas slowly withdrew. This uplift, which also figured prominently in the geologic history of Glacier, Rocky Mountain, and Grand Teton National Parks, was responsible for the formation of the Rocky Mountains.

The latest chapter in Yellowstone's geologic history began approximately seventy million years ago with the beginning of the Cenozoic Era. In early Tertiary time the Rocky Mountains

---

* C. Max Bauer, *Yellowstone, Its Underworld*, Yellowstone Park, Wyoming: Yellowstone Library and Museum Association, 1962. Written by a former Chief Naturalist of Yellowstone, this interesting booklet covers the Park's geology in some detail.

underwent renewed uplift and this region was the scene of much volcanic activity. It was during this time that extensive deposits of volcanic debris accumulated in Yellowstone, for the mountain-rimmed plateau upon which the Park is situated formed an ideal depository for the fiery products of nearby volcanoes (see below).

About one million years ago, during Pleistocene time, the Rocky Mountain region was subjected to further elevation and the climate became much colder. This marked the advent of the so-called Ice Age and the coming of the glaciers which further modified the face of the Park. As the ice plowed down the narrow, V-shaped, stream-cut canyons, these deep gorges were converted into flat-bottomed, U-shaped glacial valleys. Then, as the climate gradually warmed, the glaciers melted leaving behind accumulations of glacial drift; **Capitol Hill,** at Mammoth Hot Springs, and **Twin Buttes** and the **Porcupine Hills,** near the Lower Geyser Basin, were formed in this manner.

More recently, running water has carved numerous steep-sided gorges and the surface rocks have been attacked by various agents of weathering. Groundwater has also played an important role in the evolution of the Park's scenery. The colorful travertine terraces of the Mammoth Hot Springs area and the extensive accumulations of siliceous material found in the various geyser basins were all deposited by mineral-laden subsurface water.

Early Volcanic Activity

As mentioned earlier, the Tertiary Period was a time of considerable volcanic activity in Yellowstone. This early vulcanism appears to have been primarily of the explosive type as indicated by the presence of great quantities of *pyroclastic* ("fire-broken") *materials.* These include *tuff,* consisting of solidified volcanic ash or dust, and *volcanic breccia,* a relatively hard pyroclastic composed of angular rock fragments embedded in compacted volcanic ash.

But not all igneous activity was of the explosive type. In

places the pyroclastics contain layers of *basalt* (dark, fine-grained, solidified lava), which reveals the presence of ancient lava flows. *Rhyolite,* a light-colored, fine-grained solidified lava, and *obsidian* (a glassy, dark-colored volcanic rock) are also present. Like basalt, these igneous rocks were formed from great upwellings of lava which probably reached the surface via large fissures in the earth's crust. Exposures of rhyolite are common throughout the Park; excellent outcrops can be seen in the walls of the **Grand Canyon of the Yellowstone** and in **Gibbon Canyon** (see map).

The Glass Cliff. The best exposure of obsidian occurs at **Obsidian Cliff,** located about halfway between Mammoth Hot Springs and the Norris Geyser Basin. This is the legendary cliff of black glass reported by early mountain men and from which the Indians collected rock with which to make arrowheads and other implements. The black volcanic glass that forms this 165-foot cliff is hard but brittle and breaks with a *conchoidal* (shell-like) fracture. Surprisingly enough this glassy rock has essentially the same chemical composition as the rhyolite described above. How can two rocks look so different and yet have almost identical chemical composition? Most geologists believe this to be the result of different rates of cooling. The obsidian cooled and solidified so rapidly that its constituents did not have sufficient time to form individual mineral crystals; the rhyolite cooled more slowly, thus permitting mineral crystals to form in parts of the rock.

The relationship between cooling and *texture* (the physical appearance of a rock as indicated by the shape and size of its mineral components) may be even more clearly demonstrated by comparing obsidian and granite. It would be difficult to find two more distinct types of texture than the glassy, non-crystalline obsidian and a coarse-grained granite with its clearly observable crystals of quartz, mica, and feldspar. Yet the granite has the same basic chemical composition as obsidian and rhyolite—it simply cooled very slowly while the molten rock was still deeply buried within the crust of the earth. Thus, these rocks not only supply us with information about igneous activity of the past, they also provide some indication as to where and how fast they cooled.

Ash-Buried Forests

When the Yellowstone volcanoes were erupting during early Tertiary time, massive clouds of ash- and dust-laden vapor filled the sky and showered the land with volcanic materials. Fine pyroclastics blanketed the area like snow and in places accumulated in great thicknesses. As time passed and the eruptions continued, this material clogged and dammed many streams and completely buried large forests. Periods of volcanic violence were commonly followed by periods of quiescence during which the streams cleared their channels and the trees regained a foothold on the land. This cycle was repeated many times, for a total of twenty-seven "fossil forests" are found entombed in a thickness of about seventeen hundred feet of volcanic debris.

Although petrified forests occur in many parts of the world, the fossil forests of Yellowstone are unique in a number of respects. For example, the trees here have been preserved *in situ* (in place); they are standing upright in the same position and at the same place where they grew millions of years ago. In most petrified forests, such as Petrified Forest National Park in Arizona, the fossilized tree trunks are found in a horizontal position. This is because most of the Arizona trees have been carried into the area by ancient streams and buried in sediments deposited in broad low-lying valleys. (It should be remembered that the petrified trees in Arizona occur in Triassic rocks and are thus much older than Yellowstone's trees, which are Eocene in age.)

The Yellowstone fossil forests are also the largest known. They cover an area of more than forty square miles and in places whole successions of forests are found one on top of the other. Moreover, thousands of fossilized seeds, cones, needles, twigs, and leaves have been found associated with the fossil tree trunks, and they represent more than one hundred different types of shrubs and trees. Among these are ferns, pines, walnuts, sycamores, chestnuts, redwoods, figs, laurels, elms, willows, and many others.

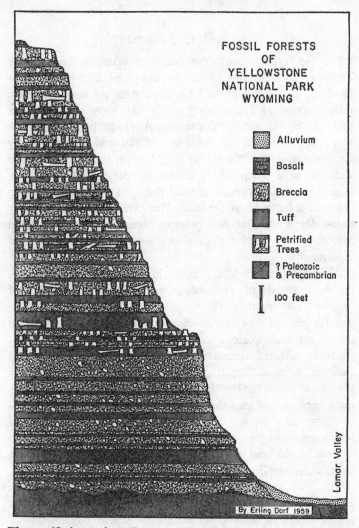

FOSSIL FORESTS
OF
YELLOWSTONE
NATIONAL PARK
WYOMING

Alluvium

Basalt

Breccia

Tuff

Petrified
Trees

? Paleozoic
& Precambrian

100 feet

Lamar Valley

By Erling Dorf 1959

The petrified trees in Yellowstone National Park are unique in that
they remain in the upright position that they grew in many millions
of years ago. REPRODUCED COURTESY OF DR. ERLING DORF

These visitors pause along the Big Room Trail to marvel at the many remarkable cave formations that adorn the largest underground chamber known to man. NEW MEXICO DEPARTMENT OF DEVELOPMENT

Lassen Peak—only active volcano in the contiguous United States—is seen here erupting in July 1915. NATIONAL PARK SERVICE

One of the world's outstanding examples of volcanic activity can be seen at Oregon's Crater Lake National Park. In the center of the photograph is Wizard Island, a cinder cone that rises more than twenty-six hundred feet from the caldera floor. The cliffs in the background form the rim of the caldera and stand almost two thousand feet above the surface of the lake. Crater Lake Lodge is seen in the lower right of the photo. OREGON STATE HIGHWAY COMMISSION

Like many lakes in the National Parks, Two Medicine Lake, in Montana's Glacier National Park, is ideally suited for boating. MONTANA HIGHWAY COMMISSION

Upheaval Dome, in Canyonlands National Park, is a feature of unusual geologic interest. Although it has been studied in detail, geologists are not in agreement as to its origin. NATIONAL PARK SERVICE PHOTO BY M. WOODBRIDGE WILLIAMS

Earthquake Lake (in foreground) was created when waters of the Madison River were impounded by the natural earthen dam created by a landslide. The slide (center of the photograph) was triggered by the Hebgen Lake earthquake, which rocked West Yellowstone, Montana, in 1959. U. S. FOREST SERVICE

The Indians called Mount McKinley Denali—"The High One." Seen here from Wonder Lake campground area, Mount McKinley's Churchill Peak rises 20,320 feet above sea level—the highest mountain on the North American continent. CHARLES J. OTT, MCKINLEY PARK, ALASKA

This typical view in Everglades National Park shows the expansive reaches of open grasslands with their "islands" of trees, called hammocks. FLORIDA STATE NEWS BUREAU

The V-shaped profile of the stream-carved Grand Canyon of the Yellowstone is obvious in this photograph. The Lower Falls—twice as high as Niagara Falls—can be seen in the distance. U. S. GEOLOGICAL SURVEY

The devastation area in Hawaii Volcanoes National Park is typical of the destruction that can be wrought by showers of pyroclastic material such as volcanic ash and cinders. The vegetation in this area was destroyed by volcanic fallout during the 1959 Kilauea Iki eruption.
NATIONAL PARK SERVICE PHOTO BY WILLIAM W. DUNMIRE

Section of Guadalupe Mountains in Guadalupe Mountains National Park, Texas. The upper sheer cliffs are rocks of the Capitan Limestone, in which the great Capitan reef is developed.

Like many of the great mountain ranges, the core of the Teton Range is composed of granite—a common rock in certain of the National Parks. NATIONAL PARK SERVICE

High atop Hurricane Ridge, in Olympic National Park, this family is having lunch in a picnic area typical of those provided in most Parks. Although they appear to be enjoying sharing their meal with their luncheon "guest," this practice is frowned upon by the National Park Service. WASHINGTON STATE DEPARTMENT OF COMMERCE AND ECONOMIC DEVELOPMENT

The process by which the trees were fossilized is much like that which preserved the Arizona trees. Briefly, mineral-bearing groundwater saturated the ash-buried trees as it percolated downward. Some of the dissolved mineral matter (especially silica) was deposited in openings within the empty cells of the wood. In this manner the durable cell walls of the wood were surrounded by silica and preserved in much their original condition.

Fossil forests occur at several localities. The most accessible of these is **Specimen Ridge** southwest of Lamar River Valley in the northern part of the Park. (Persons wishing to visit this locality can obtain directions from Park personnel or take the Fossil Forest Hike. However, an excellent example of a single petrified tree can be seen at the end of the **Petrified Tree Road,** a spur off the road from Tower Junction to Mammoth Hot Springs (see map). Here a large upright petrified stump has been partly excavated and enclosed by a high iron fence to protect it from vandalism. Many similar trees originally stood on this hillside but have long since been destroyed and removed as souvenirs.

Geology in Action

In discussing most of the National Parks, we have been primarily concerned with events of the geologic past and their role in creating the scenery of the Parks. But at Yellowstone we have a unique opportunity to observe geology in action, for many of the geologic phenomena responsible for the Park's landscape can be seen at work today. Streams unceasingly scour their channels, weathering agents steadily reduce solid rock to loose soil, and underground water continually dissolves minerals from deeply buried rocks and deposits them on the surface.

Geysers

Yellowstone is probably more famous for its geysers than for any other feature, and rightly so. There are more than a hundred active geysers and approximately three thousand noneruptive hot springs located within the Park. Indeed, this is the greatest known concentration of geysers in the world. The only other geyser regions of any consequence are located in Iceland and New Zealand, but the geysers there are neither as active nor as numerous as those of Yellowstone.

What is a geyser? Why are they found here? How do they work? These commonly asked questions are intricately related to the geology of this unusual area. The word "geyser" is derived from the Icelandic term *geysa* which means "to gush"; it may be defined as a special type of hot spring that intermittently erupts a column of steam and hot water. These gushing hot springs seldom, if ever, erupt at regular intervals— even Old Faithful does not play "every hour on the hour" but has an average eruption interval of between sixty-one and sixty-seven minutes.

The presence of geysers in Yellowstone is the result of rather special subterranean conditions which have resulted in well-defined *geyser basins*. Geyser basins appear to be located along faults or fractures that have caused one section of rock to be displaced vertically or horizontally with respect to the section adjacent to it. These fractures and other openings in the bedrock and glacial gravels permit the accumulation of groundwater and the eventual escape of steam and hot water. Within each geyser basin there is an ample supply of groundwater (derived from rainfall and melted snow) and a source of heat (probably the relic of an ancient magma).

What causes a geyser to erupt or play? The basic principle here is not unlike that of the household coffee percolator. In a geyser the "percolator" consists of the underground passageways (or plumbing) which extend downward into the hot rocks. At the bottom of this network of tubes, groundwater (under pressure exerted by the column of water above it) is "superheated" to a temperature far above the boiling point of water ($212°$ F). As the heated water in the bottom of the fissure expands, it causes some of the overlying water to overflow onto the surface; this relieves part of the pressure and allows the superheated water to flash into steam. The steam blasts out the colder water above it, causing the geyser to erupt. When most of the steam has been ejected, the geyser ceases to play and water again enters the geyser's plumbing, thus charging it for the next eruption.

The geysers of Yellowstone are as varied as they are numerous; some, such as Old Faithful, erupt at fairly regular intervals while others play only occasionally. Larger geysers may send a stream of water more than two hundred feet into the air; smaller ones produce spouts of only a few feet in height. Although many of the smaller geysers displace but a few gallons of water during an eruption, some (for example, Giant Geyser) may eject hundreds of thousands of gallons during a single outburst. Consequently the person who sees one geyser has *not* "seen them all" but should visit as many different ones as possible.

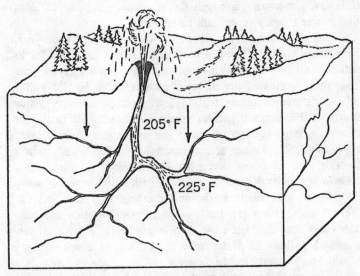

Diagrammatic cross section illustrating the mechanism of geyser eruption. Groundwater in fractures and vents below the surface is heated to the boiling point at depth, near the main source of heat, before it reaches the boiling point in the upper part of the vent. For simplicity only a few fractures are shown. From *Geology: Principles and Processes,* fifth edition, by W. H. Emmons, I. S. Allison, C. R. Stauffer, and G. A. Thiel, copyright 1960 by McGraw-Hill Book Company. Used by permission of McGraw-Hill Book Company.

Because they are so abundant and diverse, space will permit comment on only a few geysers. However, certain of the more interesting ones are briefly discussed below.

Old Faithful Geyser. Since 1870 when Old Faithful was first discovered and named, this geyser has been the "trademark" of Yellowstone National Park. Yet it is not the largest or most powerful geyser in the Park; nor is it the most beautiful. Old Faithful has become Yellowstone's most celebrated attraction because of the relative regularity with which it plays. Moreover, during the almost one hundred years that it has been under observation, Old Faithful has never failed to erupt nor has its energy diminished. But despite its regularity Old Faithful does *not* adhere to a fixed timetable; rather, its outbursts are now occurring at intervals of from sixty-one to sixty-seven minutes (although this interspace has varied from thirty-three to sixty-seven minutes).*

Of the thousands of people who visit Yellowstone each year, most come with the avowed purpose of seeing Old Faithful in action. And few are disappointed, for this famous geyser puts on a show worthy of the stellar attraction that it is. Old Faithful's performance opens with a few short preliminary water spurts as the eruption gathers strength, but within a matter of minutes a huge column of water and steam gracefully rockets skyward. After playing at maximum height for about twenty seconds, the eruption decreases in intensity and the water column begins its descent in a series of sporadic surges. Finally, after about four minutes, the water ceases to spout and the "show" ends with a few puffs of steam. During such an eruption (they may last from two to five minutes), more than ten thousand gallons of water may be hurled as much as one hundred and seventy feet in the air.

Like many other geysers, Old Faithful is situated in the center of a *mound* or *cone* of *geyserite* which was deposited by mineral-bearing thermal water. Known also as *siliceous sinter,* geyserite is a form of silicon dioxide ($SiO_2 \cdot H_2O$) and

---

* George D. Marler, *The Story of Old Faithful Geyser,* Yellowstone Park, Wyoming: Yellowstone Library and Museum Association, 1963. In this interesting booklet a Park Naturalist separates fact from fiction about Old Faithful.

its chemical composition is similar to that of quartz. The broad, oblong mound of Old Faithful is almost twelve feet high and as much as fifty feet across. Other geysers, for example, Grotto and Castle Geysers, have built similar but more ornate cones around their vents.

To learn more about the operation of Old Faithful and the other geysers in the Park, be sure to visit the displays in the **Old Faithful Museum and Visitor Center** and take the **Geyser Hill Self-guiding Nature Trail.**

Grotto Geyser. Located in the Upper Geyser Basin not far from Old Faithful, Grotto Geyser is noted for its unusual cone. It plays frequently, and during each eruption (which may last from one to five hours) water is ejected to a maximum height of thirty feet.

Castle Geyser. Castle Geyser, one of the Park's oldest, is also located in the Upper Geyser Basin. Playing to a height of sixty-five to a hundred feet, it erupts at intervals of eight to ten hours and for a period of about forty-five minutes. The geyser was named by early Park visitors who imagined its cone resembled a partially ruined feudal castle.

Riverside Geyser. One of the largest and most unusual geysers in the Upper Geyser Basin, Riverside Geyser lies on the east bank of the Firehole River. It is unique in that its water is thrown out at an angle and falls into the river. The eruptions, which occur every six to eight hours, last for about twenty minutes and reach a maximum height of approximately eighty-five feet.

Lone Star Geyser. This geyser is situated at the end of a spur which enters Grand Loop Road about one and a half miles southeast of the Old Faithful area. Characterized by a massive geyserite cone measuring ten to twelve feet in diameter and some ten feet tall, Lone Star Geyser may erupt at intervals two and a half to three hours and eject a plume of water about twenty-five feet high.

Excelsior Geyser Crater. Although the geyser is dormant, Excelsior Geyser's crater remains as a superb example of the geologic importance of thermal waters. Considered to be the largest geyser in Yellowstone, Excelsior last erupted in 1888, at which time it discharged a stream of water two hundred to three hundred feet high for from four to fifteen minutes. Today there is continual overflow from the crater atop its expansive siliceous mound; it has been estimated that six cubic feet of boiling water issue from this steaming pit each second.

Hot Springs

Yellowstone's noneruptive thermal springs are not as spectacular as the geysers but they are of considerable geologic interest. With the exception of the Mammoth Hot Springs, most of the thermal pools are in the geyser basins and derive their heat from the same source as do the geysers. Because hot water is much more active chemically than cold water, hot springs discharge water containing much dissolved mineral matter. When the heated water reaches the surface, some of the minerals are deposited, thus forming the rims, bowls, mounds, and terraces associated with many of the springs. Most such structures found in Yellowstone are composed of *siliceous sinter* (p. 438). The major exception is the group of terraces at Mammoth Hot Springs which are made of *travertine,* a form of calcium carbonate ($CaCo_3$). The terraces at Mammoth are somewhat like the travertine formations deposited by the cold-water springs at Platt National Park.

Some of the deposits formed around the hot springs are tinged with color due to the presence of minerals in the rock and algae which live in the water. These colors blend with the clear blue water to produce the rich coloration noticed in many thermal pools. A few of the more popular and interesting hot springs are briefly described below.

Mammoth Hot Springs. The hot water that feeds Mammoth Hot Springs passes through rocks which contain considerable calcium carbonate. The water dissolves limy material from these rocks and may later deposit part of it as travertine. This is the origin of travertine terraces found near Park Headquarters; they are among the largest and most beautiful in the world.

Rising like a giant staircase, the terraces occupy several levels, each "step" being somewhat smaller than the one beneath it. Some of them are brightly colored due to the presence of minerals and algae; others are snowy white—they are called "dead" terraces because the springs which created them are no longer flowing. Names have been given some of the larger and more unique terraces; among them are **Minerva, Opal, Cleopatra, Jupiter,** and **Pulpit Terraces.**

One of the more outstanding travertine deposits in this area is an enormous cone called **Liberty Cap.** Towering thirty-seven feet above the ground, this limy pillar is twenty feet in diameter at its base; it was deposited by hot water which once streamed from an opening in its top. To learn more about the terraces join the **Terrace Nature Walk** which leaves Liberty Cap several times each day.

Punch Bowl Spring. The "punch bowl" containing this spring is a bowl-shaped deposit of geyserite which accumulated as boiling mineral water spilled over the rims of the pool. It is located in the Upper Geyser Basin near Sand Basin Loop Road.

Morning Glory Pool. This unique spring has the shape and color of a huge flower. The deep blue water and delicately colored bottom make it one of the most beautiful hot pools in the Park. Unfortunately, Morning Glory Pool was almost "choked to death" by vandalistic visitors who threw more than 110 different objects into its waters. Thanks to the ingenuity of Park geologists, a system was devised to flush some of this debris from the hot spring's plumbing, thereby saving it for the enjoyment of future and—hopefully—more appreciative visitors.

Fishing Cone. In Fishing Cone (located on the western shore of Yellowstone Lake at West Thumb) we see the geological anomaly of a seething, hot spring occurring in the cold water of the lake. The spring, which is enclosed by a rather symmetrical, cone-shaped mound, was once used by fishermen to cook their freshly caught trout as they were taken from the water—a practice now prohibited by Park regulations.

Mud Pots, Mud Volcanoes, and Mud Geysers

At a number of places in the Park, there are a variety of hydrothermal phenomena associated with accumulations of viscous, steaming mud. The most common of these are *mud pots,* unique hot springs consisting of depressions filled with boiling mud containing a relatively small amount of water. Steam and sulfurous gases escaping from the mud keep it in an almost continual state of agitation. One group of mud pots is called **The Mushpots,** a name that aptly describes the sputtering mud springs which resemble a pot of mush simmering over a flame. Mud pots vary greatly in size, ranging from a few inches to as much as thirty feet in diameter. Most are relatively shallow, but in some the mud is as much as ten feet below the rim of the crater.

Certain mud pots are quite colorful due to the presence of various minerals in the mud; these are referred to as *paint pots*. Among the more vivid paint pots are those in the Lower Geyser Basin near Fountain Geyser. These are the multihued **Fountain Paint Pots,** teeming caldrons of white, orange, and pink mud. Another cluster of varicolored paint pots occurs at West Thumb near Fishing Cone.

Small *mud volcanoes* are also present; the best known of these, **Mud Volcano,** is located on a turnout on the Grand Loop Road between Fishing Bridge and Canyon Village. This "volcano" has a crater-like opening through which steam rises to the surface. Periodically the steam vent becomes choked with mud; when the mud-confined steam generates sufficient pressure, it blasts the mud out of the vent. Some mud volcanoes have built cones in this manner.

Mud geysers resemble mud volcanoes except that their craters are filled with fuming, sulfurous mud which is in constant agitation. Increasing stream pressure results in intermittent eruption similar to that of water geysers. Two mud geysers are located on the same scenic turnout as Mud Volcano. One of them, the **Black Dragon's Caldron,** is the largest and most formidable in the Park. The other, **Mud Geyser,** discharges a ten-foot column of mud from its crater every few seconds.

Fumaroles

Fumaroles, holes or vents from which steam and fumes are given off, are another type of thermal phenomenon in Yellowstone. Known also as *steam vents,* these features are found in places where there is a scarcity of groundwater. In one of these, the **Black Growler,** the steam comes out with a deafening roar and temperatures as high as 284° F have been recorded. **Hurricane Vent** (located near the Black Growler in **Porcelain Basin** of the Norris Geyser Basin) and a series of vents in the Old Faithful series of mounds are further evidence of Yellowstone's current geological activity.

The sulfurous fumes that emanate from many fumaroles are composed of hydrogen sulfide (H_2S), a gas that can be aptly described as smelling like rotten eggs. This odor is also evident near many of the hot springs and mud pots.

Hebgen Lake Earthquake

On the night of August 17, 1959, thousands of people experienced one of the more dynamic manifestations of Yellowstone's living geology. This event, now known as the Hebgen Lake Earthquake, did considerable damage in Yellowstone, and had a profound effect on the geyser basins along the Firehole River.

Although roads and buildings were damaged in the Park, the most severe damage occurred in Madison Canyon about fourteen miles west of the Park boundary. Here, more than eighty million tons of rock were shaken from the canyon wall, producing a landslide that filled the valley with several hundred feet of rock debris. This huge mass of rubble extended a mile up the valley and formed a natural dam that obstructed the flow of the Madison River, thus forming a large body of water which has since been named Earthquake Lake.

Seven miles east of the slide area, the earthen dam that impounds the Madison River to create Hebgen Lake was severely jolted and cracked. Luckily it held, thereby preventing further damage to the area downstream. However, the basin occupied by the lake was tilted: the south side of the lake was elevated and the north side was depressed, thereby submerging houses and parts of the road on the north shore. In addition, movements produced as the basin shifted, triggered a series of waves that kept the surface of the lake agitated for about eleven hours. This strong, earthquake-induced oscillation of the lake—called a *seiche* (pronounced saysh)—activated massive walls of water that inundated Hebgen Lake Dam on four different occasions.

Nine lives were lost in the slide area and at least nineteen persons are still believed to be missing. Fortunately there were no loss of life or serious injury within Park boundaries, but there was extensive road damage as hard surfaces were fractured and roads were blocked by massive rockslides. Moreover, chimneys were knocked down in the Old Faithful and Mammoth Hot Springs area and other minor property damage was inflicted by the earthquake shocks.

What caused this earthquake, and why did it occur in this area? *Seismologists,* men who study earthquakes, believe that earth tremors are initiated by a sudden jar or shock and that most such shocks are associated with movements along active fault zones. The abrupt fracture and sudden displacement of the rocks along the fault plane produce sudden wavelike motions in the rocks. The manner in which broken rocks react is best explained by the *elastic rebound theory.* According to this theory, subsurface rock masses will slowly bend and change shape if subjected to prolonged pressures from different directions. Continued stress will result in strains so great that the rocks will eventually rupture and suddenly snap back into their original unstrained state. It is the *elastic rebound,* or snapping back, that generates the *seismic* (earthquake) waves.

The *focus* (the point within the earth from which the shocks originated) of the Hebgen Lake Earthquake is believed to have been about ten miles beneath the surface below Grayling Creek along the western boundary of the Park. An active fault zone, the Red Canyon Fault, is known to be present in this area and the movement which produced the earthquake apparently took place along this rift.

Earth tremors have been recorded in Yellowstone since 1871, but the Hebgen Lake Earthquake appears to be the most severe on record. Dr. William A. Fischer made an extensive study of the earthquake area shortly after the disaster,* and

---

* William A. Fischer, *Earthquake! Yellowstone's Living Geology,* Yellowstone Park, Wyoming: Yellowstone Library and Museum Association, 1960. This is an excellent authoritative account of the Hebgen Lake Earthquake and its geologic implications.

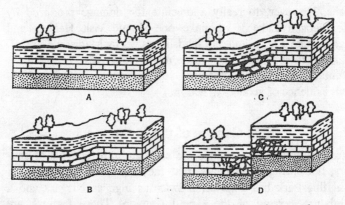

A normal section of rocks (A) may be subjected to such great stress (B and C) that it may eventually break (D), thus causing an earthquake when the rocks "snap" back into position. Reproduced by permission from *The Story of Volcanoes and Earthquakes*, by William H. Matthews III, Harvey House, Inc., Irvington-on-Hudson, N.Y. (1969).

has estimated that the total amount of energy released from this big tremor was equivalent to about two hundred atomic bombs of the type dropped on Hiroshima, Japan. The shock waves which radiated out from the focus were noticed over an area 500,000 square miles and recorded by seismograph instruments throughout the world.

As would be expected, the deep-seated shock waves produced by this earthquake had a pronounced effect on Yellowstone's hydrothermal activity. At first it was feared that the quake might have severely damaged the plumbing of the geysers and hot springs, thus diminishing the activity of these great natural features. But just the opposite occurred: geysers that had been dormant for years began to play, new hot springs appeared, earthquake-produced fissures gave rise to new fumaroles, and many geysers (including Old Faithful) began erupting at shorter intervals. There was also a general temperature increase in the water of the hot springs.

Visitors to Yellowstone can still see the effects of these changes at various points within the geyser basins; many are indicated by signs and interpretive markers along the roads and trails. However, to really appreciate the damage produced by the Hebgen Lake Earthquake one should visit Hebgen and Earthquake Lakes and the landslide area in Madison Canyon. This worthwhile side-trip can be made from West Yellowstone, Montana, in a few hours; here you will see awesome and recent evidence of geology in action.

Grand Canyon of the Yellowstone

Most visitors to Yellowstone are so filled with thoughts of geysers, hot springs, and bears, that they are astonished to learn that this Park has a waterfall twice as high as Niagara and a miniature version of the Grand Canyon. These features are embodied in the Grand Canyon of the Yellowstone River located in the north-central part of the Park.

Geologically this colorful chasm is a "textbook" example of a youthful stream-cut valley: its narrow bottom and steep-sided walls give it the V-shaped profile so typical of such valleys. Scenically it is one of Yellowstone Park's most impressive and beautiful attractions, especially as seen from Artist, Inspiration, Grandview, and Lookout Points near Canyon Village. Indeed many persons consider the view from Artist Point to be the most beautiful scene in the Park. From here one can get a tree-framed view of the thundering Lower Falls of the Yellowstone as they plunge 308 feet into the foaming water below. This is perhaps the most photogenic part of the Park, and during the peak of the visiting season it is commonly necessary to wait in line to photograph this impressive sight. If, on your visit to Artist Point, you feel that this scene is strangely familiar to you, do not be surprised—it has been a prime subject for calendar pictures for many, many years.

The Upper Falls of the Yellowstone are located a short distance upstream from the Lower Falls. This lofty 109-foot cascade marks the beginning of the Grand Canyon of the Yellowstone which continues for a distance of about twenty-four miles. Ranging from eight hundred to twelve hundred feet in depth and as much as fifteen hundred feet wide, this colorful canyon provides the ideal backdrop for the white foamy waterfalls and clear green water of the river which surges through it.

Consisting primarily of rhyolite, the sides of the canyon are predominantly yellow, but red, brown, orange, pink, and white add to the profusion of color. The colors result largely from various mineral compounds in the rocks, some of which have been altered by weathering and contact with gases and hot water from hydrothermal features in the canyon.

There are many carefully chosen and easily accessible outlooks from which to view the canyon. Visit as many of them as possible in order to appreciate the true beauty and geologic significance of this vast, multicolored gorge.

Plants and Animals of Yellowstone National Park

Because 90 per cent of Yellowstone National Park is forested, it is an ideal wildlife sanctuary for a host of interesting animals. By far the best known of these is the famous Yellowstone bear, which is common throughout the Park. Although known as the American black bear, its body may be black, brown, cinnamon red, and even platinum blond.

As a result of their abundance and public appeal, the bears are both a pleasure (to the visitor) and a problem (to Park personnel). In fact, "bear jams" (resulting from motorists stopping to watch and photograph the bears) are a common occurrence along Grand Loop Road. A mother bear followed by a couple of fuzzy cubs is certainly an appealing sight, but remember that these apparently tame creatures are *wild animals* and should always be treated as such. Park Rangers distribute special precautionary leaflets about bears and similar warnings are posted throughout the Park; for your own safety and protection please follow their instructions. Briefly, you should:

1. Never feed or molest the bears.
2. Keep car windows closed when bears are near.
3. Always keep a safe distance from bears and do not encourage them to approach.
4. Never get between an adult bear and cubs.
5. Always photograph bears from inside your automobile or at a very safe distance from them.

In addition to bears, a number of other larger mammals are found at various places within the Park. Among them are pronghorn ("antelope"), which might be seen between Mammoth Hot Springs and Gardiner, Montana; bison, especially between Old Faithful and Madison Junction; deer near Mammoth; moose which browse in shallow river beds and marshy areas (especially along Pelican Creek east of Fishing Bridge); American elk (wapiti) in certain open glades or meadows—watch for these between Mammoth and Norris and around Madison Junction and toward Old Faithful; and the coyote, which is likely to be seen in all parts of the Park. These too are wild animals and the precautions listed above also apply to these creatures.

Not all the animals are large; ground squirrel, chipmunk, marmot, and other small mammals are common. Yellowstone's bird fauna includes a variety of water fowl on Yellowstone Lake and along the Yellowstone River. Some of these are ducks, geese, sea gulls, pelicans, and, in more secluded areas, the rare trumpeter swan. Osprey, or fish hawks, can often be seen in the Grand Canyon of the Yellowstone, where they build nests on pinnacles of rock. If you are observant—and lucky—you may catch a glimpse of the white-headed, white-tailed bald eagle. Birdwatchers have reported them near the Madison, Firehole, and Yellowstone Rivers and also at Yellowstone Lake. A variety of smaller birds also occur in a variety of habitats throughout the Park.

Yellowstone's lush evergreen forests are composed primarily of lodgepole pine, but limber and white pine, alpine and Douglas fir, Engelmann spruce, juniper, and quaking aspen are also present. Cottonwood and willows grow along stream banks, and sagebrush and many wildflowers grow on the valley floor.

Tips for Tourists

America's first and largest National Park has a wide range of activities for the hundreds of thousands of visitors who go there annually. Fishing and boating are popular and the extensive wilderness areas are ideal for hiking and camping. But most of all, visitors come to marvel at the hydrothermal features, watch the antics of the ever-present bears, and enjoy the superb scenery of the incomparable Yellowstone country.

Because these attractions are so widespread and diverse, a number of visitor and accommodation centers have been established throughout the Park. These areas, which are located on the Grand Loop Road (see map), are (1) **Mammoth Hot Springs** (Park Headquarters), (2) **Norris Geyser Basin**, (3) **Madison Junction**, (4) **Old Faithful** (including Lower, Midway, and Upper Geyser Basins), (5) **West Thumb**, (6) **Lake-Fishing Bridge**, (7) **Canyon**, (8) **Tower Fall**, and (9) **Grant Village**. Each of these localities has its own contribution to make to one's total enjoyment of the Park, and ample time should be allotted to visit each of them. Most of the activities available at each place are noted below; however, the Park informational brochure (available at the Park or by mail from the Park Superintendent) or *Hayne's Guide: A Handbook of Yellowstone National Park* will provide more complete information. *Hayne's Guide* contains detailed road logs, maps, and information about the human and natural history of the area. It is available by mail from the Yellowstone Library and Museum Association, Box 117, Yellowstone National Park, Wyoming 82190, or Haynes, Inc., 801 North Wallace Avenue, Bozeman, Montana 59715.

Yellowstone's colorful paint pots, hot springs, plant life, and geologic formations are ideal subjects for the photographer. Erupting geysers and the bears are perhaps the most commonly photographed objects in the Park, but extreme care should be exercised when around them. Use your light meter if you have one, and a haze filter will generally improve the clarity of most of your pictures.

Your enjoyment of this unusual area will be heightened if you observe posted regulations pertaining to the Park. Briefly, do not disturb natural features, do not feed, tease, or approach the bears, keep away from the edges of thermal pools and geysers, and stay on the constructed walks in the thermal areas.

Yellowstone National Park at a Glance

Address: Yellowstone National Park, Wyoming 82190.

Area: 2,221,772 acres.

Major Attractions: World's greatest geyser area, with about three thousand geysers, hot springs, and other hydrothermal features; spectacular falls and canyon of the Yellowstone River; one of the world's greatest wildlife sanctuaries.

Season: May 1 through October 31.

Accommodations: Cabins, campgrounds, group campsites, hotels, motels, and trailer sites. *For reservations contact:* Yellowstone Park Co., Yellowstone National Park, Wyoming 82190.

Activities: Boating, boat rides, camping, fishing, guided tours, hiking, horseback riding, nature walks, picnicking, and scenic drives.

Services: Boating facilities, boat rentals, food service, gift shop, health service, laundry, nursery, photo shops, post office, religious services, service station, telegraph, telephone, transportation, picnic tables, rest rooms, and general store.

Interpretive Program: Campfire programs, museum, nature walks, roadside exhibits, self-guiding trails, and trailside exhibits.

Natural Features: Canyons, erosional features, forests, fossils, fumaroles, geologic formations, geysers, hot springs, lakes, mountains, mud pots, mud volcanoes, rivers, rocks and minerals, unusual birds, unusual plants, volcanic features, waterfalls, wilderness areas, and wildlife.

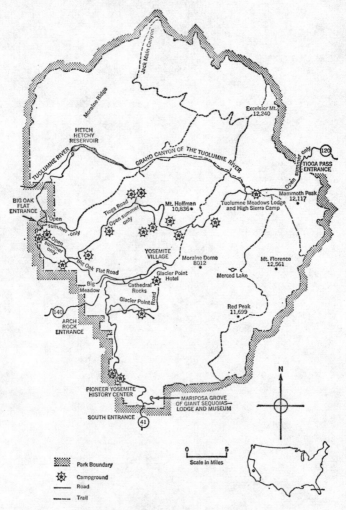

Excelsior Mt.
12,240

Jack Main Canyon

Moraine Ridge

HETCH
HETCHY
RESERVOIR

TUOLUMNE RIVER

GRAND CANYON OF THE TUOLUMNE RIVER

TIOGA PASS
ENTRANCE

Open summer only

120

BIG OAK
FLAT
ENTRANCE

Open
summer only

Tioga Road

Open summer only

Mt. Hoffman
10,836

Mammoth Peak
12,117

Tuolumne Meadows Lodge
and High Sierra Camp

Open
summer only

Big Oak Flat Road

YOSEMITE
VILLAGE

Moraine Dome
8012

Mt. Florence
12,561

Glacier Point
Hotel

Merced Lake

Big
Meadow

Cathedral
Rocks

Glacier Point Road

Glacier Point

Red Peak
11,699

140

ARCH
ROCK
ENTRANCE

N

PIONEER YOSEMITE
HISTORY CENTER

MARIPOSA GROVE
OF GIANT SEQUOIAS
LODGE AND MUSEUM

SOUTH ENTRANCE

41

Park Boundary

Campground

Road

Trail

0   5
Scale in Miles

Yosemite National Park, California.

37. Yosemite National Park, California

Sierra Wonderland

Located high on the western flank of the Sierra Nevada is Yosemite National Park, a 1200-square-mile wonderland of granite domes, foamy waterfalls, and lofty giant sequoias. Here is some of the world's most spectacular and accessible glaciated scenery, and here is Yosemite Valley, an incomparable ice-carved trough that is world famous for its rugged grandeur. Although it represents only seven square miles of the Park's total area, nature has been generous to Yosemite Valley. Its broad level floor contains wooded groves and meadows flanked by sheer granite walls which, in places, tower more than four thousand feet above the level, stream-watered valley floor. At various points along the rim, water tumbles from hanging valleys, thereby producing waterfalls of unusual height and beauty. Elsewhere, the valley walls have been sculptured into a host of interesting shapes including bulbous domes, slender spires, and graceful arches.

Since 1855, when the first tourist party rode horses into Yosemite Valley, tens of millions of visitors have succumbed to the magic splendor of its rocks and waters. It is probable that the early visitors, like those of today, wondered how nature could bestow such lavish beauty on so small an area and what natural agents could have shaped the valley walls into such an unusual array of geometrical forms.

Behind the Scenery

The geologic story of Yosemite is much like that of Kings Canyon and Sequoia National Parks which lie to the southeast. All three parks lie across the heart of the Sierra Nevada in east-central California and all are distinguished by similar granite mountains, steep-walled chasms, and magnificent forests. The development of the Sierra Nevada and the early geologic history of the Yosemite-Kings Canyon-Sequoia region has already been discussed elsewhere, so let us now direct our attention to Yosemite Valley and the way in which it was formed.

A Valley Is Born

The sculptured walls of Yosemite Valley pay silent tribute to the geologic efficiency of natural forces, for ice, running water, and the agents of weathering have played key roles in shaping this unusual gorge. But equally important is the character of the bedrock, for in places the granitic rocks are riddled with fractures which render them more susceptible to weathering and facilitate glacial erosion. Other exposures consist of massive unfractured granites which have more effectively withstood the onslaught of weather and ice.

As mentioned earlier (p. 378), the rocks of the Sierra Nevada date far back into the geologic past, but the valley itself was carved within relatively recent geologic time. During early Tertiary time, perhaps during the Eocene or Oligocene Epochs, the area now occupied by the Sierra Nevada began to acquire its characteristic slant to the southwest. It was then that the Merced River (which nows flows along the valley floor) came into existence and began to flow southwestward to empty into the ancient sea which then occupied most of the Great Valley of California. Throughout early Miocene time the Merced meandered sluggishly through the area, gradually developing a wide, level, valley floor.

During later Tertiary time, the Sierra region was subjected to renewed uplift which further steepened the western slope and raised the eastern margin several thousand feet. So strong were these movements that a series of great faults developed along the eastern border, causing the land beyond to subside or remain stable. Thus, the Sierra Nevada was eventually produced into a massive, tilted fault-block mountain range similar to the Teton Range in Grand Teton National Park. The further steepening of the block accelerated the flow of the Merced River, thereby increasing its entrenching, or down-cutting, ability. This enabled the rejuvenated stream to carve a new inner gorge on the old valley floor. Throughout Pliocene time the Sierra Block remained relatively stable, but the Merced continued to deepen its gorge until it eventually produced a sharply incised valley more than one thousand feet deep.

What appears to have been the final elevation of the Sierra Block occurred about two million years ago near the end of the Pliocene Epoch and culminated in a series of uplifts that raised the Sierra Nevada to its present height of more than fourteen thousand feet. Concurrently the eastern lowlands such as Owens Valley were depressed or remained stable, thereby producing the impressive rocky face that marks the eastern Sierra escarpment today. Once more the Merced experienced renewed vigor as it plunged down the steepened southwestern slope and once more it incised the granitic valley floor, this time to a depth of fifteen hundred feet.

Ice—The Master Sculptor

But this early valley was not the Yosemite that we see today. With the advent of the Ice Age, great thicknesses of snow began to accumulate in the wintry heights of the High Sierra. At the heads of the deeper valleys the snow became compacted into dense masses of glacial ice; it was then that the powerful glaciers began their slow, relentless descent into the stream-cut valleys many miles below. Geologic evidence indicates that Yosemite underwent not one but at least three distinct periods of glaciation. Each of these *glacial stages* lasted for thousands of years and each was followed by a warmer (but equally long) *interglacial stage* during which time the ice melted and receded.

Glaciation in the Valleys. The first two ice advances were by far the most prolonged and erosive. During this time ice filled the valley from rim to rim and Glacier Point, which now looms thirty-two hundred feet above the valley floor, was covered by five hundred feet of ice. Only the more lofty prominences such as El Capitan, Half Dome, Eagle Peak, and Sentinel Dome, were not overridden; their towering summits appeared as ice-surrounded rock "islands" called *nunataks*. (Nunataks are present today in certain parts of the great ice sheets of the Arctic and Antarctic.) The first two glaciers reached as far southwest as El Portal; here at an elevation of two thousand feet above sea level, the climate was warmer and the glaciers began to melt.

During the first two glacial invasions, a great tongue of ice (called a *trunk glacier*) was formed by two smaller glaciers which entered Yosemite Valley from the upper Merced and Tenaya Canyons. As the trunk glacier moved slowly down the valley, the twisting, river-cut, V-shaped inner gorge was gradually transformed into a slightly sinuous, U-shaped glacial trough, and it was deepened as much as two thousand feet. The smaller tributary valleys, such as those of Yosemite and Bridalveil Creeks, contained only small bodies of ice and were not eroded downward into the Yosemite Valley. When the ice receded, these were left as hanging valleys from which drop the waterfalls that we see in the Park today.

Although most of the ice erosion was caused by the Yosemite Glacier, other ice streams which flowed from Little Yosemite Valley and Tenaya Canyon also were effective agents of erosion. Their ability to erode was greatly facilitated by the highly fractured bedrock over which they passed. These vertical fractures, or *joints,* divide the granite into natural blocks which may become frozen into the base of the glacier. As the glaciers moved downslope, some of these blocks were pulled from the valley floor—a process known as *glacial quarrying* or *plucking*—and transported downstream by the flowing ice. Thus, the Tenaya Glacier, which moved parallel to one set of

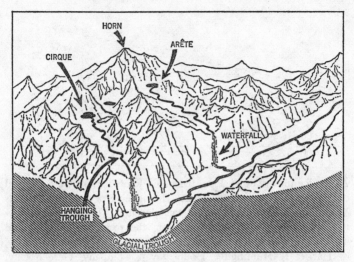

Sketch showing a main glacial trough and hanging valleys left sus-
pended after the ice disappeared. From *Geology: Principles and
Processes,* fifth edition, by W. H. Emmons, I. S. Allison, C. R.
Stauffer, and G. A. Thiel, copyright 1960 by McGraw-Hill Book
Company. Used by permission of McGraw-Hill Book Company.

joints, was able to erode a canyon with a gently sloping floor.
But the glacier which flowed into Merced Canyon from Little
Yosemite Valley moved at an angle across the trend of the
joints. Here the ice excavated with varying degrees of efficiency; in places it removed large blocks from the fractured
granite, elsewhere it could only scour and polish the massive
unfractured rock. This *selective quarrying* resulted in the formation of a glaciated valley whose floor rises in a series of
irregular steplike benches literally forming a giant stairway.
Parts of this long *glacial stairway* are now occupied by Nevada
and Vernal Falls (p. 467).

The third and final glacial invasion filled the valley to only
one-third of its depth and the ice extended but a short distance
below El Capitan. There, the glacier began to melt, depositing
its load of glacial debris in a series of arcuate, moundlike, *recessional moraines,* each of which represents a temporary de-

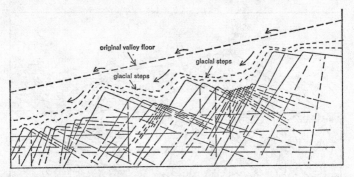

As the ice moved downslope, the much jointed rocks were plucked out, leaving the less jointed rocks. The upper broken line represents Yosemite's valley floor before it was glaciated. The thin lines below it show the stages of glacial erosion, the result of plucking and abrasion. Reproduced by permission from *The Yosemite Story*, by Harriet E. Huntington, Doubleday & Company, Garden City, N.Y. (1966).

crease in the rate of glacial retreat. One of these glacier-deposited ridges, the El Capitan Bridge moraine, near Bridalveil Fall, formed a natural dam that impounded the melt water from the last glacier. This lake—called Ancient Lake Yosemite—occupied a basin scooped from the valley floor by the old Yosemite Glacier and is believed to have extended to the head of the valley, a distance of about five and a half miles. But like most lakes, Ancient Lake Yosemite lasted only a short time, geologically speaking; it soon became filled with sediment deposited by the many postglacial streams which emptied into it. Thus, Yosemite Valley was eventually filled with thousands of feet of silt, sand, and gravel which form the broad, level valley floor that we see today.

Glaciation in the High Sierra. Although the visitor usually sees the evidence of glaciation that has taken place in the valleys, the effects of the Ice Age are even more obvious in the higher elevations of the Park. Here, the glacial scenery assumes an entirely different aspect, for in these areas the glaciers were not confined between narrow valley walls; rather, the broad rolling topography permitted the ice to overrun the region and

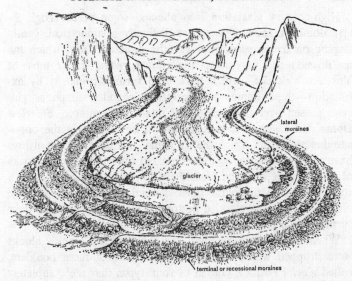

lateral moraines

glacier

terminal or recessional moraines

When the ice withdrew from Yosemite's valleys, it left behind a series of moraines. The moraines are the ridges of boulders, rocks, and soil left by the glacier. The melting water from the glacier has eroded two openings in the moraines. The ice at the end of a glacier is not so high as in the main body of the glacier. Reproduced by permission from *The Yosemite Story*, by Harriet E. Huntington, Doubleday & Company, Garden City, N.Y. (1966).

to cover most of the land surface. In Tuolumne Basin, the area now occupied by beautiful **Tuolumne Meadows** (see map), a broad ice sheet four and a half miles wide spread out over the landscape. So thick was this glacier that many of the mountain peaks in the area were overridden by an ice mass more than fourteen hundred feet thick; even mighty **Fairview Dome,** which now rises twelve hundred feet above the surrounding meadows, was buried beneath eight hundred feet of ice.

Today we see Tuolumne Meadows as a lovely mountain garden; the Tuolumne River lazily meanders over its grassy surface and in places large granite domes project above the basin floor. Mountain meadows such as Tuolumne and Big Meadow appear to be of glacial origin for they are believed to have formed from the filling of shallow, interconnected, glacial lakes.

High-country glaciation also affected some of the High Sierra domes; certain of these are distinctly asymmetrical, gently sloping on their *upstream* sides (the direction from which the ice flowed) and very steep-sided *downstream*. Unlike most of Yosemite's domes—which have been formed primarily by exfoliation—these domes assumed their rounded shape as glaciers passed over them. Some of these, for example, **Fairview Dome** and **Lembert Dome,** appear to be leaning in the opposite direction from which the ice came. But they did not always have this appearance; as the ice passed over them, great blocks of granite were quarried from their downstream face to leave a steep cliff. The upstream sides resisted the removal of large blocks but the rock-studded bottom of the glaciers scoured their surface into a gentler, more streamlined slope.

When the glaciers receded, the ice-quarried granite blocks were dropped where the ice melted. Some of these boulders, called *glacial erratics,* consist of rock types that are completely foreign to the area in which they now rest, thus indicating that they were picked up elsewhere and carried many miles downstream.

As the glaciers ground over the bedrock, they produced a smooth, rather lustrous surface called *glacial polish.* Although evident at various localities throughout the Park, glacial polish is especially well developed on the upstream flanks of **Fairview, Lembert,** and **Pothole Domes.** The latter dome, a low easily climbed feature located just north of where the Tioga Road enters Tuolumne Meadows, is a good place to get a close look at glacial polish. Although these mirror-like surfaces have been weathered away in most places, it may still be found in irregular patches, especially under glacial erratics which have protected the underlying bedrock from the elements. Ice-polished rock surfaces are also common on the shores of **Tenaya Lake** and in **Tenaya Canyon.** Interestingly enough, even the Indians noticed the abundance of glacial polish in this area; they referred to the body of water now called Tenaya Lake as Pywe-ak, "the lake of shining rocks."

Many glacially polished rock surfaces contain rounded cavities called *weather pits.* Although somewhat similar to potholes (which are produced by running water), these pits are the results of strongly localized weathering. Look for weather pits on **Glacier Point, Pothole Dome,** and in other areas throughout the Park.

Most of Yosemite's glacial ice melted thousands of years ago but a few small glaciers and icefields can still be seen in well-protected valleys on the flanks of the higher mountains. The largest of these is **Lyell Glacier,** an ice mass about a half mile long and a mile in width; it lies near the summit of 13,114-foot Mount Lyell, the highest point within the Park. Nearby **Maclure Glacier** measures about a half mile in length and clings to the side of Mount Maclure (13,005 feet in elevation). Although not within Park boundaries, **Dana Glacier,** a smaller body of ice, is situated in a shaded cirque north of the 13,053-foot summit of Mount Dana, which is located about ten miles northeast of Lyell and Maclure Glaciers. But as interesting as they are, these small ice bodies are quite insignificant when compared to the massive Pleistocene glaciers which shaped the face of the High Sierra.

Shaping of the Landscape

Most visitors to Yosemite soon notice the irregular but distinctive outline of the valley rim: bold cliffs, angular spires, massive pillars, and rounded domes rise abruptly above the broad valley floor, sharply accentuating the valley's rock-ribbed walls. Hopefully, the more inquisitive visitor will wonder how and why these fantastic rock structures evolved. Why, for example, do the walls exhibit such an incredible variety of rock structure? And how could such dissimilar geometric forms as, say, a dome and a pinnacle develop side by side from a seemingly homogeneous rock mass? Finally, what combination of geologic processes joined forces to produce so many varied and diverse landforms in such a small area?

To answer the above questions we must carefully examine Yosemite's rocks, for in the final analysis it is the structure and composition of the rock as well as the erosive agent that determines what forms shall be developed.

Cliffs, Pinnacles, and Pillars. To the casual observer, Yosemite's landforms appear to have all been shaped from exactly the same kind of rock. But the rocks are not identical; in fact, approximately twelve different types of granitic rocks have been identified in the Yosemite area. Despite the fact that these are all granite-like rocks, they vary greatly in color, texture, and durability and these variations played an important role in determining how the rocks will react to various erosional processes.

For example, some of the rocks consist of massive, unfractured granite; such rock is highly resistant to erosion and has resisted destruction by ice or weathering. The majestic **El Capitan,** a vast monolith whose almost vertical cliff face rises three thousand feet above the valley floor, is a classic example of the type of landform developed in unjointed granite. Rock of this type is most resistant to erosion—it can even withstand the rigors of glaciation. Thus, **Mount Broderick** and **Liberty Cap** were only rounded and polished by the glaciers, whereas similar bodies composed of fractured granite were completely demolished as the ice moved over them. But, in general, most of Yosemite's rocks have undergone varying degrees of jointing and this has facilitated their destruction. Some have split readily along rather widely spaced vertical joints; these rocks have given rise to most of the perpendicular walls and smooth cliffs that characterize the valley. The cliff face of **Half Dome** is a good example of structures controlled by this kind of jointing. In addition, the main cliff at **Glacier Point,** the precipice of **Upper Yosemite Fall,** and the sheer cliffs adjacent to **Ribbon Fall** are also associated with this type of rock. Elsewhere, rocks containing more closely spaced vertical joints have been sculptured into angular, multifaced structures such as obelisks, columns, and monuments. **Washington Column,** a massive natural pillar seventeen hundred feet tall, and **Sentinel Rock,** a colossal obelisk with a flat front and sharp, splintered crest, exemplify structures originating in this type of rock.

Other rock forms have developed in rocks containing sets of master joints trending in several different directions. These include **Cathedral Rocks,** whose three summits loom 1650, 2590, and 2680 feet respectively above the valley floor, and the **Three Brothers,** an asymmetrical, gabled structure whose slanting "roofs" have formed along westward-sloping joint planes. In some places the rocks are crosscut with a bewildering array of intricate, intersecting joints; such formations have given rise to some finely carved, castlelike, columnar structures.

But not all the landforms have developed from joints. A few prominent landmarks on the valley rim are composed of granite intrusions which have been injected into the surrounding rocks.

Domes and Arches. Yosemite possesses a variety of scenic wonders, but none are as distinctive as the massive granite domes that dot its spectacular landscape. Because of the region's icy past, it was originally thought that all of these helmet-shaped monoliths were upheaval domes or that they had been rounded and polished by overriding glaciers. It is now known, however, that the Sierra Nevada was never completely ice-covered; instead, the glaciers were local in extent and the higher elevations were never buried by ice. For example, such well-rounded features as **Half Dome** and **Sentinel Dome,** could not possibly have been shaped by ice, for their summits were never subjected to glaciation. Moreover, one of the world's most perfectly formed granite domes—Stone Mountain near Atlanta, Georgia—is located several hundred miles beyond the known southern limits of the Pleistocene glaciers.

But if the domes were not formed by glaciation, how were they formed? Geologic field studies indicate that their smooth, rounded surfaces were produced by *exfoliation,* a special type of weathering in which curved sheets or plates of rocks (called *shells*) are stripped from a larger rock mass. You will notice that the surface of every dome is covered with these curving shells, or *spalls* as they are also called. Closer observation reveals that these great sheets of granite are arranged in concentric layers—like the rings of an onion. One should not, however, think of the domes as huge "rock onions." Unlike onions, their concentric shells are confined to the exterior of the dome. The interior is composed of solid rock.

As weathering progresses, the outer spalls—which range in thickness from a few inches to tens of feet—become loosened and drop off, exposing a fresh rock surface to the elements. In time, this too will become separated and fall away; thus, as exfoliation proceeds, successive spalls are removed and the angularities of the original rock are replaced by smooth curves.

That the domes were produced by exfoliation rather than glaciation has been conclusively established, but the process by which the massive, stony slabs are peeled from the parent rock is still not clearly understood. Most geologists agree that exfoliation is a type of *sheeting,* a form of rock rupture similar to jointing. Sheeting develops along fractures that have slightly curved surfaces and lie essentially parallel to the topographic surface, and although the rupture separating each sheet can be seen at the surface, visible sheeting disappears with depth. However, invisible zones of weakness parallel to the sheeting are still present in the rock.

The effects of exfoliation are clearly observable, but its cause is not so obvious. We are not sure, for example, of the origin of the expansive stresses that produce sheeting fractures. It has been suggested that expansion was caused by *hydration* as water combined chemically with the minerals in the granite thus causing it to swell. There is also evidence to indicate that external solar heating may cause the rock to expand. Still another theory proposes that sheeting fractures develop from tension cracks produced during the cooling and crystallization of the rock. Research indicates, however, that solar heating, hydration, or crystallization are not in themselves capable of generating sufficient stress to produce exfoliation. The most accepted belief is that the ruptures develop due to release of stress as the overlying rocks are stripped away by erosion. As confining pressures are relieved, the underlying rock mass gradually swells, producing expansion fractures which lie approximately parallel to the exposed surface. These partings mark off the slabs which, upon being exposed to weathering, will eventually disintegrate and drop off the dome.

Sheeting is best developed in granite-like rocks such as those exposed in the core of the Sierra Nevada and is strikingly illustrated by the scaly granite domes of Kings Canyon, Sequoia, and Yosemite National Parks. For example, such famous

Yosemite features as **Half Dome** and **Sentinel Dome** were produced by exfoliation, and similar exfoliation domes are common throughout the Park. Half Dome, which might be considered Yosemite's "trademark," is by far the most unusual dome
in the Park, if not in the world. Visitors commonly ask: "What
happened to the other half of Half Dome?" The answer to this
question is simple: "It was never there." Contrary to earlier belief, the "missing half" of the dome was not sheared off by an
ancient glacier. Neither the Tenaya Glacier nor the normal
agents of erosion could have effectively done away with this much
rock material during the amount of time involved. Instead, the
stark, precipitous front of Half Dome was developed relatively
rapidly as the result of sheeting along a master set of vertical
joints. As exfoliation proceeded, great slabs of rock were displaced in a plane parallel to the set of joints, thereby producing zones of weakness that rendered the rocks more susceptible
to the processes of glacial erosion. Thus, as these rocky scales
dropped away or were plucked from the dome by glaciers, the
familiar sheer face of this famous monolith was gradually developed. The smoothly rounded back of Half Dome evolved
more slowly; it appears to be the product of exfoliation along
convex fractures whose planes lie essentially parallel to the
gently curving surface of the dome.

Sentinel Dome is easily accessible and is a good place for a
closer look at the effects of exfoliation. It is reached by a spur
road off the Glacier Point Road near Washburn Point and by
means of a short but steep scramble you can reach its 8122-
foot summit. The concentric shell structure of Sentinel Dome
can readily be observed, and its surface is covered with exfoliation spalls which range from six inches to several feet in thickness. In addition, the various Park roads pass by and over a
number of other exfoliation domes. Note especially the great
slabs which litter the surface of **Turtleback Dome;** you will see
these on the south side of Wawona Road about one-half mile
west of Wawona Tunnel.

But not all of Yosemite's domes are the exclusive products of
exfoliation; as noted earlier some have been shaped by glaciers.
Two of these, **Liberty Cap** and **Mount Broderick,** were sculp-

tured by the Merced Glacier. Their curving backs and smooth summits were ground and polished as the ice moved downslope; their steep, craggy fronts face in the direction of ice flow and were subjected to the quarrying action of the glacier. Asymmetrical, ice-hewn landforms of this type are called *roches moutonnés,* a term which literally means "sheep rocks." They were so named because when viewed from a distance their rounded polished forms are somewhat reminiscent of the backs of grazing sheep. Domes such as Liberty Cap and Mount Broderick are composed of massive, unyielding granite and have successfully withstood demolition by the Merced Glacier.

Among the more remarkable geological features seen in Yosemite are the unusual *arches* which sometimes develop in exfoliating granite. Consisting of a series of vaulted, sculptured arcs recessed one within another, these graceful structures are formed by the collapse of the lower portion of exfoliation spalls; the remaining portion of each spall tends to assume the shape of an arch. Although imperfectly formed arches can be seen on a number of domes and cliff faces, the **Royal Arches** are a classic example of this type of weathering phenomenon. Located on the north wall near the head of Yosemite Valley, these arches are carved on the face of a slanting cliff that rises fifteen hundred feet above the valley floor. The arches are of gigantic proportions—the largest has a span of eighteen hundred feet and its underside looms one thousand feet above the base of the cliff. The massive shells which frame the arches range from ten to eighty feet thick, but they join to form a ponderous sheet two hundred feet thick near the top of the main arch. Glacial erosion must also have played an important part in the sculpture of the arches, for during the last stage of glaciation the Yosemite Glacier probably quarried away the lower portions of the shells which had previously been loosened by sheeting.

The Waterfalls

Yosemite's waterfalls, like its cliffs and domes, reflect the influence of rock structure on the scenery. This is especially true in the development of Yosemite's spectacular free-leaping waterfalls, for these plunge from hanging valleys and are associated with sheer cliff faces developed along vertical or steeply inclined master joints. The waters of such falls pour over sheer precipices and, whipped by the wind, fall gracefully downward to pound the rocks below. An alcove—the product of the action of exfoliation by hydration—is eventually produced at the base of the falls.

Yosemite's waterfalls are noted for their height. For example, **Upper Yosemite Fall** plummets 1430 feet—a height equal to nine Niagara Falls piled one on top of the other. After hitting the base of the Upper Falls, the water cascades and falls an additional 675 feet before pitching over the lip of **Lower Yosemite Fall** to descend yet another 320 feet. The combined distance of the Upper and Lower Yosemite Falls plus the intermediate cascades is 2425 feet, a drop which makes Yosemite Falls one of the highest waterfalls in existence. Other examples of free-leaping waterfalls include **Bridalveil Fall,** with an untrammeled vertical descent of 620 feet and **Illilouette Fall,** which drops 370 feet from the hanging valley of Illilouette Creek.

Ribbon Fall, the highest in Yosemite, pours over the north rim of Yosemite Valley almost directly opposite Bridalveil Fall. Here, the waters of Ribbon Creek drop a sheer 1612 feet, a distance greater than the height of the Empire State Building.

Not all of Yosemite's waterfalls are associated with hanging valleys. **Vernal** and **Nevada Falls** tumble down "steps" of the giant stairway carved by the Merced Glacier. In this part of the canyon, the Merced River drops two thousand feet in a distance of one and a half miles. In the lower part of this stretch the river descends in series of tumultuous cascades and rapids, but in its upper reaches, the river makes a more rapid descent via Vernal and Nevada Falls. **Nevada Fall,** at the upper step, drops 594 feet and one-half mile downstream **Vernal Fall** has a drop of 320 feet, a height almost twice that of Niagara Falls.

To see the waterfalls in full splendor, visit Yosemite early in the season; they are at their fullest in May and June while the winter snows are melting, but their volume decreases rapidly during July. Although a few falls run all year, in dry years some have no visible water after the middle of August.

Plants and Animals of Yosemite National Park

As in most of the mountain National Parks, the majority of Yosemite's plants and animals are confined to definite life belts that are related to climate and altitude. There are five life zones in Yosemite National Park and they range from two thousand feet above sea level at Arch Rock to 13,114 feet on Mount Lyell, the highest point in the Park. Each of these zones supports its own characteristic assemblage of plants and animals and it is interesting to watch the fauna and flora change as you ascend to the higher elevations in the Park. The change is more noticeable in the vegetation, for certain of the animals migrate from one zone to the next according to the season. Plants characteristic of the warmer, drier slopes, the Upper Sonoran Zone at about two thousand feet above sea level, are brushy plants like the buckthorn and manzanita with an occasional stand of redbud and Digger pine. Animals that occupy this zone include California mule deer, gray fox, ringtail, the scrub jay, and the thrasher.

At four thousand feet, Yosemite Valley is in the *Transition Zone* which ranges from thirty-five hundred to about sixty-five hundred feet elevation. Living here are ponderosa pine, incense cedar, white fir, and canyon live oak and black oak. California mule deer, gray and Douglas squirrel, and the chipmunk are common mammals. Steller's jay is the bird that you are most likely to see in this zone.

Coniferous trees increase with altitude and predominate in the *Canadian Zone,* which begins at about six thousand feet. Prominent in the extensive evergreen forests are California red fir, Jeffrey, sugar, lodgepole, and western white pines. Animals that may be seen include the golden-mantled ground squirrel, Townsend's solitaire, blue grouse, western bluebird, western tanager, mountain quail, and fox sparrow.

Still higher is the *Hudsonian Zone,* which is encountered between eight thousand and ten thousand feet above sea level. This life belt supports a fauna and a flora that resemble those of the Hudson Bay area of Canada. Lodgepole pine and mountain hemlock are the most common trees; Belding's ground squirrel, marmot, Clark's nutcracker, white-crowned sparrow, and the mountain bluebird also live here.

Above about eleven thousand feet lies the treeless, barren *Arctic-Alpine Zone.* Life is sparse here; trees are stunted and deformed by wind and weather and few animals can withstand the rigors of this inhospitable area.

The Big Trees

Worthy of special note are the famous giant sequoia trees of **Mariposa Grove** in the extreme southern part of the Park. These huge trees, like those of Sequoia National Park (p. 384), are among the oldest and largest living things on earth. The largest tree in the grove is the **Grizzly Giant** with a base diameter of 30 feet, girth of 94.2 feet, and a height of 200 feet. Although it is not possible to determine the exact age of this monstrous tree, its size and gnarled appearance suggest that it is almost three thousand years old. The Grizzly Giant is the largest and oldest tree in Mariposa Grove, but the best-known and most photographed sequoia is **Wawona Tunnel Tree,** through which passed the Mariposa Grove Road. Cut in 1881, the tunnel was 8 feet wide, 10 feet high, and 26 feet long. Unfortunately, this 234-foot giant crashed to the ground during savage winter storms of 1969. Today, this two thousand-year-old tree rests on the forest floor in three segments estimated to contain three hundred thousand board feet of pastel-redwood. Other interesting trees to be seen here include the **California Tree,** which also has a tunnel in it, and the 286-foot-high **Columbia Tree,** which is the tallest in the grove. In addition to Mariposa Grove, giant sequoias grow in **Merced** and **Tuolumne Groves** southeast of Big Oak Flat Entrance in the western part of the Park.

Tips for Tourists

One of Yosemite's more pleasant features is that you can enjoy many of its attractions without having to leave your automobile, or by walking only a very short distance. And this Park has something for everyone. Unlike most National Parks, Yosemite has tennis courts, golf courses, and a swimming pool; and in season you can hike, ride a horse, or rent a bicycle. Most of the visitor activity is confined to Yosemite Valley for this eight-square-mile area is the "heart" of the Park. But minutes away from the crowded valley floor is the back country, whose wilderness areas are the goal of the hiker, the camper, the fisherman, and the lover of nature. Another feature of Yosemite National Park is its especially well-developed and diversified naturalist program. And this is as it should be, for Yosemite was the first National Park in which naturalist programs were offered.

As in all National Parks, your visit will be more enjoyable and you will understand more of what you see if you start your visit at one of the museums. The best place to start here is in the **Yosemite Visitor Center** at Yosemite Village. There are educational exhibits which will explain the geology of the Park and the geologic history of the Sierra Nevada. In addition, there are exhibits which will acquaint you with Yosemite's plants and animals and provide interesting information about the Indians and early human history in the area. Of special interest is the **Indian Circle** located behind the museum; during the summer there are daily explanations of how the Indians hunted, cooked, dressed, and lived. You can also see Indian basketry demonstrations here during the summer.

Yosemite Visitor Center is the headquarters for all interpretive programs in the Park and a schedule of such activities is usually posted. Here, too, there is a book counter where you can purchase books and maps treating subjects of interest to Park visitors.

The **Mariposa Grove Museum** is situated in the "Big Trees" area in the southern end of the Park. Housed in a replica of a log cabin somewhat like the one which occupied this site more than a hundred years ago, the exhibits here deal primarily with the giant sequoias. During the summer, there are daily talks by Ranger-Naturalists.

The **Pioneer Yosemite History Center** at Wawona is devoted exclusively to telling the story of Yosemite's early settlers in the age of the horse. Located nearby is the reconstructed covered bridge over the South Fork of the Merced River; this is the only covered bridge in the entire National Park System. Completely furnished historic cabins and an excellent collection of stagecoaches can be seen here. At El Portal, on the highway from Merced, is the **Pioneer Yosemite Transportation Center.** Its theme of the age of steam and early gasoline vehicles is supported by a number of interesting old vehicles.

Although much can be seen from an automobile, those who would *know* Yosemite must take to its trails. Hiking brings you into close contact with nature and you will be surprised to find how many things you notice while hiking that you would not otherwise see. There are more than seven hundred and fifty miles of well-marked trails radiating from Yosemite Valley to all sections of the Park, and there are several camps, lodges, or hotels situated within an easy day's walking distance of each other. Thus, the hiker may travel light, depending upon the lodges and hotels for accommodations, or he may carry his equipment on his back or by pack animal and thereby be totally independent.

Yosemite's trails are numerous and varied, permitting short, easy trips that require only a few hours, and longer, more difficult ones that may last for several weeks. Among the more popular trails that originate in the valley are those to **Vernal Fall** (two miles round trip from Happy Isles to base of the fall) and the longer **Nevada Fall** trail, a round-trip distance of about six and a half miles from Happy Isles to the top of the fall. More demanding are the trails which involve climbs from the valley floor to the valley rim. These include the **Yosemite Falls Trail,** from Camp 4 to the top of the falls (a round trip of almost seven miles) and the trail to **Glacier Point,** a strenuous 9½-mile round trip from the base of Sentinel Rock. These are but a few of the valley trails, for there are dozens of other trails leading from Wawona and Tuolumne Meadows. You can get additional information at the Visitor Center or any ranger station.

A long-time favorite of veteran hikers is the **High Sierra Loop,** a 53-mile round trip from the valley via Tuolumne Meadows. You can "pack in" if you wish, or you may obtain food and lodging at one of the six High Sierra Camps, most of which are about ten miles apart on the Loop Trail. One of the longest Sierra trails is the **John Muir Trail,** which covers two hundred and twelve miles of High Country wilderness. Named after John Muir, famous Scottish-born naturalist who wrote extensively on Yosemite Valley and the High Sierra, the trail starts in Yosemite Valley, climbs to Tuolumne Meadows, and ends at Mount Whitney in Sequoia National Park.

Persons intending to take longer hikes will find the U. S. Geological Survey topographic maps to be most helpful. They can be purchased at Yosemite Visitor Center at any time of the year and at Tuolumne Meadows and Wawona during the summer. But before starting overnight trips, be sure to stop at ranger stations to check trail conditions and obtain camping information and a fire permit.

The **Badger Pass Ski Area** (p. 452), twenty miles from Yosemite Valley on Glacier Point Road, is headquarters for the winter sports season, which usually lasts from about mid-December to mid-April, depending on the weather. There are ski slopes for most degrees of skill at Badger Pass and all have T-bar lifts and there is also one chair lift. In addition, there are marked ski trails through the nearby woods; these are maintained by the National Park Service. Ski equipment can be rented and lessons are available. You may also arrange an all-expense ski-tour through Yosemite Park and Curry Co., Yosemite National Park, California 95389. In the valley, there is an ice-skating rink at Camp Curry; skates and sleds can be rented there.

Yosemite has been called a photographer's paradise, for the unusual rock formations are as striking in black and white as they are in color. Even the rankest amateur can produce interesting photographs with such outstanding subjects as Yosemite and Bridalveil Falls, Sentinel Rock, El Capitan, Royal Arches, and Half Dome.

Because of the distance involved and the high altitude of the Park, a haze filter should be used on long-distance shots. And when photographing the waterfalls get far enough away to include both the top and base of the fall.

Tunnel View and Glacier Point are especially good vantage points from which to take pictures, but the entire Park offers unusual photographic opportunities at every turn.

Yosemite National Park at a Glance

Address: Superintendent, Box 577, Yosemite National Park, California 95389.

Area: 761,320 acres.

Major Attractions: Mountainous region of unusual beauty; Yosemite and other inspiring gorges with sheer granite cliffs; spectacular waterfalls; three groves of giant sequoias.

Season: Year-round.

Accommodations: Cabins, campgrounds, group campsites, hotels, tents, and trailer sites. For reservations contact: Yosemite Park and Curry Co., Yosemite National Park, California 95389

Activities: Boating, camping, fishing (license required), guided tours, hiking, horseback riding, mountain climbing, nature walks, picnicking, scenic drives, swimming, water sports, and winter sports.

Services: Food service, gift shop, guide service, health service, kennel, laundry, post office, religious services, service station, shuttle bus, ski rental, ski town, ski trails, telegraph, telephone, transportation, picnic tables, rest rooms, and general store.

Interpretive Program: Campfire programs, ecology water float trip, museum, night prowls, nature trails, "rap" sessions, roadside exhibits, self-guiding trails, trailside exhibits.

Natural Features: Canyons, caves, erosional features, forests, geologic formations, glaciation, glaciers, lakes, mountains, rivers, rocks and minerals, soda springs, unusual birds, unusual plants, waterfalls, wilderness area, wildlife, and early pioneer features.

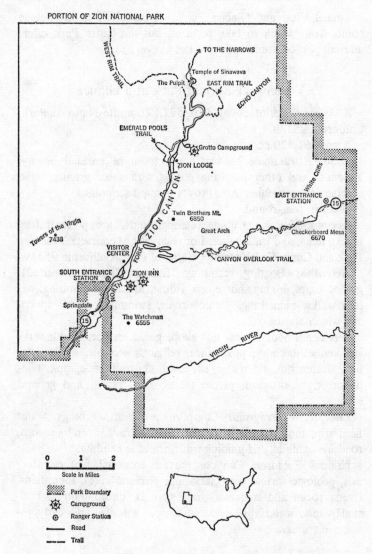

Zion National Park, Utah.

38. Zion National Park, Utah

Heavenly City of God

Like the early Mormon settler who named Zion Canyon, to-day's visitor to this area is likely to view the multihued canyon walls and towering, temple-like monoliths with awe. Indeed, it was the cathedral-like quality of this remarkable region that inspired an early inhabitant to call the area Zion, "the heavenly city of God."

The Geologic Story

Zion National Park—like Bryce Canyon and Canyonlands National Parks—is located in the arid plateau country of Utah, a region long famous for its scenic wonders. Here the relentless agents of erosion—wind, frost, rain, and running water—have carved the land into steep-sided gorges and precipitous cliffs, leaving behind massive erosional remnants resembling lofty temples of stone. But this spectacular desert-canyon landscape reveals only the latest chapter in the geologic history of Zion National Park, for the weather-scarred sedimentary rocks disclose a geologic chronicle that began more than 225 million years ago.

Mesozoic Lands and Seas

Now a desert region several thousands of feet above sea level, the Zion area has not always been so. At various times during the early part of the Triassic Period, this portion of Utah was flooded by a widespread and probably rather shallow area. It was in this chapter of geologic history that the sediments which gave rise to the *Moenkopi Formation* were deposited. Although the Moenkopi consists chiefly of thousands of thin beds of shale and sandstone, thicker layers of gypsum and limestone are found in some parts of the formation. Many of the sediments from which these rocks were formed appear to be of marine origin, but some of them represent continental sediments that were laid down by inland streams and others along a coastal plain. The Moenkopi Formation is best displayed in the southwestern corner of the Park, where approximately eighteen hundred feet of red, brown, pink, and gray rocks are exposed along the Virgin River.

Following the deposition of the Moenkopi sediments, the Zion area was elevated and the land was exposed to the agents of weathering. This made possible the erosion of the uppermost Moenkopi rocks and produced the uneven weathered surface upon which the *Shinarump Conglomerate* was deposited. Although originally believed to have been a separate geologic formation, the Shinarump is now generally considered to be a basal phase of sedimentation of the *Chinle Formation* (see below). As seen in the Zion region, the Shinarump consists primarily of beds of brown, gray, and white sands and gravel which have been cemented together to form course sandstones and conglomerate. It is less than a hundred feet thick within the Park boundaries, but because it is a cliff-forming unit, it is fairly conspicuous as a vertical gray cliff between the more colorful Chinle and Moenkopi strata.

The Shinarump Conglomerate is overlain by the brightly colored shales, sandstones, and conglomerates of the *Chinle Formation* which in this area is about one thousand feet thick. These rocks suggest that the Zion region was a flat lowland during Late Triassic time, for the rocks of the Chinle are believed to have been formed from sands, muds, and gravels that were deposited by streams and in shallow bodies of fresh water. In places there are also layers of volcanic ash interbedded with the sediments; this material probably drifted into the area from nearby volcanoes which erupted during the latter part of the Triassic Period. The postulated continental origin of the Chinle is further substantiated by its fossils, which consist mostly of reptile bones and tracks, the shells of freshwater clams, and large concentrations of fossil wood. The latter is especially abundant in the Petrified Forest and Painted Desert areas of eastern Arizona (p. 331). The Chinle Formation is also remarkably displayed in Utah's brilliant multicolored vermilion cliffs.

The *Moenave Formation*—formerly the Springdale Sandstone Member of the Chinle Formation—lies above the Chinle. Probably Triassic in age, the Moenave appears to be of *fluvial* origin; that is, it was deposited by an ancient river. The formation consists primarily of sandstone and siltstones and the presence of primitive crocodile remains would seem to indicate a tropical or subtropical climate. In addition, the general lack of fossilized land plants and the presence of tongues and lenses of wind-blown sandstone suggest sparseness of vegetation for this area during this segment of geologic time.

The *Kayenta Formation,* a series of stream-deposited purple and maroon sandstones, overlies the Moenave and in places forms a shelving slope at the base of the steep-walled Navajo Sandstone. You will see a good outcrop of the Kayenta Formation at Stop No. 12 on the Weeping Rock Self-guiding Nature Trail. Although the exact age of this formation is not known, it has been variously dated as Late Triassic and Early Jurassic, but it is probably the latter.

Ancient Sand Dunes

Near the end of Triassic time or at the beginning of the Jurassic Period—we cannot be sure which—the Zion region underwent a rather drastic physical change. The climate became much more arid, thereby producing a decidedly desert-like environment. As time passed, strong winds blew the surface sands into great dunes like those of the Sahara Desert, and these ancient "frozen sand dunes" constitute a large part of the *Navajo Sandstone,* one of Utah's most spectacular rock units.

Forming virtually unscalable cliffs several hundred feet high, the Navajo is composed largely of white, fine-grained quartz sandstone, but in places the rocks may be tan, brown, or red. From it have been carved the mighty sandstone cliffs of such Zion landmarks as Checkerboard Mesa, the Great White Throne, and the upper part of the West Temple. Probably the most striking features of the Navajo Sandstone is its unusually well-developed *cross-bedding,* a special type of rock layering or stratification. As noted earlier, most sedimentary rocks are composed of layers which were deposited in a nearly horizontal position; however, bedding planes in parts of the Navajo Sandstone are found to lie at angles that differ greatly from the true horizontal planes that mark the bottom and top of the formation. This type of stratification is known to be developing in sand dunes today, and the cross-bedding in the Navajo is thought to have formed under similar conditions. It is believed, for example, that the steeply inclined beds originated as tilted layers of sand on the *slip face* or leeward slope of a sand dune. Those cross-beds that are tilted in different directions and at different angles were apparently formed when the wind changed its direction and/or its velocity. Thus, the constantly changing wind currents eventually produced the intricate pattern of cross-stratification which was preserved when the wind-blown sand later became cemented into solid sandstone.

The cross-bedding in Zion's ancient sand dunes has not only added to the beauty of the Navajo Sandstone, it has been most valuable in reconstructing the geologic history of the region. For example, the orientation and alignment of the cross-beds indicate that the sands were deposited by winds which blew from the north and northwest. Moreover, the *eolian* (wind-blown) origin of the Navajo is also supported by the presence of innumerable well-sorted, almost perfectly rounded quartz sand grains, many of which have a characteristic "frosted" appearance. The "frosted" or "ground-glass" surface of these translucent grains is similar to the exterior of sand grains that make up typical dune sands in many parts of the world today.

But like most modern deserts, the Navajo desert was not perpetually dry. Occasional rainstorms caused streams to flow and produced small lakes and ponds in which freshwater sediments were deposited. There is little doubt, however, that wind was the principal geologic agent responsible for the creation of this massive sandstone formation.

The Sea Returns

Near the latter part of the Jurassic Period, the sea returned to southwestern Utah and the eroded Navajo desert sands were covered with marine sediments. These sediments later hardened into the sedimentary rocks of the *Carmel Formation:* a series of thin, even-bedded, cream-colored limestones containing many marine fossils characteristic of Late Jurassic time. These rocks—which are perhaps 130 million years old—are the youngest sedimentary rocks in the Park.

With the exception of the previously mentioned layers of volcanic ash, the only igneous rocks within the Park are rather recently extruded lava flows. These basaltic flows and pyroclastics (fragmental volcanic rocks) can be traced to craters which developed within relatively recent geologic time.

A Landscape Is Born

Although the rocks that are exposed in Zion National Park are extremely old, not so the landscape. Zion's famed gorges and sandstone monoliths have been created quite recently, geologically speaking, and they are, in fact, still being shaped by the restless agents of erosion.

The reader, like most visitors to Zion National Park, may wonder what forces are responsible for carving the profound gorge that is Zion Canyon and why was it formed in this particular area. Early visitors to Zion were equally puzzled by this deep chasm and it is not surprising to learn that they considered it to be a great fissure formed where the earth's surface had literally been rent asunder. But although it is possible for chasms of this sort to be created by fractures along faults, there is no evidence that faulting of this magnitude occurred in the canyon. Moreover, the rocks on either side of the canyon walls correlate (match) perfectly, providing proof that this great fissure was not produced by displacement of the rocks along a fracture. What about glaciation? Could Zion Canyon—like incomparable Yosemite Valley—have been gouged from the land by a massive "river" of ice? Not if the rock record is accurate, for no proof of glacial erosion has yet been found. Nor was the canyon formed by continued wind action; geologic evidence to support an eolian origin is also lacking.

How, then, was the sheer-walled, flat-bottomed canyon created? Although it may appear unlikely, Zion Canyon was carved primarily by the running waters of the Virgin River. But the Virgin was not unassisted in its work. The structural framework and composition of the rocks, and geologic agents such as wind, rain, frost, gravity—even plant life—facilitated the cutting of the canyon.

The first geologic event that favored the canyon's development took place many millions of years ago near the end of the Tertiary Period. During this time the Zion National Park region was subjected to a series of intermittent uplifts that gradually raised the area from near sea level to an elevation of more than ten thousand feet. The geologic significance of this uplift was twofold. First, the vertical movements broke the earth's crust into massive fault blocks covering hundreds of square miles. One of these—the *Markagunt fault block*—is the site of the Markagunt Plateau upon whose surface Zion Canyon is incised. Second, the uplift of the area rejuvenated the streams and gave them greater capacity to erode. Their erosive power was, moreover, greatly enhanced by the countless fractures that had been produced by the stress of the structural uplift. Water entered these cracks, cutting them deeper and grinding up and carrying away tons of rock. Thus, the combination of fast-flowing, rejuvenated Virgin River and the uplifted, fracture-riddled Markagunt fault block set the scene for the creation of Zion Canyon.

One's first impression of the usually placid Virgin River is that this rather unimposing stream could not possibly have produced the deep, vertical-walled chasm of Zion Canyon. But the Virgin is, as one geologist has said, "a moving ribbon of sandpaper," whose load of sand and silt has cut and scoured the canyon walls and floor for hundreds of thousands of years. So great is this load of sediment that an estimated three hundred thousand tons of rock debris is removed from the Park each year. The cutting ability of the Virgin River is further enhanced by the steep gradient of the stream, which drops about fifty to seventy feet per mile. This gradient is much steeper than that of most rivers, and greatly increases the velocity of the stream. The Virgin is also aided by water and sediment which it receives from its tributaries; this added water substantially increases the volume and velocity of the river. Although most of the tributary streams are dry throughout much of the year, they may become raging torrents during heavy rainstorms. Moreover, many of these streams flow on bare rock which is not protected by soil or vegetation and their waters are quickly transported to the master stream. Thus, the volume and velocity of the Virgin River make it possible—especially during flood periods—for the river to carry a large load of rock particles which effectively erode the stream channel.

Most of the energy of the stream has been expended in downcutting, for the canyon has been deepened much more rapidly than it has been widened. But as the stream gouged its channel deeper into the solid bedrock, an ever-increasing expanse of canyon wall was exposed to other agents of erosion. It is evident that the processes responsible for widening Zion Canyon have not acted as rapidly as have the process of downcutting, yet slowly—almost imperceptibly—the walls of the gorge have been eroded backward. Rain pounding against the cliff faces, frost wedging in rock crevices, acids formed in air and water, and the prying action of plant roots have joined forces to gradually wear away the rocks of the canyon walls. The disintegration and decomposition of the walls has also been advanced because of the composition and texture of the rocks of which they are composed. The sandstone is quite *friable* (easily crumbled) and consists primarily of weakly cemented rounded quartz grains. These grains are generally cemented together by calcite, a mineral easily decomposed by weak acids. Present also are varying amounts of iron oxides; their presence is reflected in the variegated color of the rock. Side-wall erosion is further accelerated by the large number of joints and bedding planes that lace the canyon walls. Rain that beats on the walls and water that seeps through the rock gradually dissolve the calcareous cement, thereby loosening the individual sand grains and permitting them to fall to the canyon floor or hastening their removal by wind and rain. Thus, the sandstone is reduced to abrasive particles of loose sand that are removed by the river and which assist in the further cutting of the stream channel. Fresh scars on cliff faces and relatively unweathered sandstone slabs on the canyon floor offer proof that this process continues today just as it has in ages past.

Probably the best place to appreciate fully the height-width relationship of Zion Canyon is at the Narrows. Here, the canyon is only a few feet in width but the perpendicular canyon walls loom some fifteen hundred feet above the stream bed. However, you should not plan your visit so as to be there during a heavy rain, for the Virgin River has been known to rise as much as twenty-five feet in as little as fifteen minutes. Flash floods of this type further scour the stream channel and remove tremendous quantities of rock material. Thus, the Virgin River, with an able assist from wind, rain, and frost, continues to mold and reshape the landscape of the Park. And if the geologic cycle of erosion continues as it has in the past, the superb desert-canyon terrain will—in future millions of years—be again reduced to low-lying plains at or near sea level.

Plants and Animals of Zion National Park

The unique desert setting of Zion National Park is considerably enhanced by the plant and animal life in the Park. A variety of plants have adapted themselves to life in the area, and most are somewhat restricted as to where they can grow. In the lower more-arid parts of the canyon, there are many yuccas, cacti, mesquite, and other desert-dwelling plants; piñon and juniper trees thrive near certain of the canyon mouths. The moister places, such as along the riverbanks and around springs, are inhabited by maple and box elder. A colorful array of wildflowers are native to Zion and these include the white evening primrose, the sacred datura or "Zion Moonflower," golden and red-spurred columbine, crimson monkey flower, and the dark-throat shooting star.

An interesting assemblage of animals also called Zion their home. Among the more common small mammals are squirrels, chipmunks, coyotes, ringtails, foxes, skunks, and weasels. Probably the only large animal that you will see is the Rocky Mountain mule deer: look for them in the meadows in late evening. A large variety of snakes, lizards, and frogs also inhabit the Park, but only the rattlesnake is venomous.

Although you are more apt to hear rather than see them, a number of birds live in the Park. Wooded areas at lower elevations frequently attract finches, western tanagers, towhees, vireos, grosbeaks, and nuthatches. At the higher elevations, especially atop cliffs and pillars, you may get a glimpse of a hawk or a golden eagle.

Tips for Tourists

There is much to do and see at Zion National Park, and the area lends itself well to a visit of one day or one week. Free interpretive services are planned to help you to appreciate more fully the natural history of the area, and recreational facilities are designed with the vacationer in mind.

The logical place to start your visit is the Visitor Center, which is located near the South Entrance to the Park. Plan first to see the fourteen-minute illustrated orientation program and then browse the fine exhibits that portray the geology, plants and animals, and human history of this great scenic area. The Visitor Center is open daily throughout the year and uniformed National Park Service personnel stand ready to answer your questions and to help you plan your visit.

Road building in Zion is not an easy task; nevertheless, the National Park Service has constructed some twenty miles of improved roads within the Park. The most spectacular of these —and a Zion "must"—is the Zion-Mt. Carmel Highway

(Utah 15) which extends about eleven miles from the eastern Park boundary to Zion Canyon Junction (see map). Shortly after leaving the East Entrance Station you will see **Checkerboard Mesa,** a Zion landmark. Here you can observe a classic example of the cross-bedding that characterizes the upper part of the Navajo Sandstone. The cross-beds are intersected by many vertical joints, thus forming the checkered pattern from which this distinctive feature derives its name.

About six miles from the Entrance Station, and after descending approximately eight hundred feet by means of a series of six sharp turns, you will encounter the **Zion-Mt. Carmel Tunnel.** This mile-long tunnel took three years to blast through the mountainside and was completed in 1930 after an expenditure of some one million dollars. But this is not just another dark ride through a typical mountain tunnel, for at well-spaced intervals large viewing galleries have been cut through the side of the tunnel and these offer excellent views of Pine Creek Canyon and geologic attractions such as the **East Temple, Sentinel Mountain,** the **Great West Wall,** and the **Great Arch.** The latter has been developed on the south wall of Pine Creek Canyon and is a good example of one of the more important processes that have been responsible for the widening of Zion's steep-sided canyons. At the site of the Great Arch water flowing from seepage springs in the Navajo Sandstone has dissolved the limy cement that binds the sand grains, thus allowing the sand grains to fall apart. As the disintegration of the sandstone progresses, the fabric of the rock is weakened and large, arcuate slabs have become loosened and dropped to the floor of the canyon. Similar arches, alcoves, and undercut areas occur elsewhere in the Park, but, as its name implies, the Great Arch is the largest of these. While driving in the tunnel keep your headlights on at all times, and stop only at officially designated parking areas in the observation galleries.

After leaving the Lower Entrance of the tunnel—so named because it is 287 feet lower than the upper entrance—the road descends into the valley by another series of hairpin turns. Driving along the canyon floor, the road leads past many fine examples of Zion's unusual geology and provides excellent views of the **East Temple, Mount Spry, Sentinel Mountain,** and the domed mass called the **Beehive.** About one-half mile before arriving at Zion Canyon Junction, the highway crosses **Pine Creek,** near which there are good exposures of the colored sandstones and shales of the Chinle Formation.

The other major road in the Park is the **Zion Canyon Road,** which extends for eight miles from the South Entrance to the Temple of Sinawava (see map). The first part of this road is Utah 15, which enters the Park about one mile northeast of Springdale and leads to the South Entrance Station. Approximately one-half mile from the entrance, a side road loops eastward to **South Campground** and **Zion Inn** and rejoins the highway near the **Visitor Center.** At **Zion Canyon Junction,** Utah 15 continues to the east, where it is known as the Zion-Mt. Carmel Highway, which was discussed earlier. Zion Canyon Road proceeds up-canyon past a spectacular display of cliffs and mountains of many shapes, sizes, and colors; there, too, are many of the geologic features for which this Park is famous. On the right (east) side of the road look for the **East Temple,** the paired peaks of the **Twin Brothers** (one "twin" is taller than the other), sharp-pointed **Mount Spry,** the **Mountain of the Sun, Red Arch Mountain,** and finally, towering some 2216 feet above the canyon floor, Zion's trademark—the **Great White Throne.** There is also much to see on the left (west) side of the road. After leaving the Visitor Center look for the **Towers of the Virgin,** the **Beehives,** 7157-foot, flat-topped **Sentinel Mountain,** the **Three Patriarchs, Angels Landing,** and **Majestic Mountain.** The **Zion Lodge Loop** and the road to **Grotto Campground** and picnic area are also located on this section of the road about three miles from Zion Canyon Junction.

Continuing up-canyon past the Great White Throne, you will see the road that leads to the **Weeping Rock** parking area, start of the Weeping Rock Self-guiding Nature Trail. The road then follows the Virgin as it makes its big bend around **Angels Landing,** the rather flat-topped peak which looms fifteen hundred feet above the river on your left. Near here on the east (right) side of the road is 6496-foot **Cable Mountain.**

At the turn of the century, before this area was a National Park, early settlers rigged a 2136-foot cable from the top of this mountain and this was used to lower lumber into the valley below. The sharp-eyed visitor may still be able to see the remains of the old cable-works on the top of the mountain.

After passing **Cathedral Mountain** (elevation 6900 feet) on the left and **Observation Point** on the right, the road ends at the **Temple of Sinawava,** the cool, green, valley oasis that is the climax of this 6½-mile drive. This area is also the starting point of the ever-popular, mile-long trail to the **Narrows** and the **Hanging Gardens of Zion.**

The perpendicular walls of Zion's canyons and monoliths beckon the mountaineer as well as the hiker. But, because of the friable nature of the weathering sandstones, climbing alone is not permitted and all climbers must register at the Visitor Center before starting their ascent.

"Picture taking unlimited" might well describe the photographic opportunities at Zion National Park. Because the steep, vertical canyon walls control the amount of available light, you will probably want to photograph features on the east wall in the afternoon and the west wall in the morning. The Temple of Sinawava and the Great White Throne are best photographed in the early afternoon; the noonday sun provides the most satisfactory light for shooting in the Narrows. Hikers and horseback riders will view an ever-changing panorama which will provide many scenes of unusual beauty. This is equally true of motorists on the Zion-Mt. Carmel Highway from where particularly dramatic shots can be taken from the tunnel's viewing galleries—especially if your subject is carefully framed by the walls of the tunnel window.

Zion National Park at a Glance

Address: Superintendent, Springdale, Utah 84767.

Area: 147,035 acres.

Major Attractions: Outstanding colorful canyon and mesa

scenery; erosion and faulting patterns that create phenomenal shapes and landscapes; former volcanic activity.

Season: Year-round.

Accommodations: Cabins, campgrounds, group campsites, and lodge. *For reservations contact:* Utah Parks Company, Box 400, Cedar City, Utah 84720.

Activities: Camping, fishing (license required), guided tours, hiking, horseback riding, mountain climbing, nature walks, picnicking, scenic drives, swimming.

Services: Food service, gift shop, post office, religious services, service station, telephone, transportation, picnic tables, swimming pool, and rest rooms.

Interpretive Program: Campfire programs, museum, nature trails, roadside exhibits, self-guiding trails, and trailside exhibits.

Natural Features: Canyons, deserts, erosional features, forests, fossils, geologic formations, rivers, rocks and minerals, unusual birds, unusual plants, waterfalls, wilderness area, wildlife, Indian ruins, and hanging gardens.

Glossary

Aa—Hawaiian term for rough, clinkery lava. (Pronounced *ah-ah*).

Abrasion—The wearing away of rocks by rubbing or grinding, chiefly by small grains of silt and sand carried by water or air currents and by glaciers.

Agate—A variety of chalcedony with alternating layers of chalcedony and opal.

Algae—Simple forms of plants, most of which grow in water. Seaweeds are the most common forms found as fossils.

Alluvial Plain—A plain formed by the deposition of materials from rivers and streams.

Alpine Glacier—A glacier confined to a stream valley; usually fed from a cirque. Also called valley glacier or mountain glacier.

Amphibian—A cold-blooded animal that breathes with gills in early stages of life and with lungs in later stages. Intermediate between fish and reptiles.

Amygdaloid—A general name for volcanic rocks that contain numerous gas cavities (vesicles) filled with secondary minerals.

Anticline—An arch, or upfold, of rock strata, with the flanks dipping in opposite directions from its axis.

Anticlinorium—A series of anticlines and synclines so arranged structurally that together they form a general arch or anticline.

Appalachian Revolution—The closing event of the Paleozoic Era; the time when the Appalachian Mountains were originally formed by buckling and folding.

Aquifer—Porous, permeable, water-bearing layer of rock, sand, or gravel capable of supplying water to wells or springs.

Archeozoic—The earliest era of geologic time during which the first known rocks were formed, known also as the Early Precambrian.

Arête—Sharp crest of a mountain ridge between two cirques or two glaciated valleys.

Artifacts—Structures or implements made by man.

Ash—Fine-grained material ejected from a volcano.

Basalt—A common extrusive igneous rock, usually occurring as lava flows and typically black or dark gray in color.

Base Level—The lowest level to which land can be eroded by running water; equivalent to sea level for the continents as a whole.

Basin—Applied to a basin-shaped feature which may be either structural, with rocks dipping inwards, or purely topographical.

Batholith—A huge mass of crystalline igneous rock originating within the earth's crust and extending to great depths.

Bed—The smallest division of a stratified rock series.

Bed Load—Material in movement along a stream bottom or, if wind is the transporting agency, along the surface.

Bedding Planes—Surfaces along which rock layers part readily, by which one layer may be distinguished from another.

Bedrock—The unweathered solid rock of the earth's crust.

Bergschrund—The gap between glacier ice and the headwall of a cirque.

Biochemical Rock—A sedimentary rock composed of deposits resulting directly or indirectly from the life processes of organisms.

Block Mountains—Mountains that result from faulting.

Blue Ridge—The easternmost range of the Appalachian Mountain System, composed largely of very ancient Archeozoic and Proterozoic rocks.

Bomb, Volcanic—A mass of lava ejected from a volcanic vent in a plastic condition and then shaped in flight or as it hits the ground. Larger than one and a half inches across.

Boulder—Large, water-worn, and rounded blocks of stone, most commonly found in stream beds, on beaches, or in glaciated areas.

Braided Stream—A stream whose channel is filled with deposits that split it into many small channels.

Breaker—A wave breaking into foam in the shallow water near the shore.

Breccia—A rock made up of coarse angular fragments of pre-existing rock which has been broken and the pieces recemented together.

Butte—A flat-topped, steep-walled hill; usually a remnant of horizontal beds and smaller and narrower than a mesa.

Calcareous—Composed of calcium carbonate.

Calcareous Algae—Algae that form deposits of calcium carbonate, fossils of which are found in the United States.

Calcite—A mineral composed of calcium carbonate, $CaCO_3$.

Caldera—A large basin-shaped volcanic depression.

Cambrian—The earliest period of the Paleozoic Era or the system of rocks formed in this period.

Carboniferous—Composed largely of carbon. Also, a former period of the Paleozoic Era, now divided into the Mississippian and Pennsylvanian Periods, so-called because it contained the world's greatest coal deposits.

Cementation—The process whereby loose grains, such as silt, sand, or gravels, are bound together by precipitation of mineral matter between them to produce firm rock beds.

Cenozoic—The latest of geologic time, containing the Tertiary and Quaternary Periods, and continuing to the present time.

Central Vent—An opening in the earth's crust, roughly circular, from which magmatic products are extruded. A volcano is an accumulation of igneous material around a central vent.

Chalcedony—The noncrystalline forms of quartz, such as chert, flint, and agate.

Chemical Weathering—The weathering of rock material by chemical processes whereby the original material is transformed into new chemical combinations.

Cinder Cone—Cone formed by the explosive type of volcanic eruption; it has a narrow base and steep, symmetrical slopes of interlocking, angular cinders.

Cinder, Volcanic—A fragment of lava, generally less than an inch in diameter, ejected from a volcanic vent.

Cirque—Steep-walled basin high on a mountain, produced by glacial erosion and commonly forming the head of a valley.

Clastic Rock—Those rocks composed largely of fragments derived from pre-existing rocks and transported mechanically to its place of deposition, such as shales, siltstones, sandstones, and conglomerates.

Clastic Texture—Texture shown by sedimentary rocks formed from deposits of mineral and rock fragments. *See* Clastic Rock.

Clay—The finest type of soil or clastic fragments; having high plasticity when wet, and consisting mainly of aluminum and silica.

Coal—A black, compact sedimentary rock, containing 60 to 100 per cent of organic material, primarily of plant origin.

Coastal Plain—An exposed part of the sea floor, normally consisting of stream- or wave-deposited sediments.

Col—A pass through a mountain ridge. Formed by the enlargement of two cirques on opposite sides of the ridge until their headwalls meet and are broken down.

Column—A column or post of dripstone joining the floor and roof of a cave; the result of joining of a stalactite and a stalagmite.

Columnar Jointing—A pattern of jointing that blocks out columns of rock. Characteristic of tabular basalt flows or sills.

Complex Mountains—Mountains that result from a combination of faulting, folding, and volcanic action.

Composite Cone—Cone formed by intermediate type of volcanic eruption, consisting of alternate layers of cinders and lava; also called a strato-volcano.

Conchoidal—A characteristic break or fracture of a mineral or rock resulting in a smooth, curved surface. Typical of glass, quartz, and obsidian.

Concordant Pluton—An intrusive igneous body with contacts parallel to the layering or foliation surfaces of the rocks into which it was intruded.

Concretion—A nodular or irregularly shaped structure which has grown by mineral concentration around a nucleus, such as siderite concretions or oölitic hematite.

Conglomerate—Water-worn pebbles cemented together; the pebbles are usually of mixed sizes.

Contact Metamorphism—Alteration of rocks caused contact with igneous intrusions.

Continental Glacier—An ice sheet that obscures mountains and plains of a large section of a continent. Existing continental glaciers are on Greenland and Antarctica.

Continental Shelf—The relatively shallow ocean floor bordering a continental landmass.

Contour Lines—Lines of a map joining points on the earth having the same elevation.

Coquina—A coarse-grained, porous variety of clastic limestone composed mostly of fragments of shells.

Correlation—The process of establishing the contemporaneity of rocks or events in another area.

Crater—A bowl-shaped depression, generally in the top of a volcanic cone.

Creep—The slow, imperceptible movement of soil or broken rock from higher to lower levels.

Cretaceous—The latest period of the Mesozoic Era of geologic time.

Crevasse—A deep crack in a glacier.

Crust—The outer zone of the earth, composed of solid rock between twenty and thirty miles thick. Rests on the mantle, and may be covered by sediments.

Crystal—The form of a mineral occurring in a geometric shape with flat or smooth faces meeting each other in definite angles.

Crystalline—Pertaining to the nature of a crystal, such as a rock composed of crystals or crystal grains; often glassy in appearance.

Decomposition—Synonymous with chemical weathering.

Deflation—The removal of material from a land surface by wind action.

Deformation—The result of diastrophism as shown in the tilting, bending, or breaking of rock layer.

Delta—A deposit of sediment built at the mouth of a stream as it enters a larger, quieter body of water, such as the sea, a lake, or sometimes a larger, more slowly flowing stream.

Deposition—The laying down of material which may later become a rock or mineral deposit.

Detrital Sedimentary Rocks—Rocks formed from accumulations of minerals and rocks derived either from erosion of previously existing rock, or from the weathered products of these rocks.

Devonian—The fourth period of the Paleozoic Era.

Diastrophism—The process by which the earth's crust is deformed, producing folds and faults, rising or sinking of the lands and sea bottom, and the building of mountains.

Differential Weathering—The process by which different sections of a rock mass weather at different rates. Caused primarily by variations in composition of the rock itself and also by differences in intensity of weathering from one section to another in the same rock.

Dike—Wall of intrusive igneous rock cutting across the structure of other rocks.

Diorite—A coarse-grained igneous rock with the composition of andesite (no quartz or orthoclase), composed of about 75 per cent plagioclase feldspars and the balance ferromagnesian silicates.

Dip—The slope of a bed of rock relative to the horizontal.

Disintegration—Synonymous with mechanical weathering.

Disturbance—Regional mountain-building event in earth history; commonly separating two periods.

Divide—The ridges or regions of high ground that separate the drainage basins of streams.

Dome—An upfolded area from which the rocks dip outwards in all directions.

Drainage Basin—The area from which a given stream and its tributaries receive their water.

Drift—General term for glacial deposits.

Dripstone—A deposit, usually of limestone, made by dripping water, such as stalactites and stalagmites in caverns.

Drumlin—Oval-shaped hill composed of glacial drift, with its long axis parallel to the direction of movement of a former ice sheet.

Dune—A mound or ridge of wind-deposited sand.

Earthquake—The shaking of the ground as a result of movements within the earth, most commonly associated with movement along faults.

End Moraine—A ridge or belt of till marking the fartherest advance of a glacier; also called a terminal moraine.

Environment—Everything around a plant or animal which may affect it.

Eocene—Second oldest epoch of the Tertiary Period of the Cenozoic Era.

Eolian—Pertaining to the erosion and the deposits resulting from wind action and to sedimentary rocks composed of wind-transported material.

Epoch—A subdivision of a geologic period, such as the Pleistocene Epoch of the Quaternary Period.

Era—A major division of geologic time. All geologic time is divided into five eras: the Archeozoic, Proterozoic, Paleozoic, Mesozoic, and Cenozoic Eras.

Erosion—The process whereby loosened or dissolved materials of the earth are moved from place to place by the action of water, wind, or ice.

Erratic—A large boulder, deposited by glacial action, whose composition is different from that of the native bedrock.

Escarpment—*See* Scarp.

Exfoliation—The scaling or flaking-off of concentric sheets from bare rock surfaces, much like the peeling of onion layers.

Exfoliation Dome—A large, rounded, domal feature produced in homogenous coarse-grained igneous rocks (and sometimes in conglomerates) by the process of exfoliation.

Extrusive—As applied to igneous rocks, rocks formed from materials ejected or poured out upon the earth's surface, such as volcanic rocks.

Extrusive Rock—A rock that has solidified from a mass of magma that poured or was blown out upon the earth's surface.

Fault—A fracture in a rock surface, along which there is displacement of the broken surfaces.

Fault-block Mountain—A mountain bounded by one or more faults.

Faulting—The movement of rock layers along a break.

Fault Scarp—A cliff formed at the surface of a fault.

Fauna—The forms of animal life of a particular region or time period.

Firn—Granular ice formed by the recrystallization of snow. Intermediate between snow and glacial ice; also called *névé*.

Fissure—An open fracture in a rock surface.

Fissure Eruption—Extrusion of lava from a fissure in the earth's crust.

Fjord—A drowned glacial valley.

Flood Plain—The part of a stream valley which is covered with water during flood stage.

Flora—The forms of plant life of a particular region or time period.

Flowstone—A sedimentary rock, usually of limestone, formed by flowing water, most commonly in caverns.

Fold—A bend in rock layers, such as an anticline or syncline.

Folded Mountains—Mountains that result from the folding of rocks.

Foliation—An extremely thin layering or laminated structure in rocks or minerals, often so pronounced as to permit separation or cleavage into thin sheets.

Formation—Any assemblage of rocks having some character in common, whether of origin, age, or composition. Also, anything that has been naturally formed or brought into its present shape, such as dripstones in caverns.

Fossil—Any remains or traces of plants or animals that have been naturally preserved in deposits of a past geologic age.

Fossiliferous—As applied to rocks, any rock containing fossils.

Fossiliferous Limestone—Limestone made from the skeletons of fossilized sea animals.

Friction—The resistance due to surface rubbing.

Frost Action—Process of mechanical weathering caused by repeated cycles of freezing and thawing. Expansion of water during the freezing cycle provides the energy for the process.

Frost Wedging—Prying off of fragments of rock by expansion of freezing water in crevices.

Fumaroles—Fissures or holes in volcanic regions, from which steam and other volcanic gases are emitted.

Geanticline—Very broad upfold in the earth's crust, extending for hundreds of miles.

Geode—A hollow stone, usually lined or filled with mineral matter, formed by deposition in a rock cavity.

Geologic Column—A chronologic arrangement of rock units in columnar form with the oldest units at the bottom and the youngest at the top.

Geologic Revolutions—Periods of marked crustal movement separating one geologic era from another.

Geologic Time—All time which has elapsed since the first known rocks were formed and continuing until recent, or modern, time when the glaciers of the last ice age retreated.

Geologic Time Scale—A chronologic sequence of units of earth time.

Geological Cycle—A period in which mountains are born, and rise above the sea and are again eroded.

Geologist—A person engaged in geological work, study, or investigation.

Geology—The science which deals with the origin and nature of the earth and the development of life upon it.

Geophysics—The physics of the earth.

Geosyncline—A great elongated downfold in which great thicknesses of sediments accumulate over a long period of time.

Geyser—A hot spring which periodically erupts steam and hot water.

Glacial Drift—Boulders, till, gravel, sand, or clay transported by a glacier or its meltwater.

Glaciation—A major advance of ice sheets over a large part of the earth's surface.

Glacier—A body of ice compacted from snow, which moves under its own weight, and persists from season to season.

Glaciofluvial—Pertaining to streams flowing from glaciers and their deposits.

Gneiss—A metamorphic rock, usually coarse-grained, having its mineral grains aligned in bands or foliations.

Graben—A trough developed when parallel faults allow the blocks between them to sink, forming broad valleys flanked on each side by steep fault scarps; also called *rift valley*.

Gradient—The difference in elevation between the head and mouth of a stream.

Granite—An intrusive igneous rock composed of orthoclase feldspar and quartz, and may contain additional minerals, most commonly mica.

Granitization—The process of alteration of other rocks into granite without actual melting.

Granodiorite—A coarse-grained igneous rock intermediate in composition between granite and diorite.

Gravel—A loose deposit of rounded, water-worn pebbles, mostly ranging in size from that of a pea to a hen's egg, and often mixed with sand.

Greenstone or Greenschist—A metamorphosed basaltic rock having a greenish black color.

Ground Moraine—Till deposited from a glacier as a veneer over the landscape and forming a gently rolling surface.

Hanging Valley—A tributary valley which terminates high above the floor of the main valley due to the deeper erosion of the latter; commonly by glaciation.

Headward Erosion—The process whereby streams lengthen their valleys at the upper end by cutting of the water which flows in at the head of the valley.

Historical Geology—The branch of geology that deals with the history of the earth, including a record of life on the earth as well as physical changes in the earth itself.

Horn—A spire of bedrock left where cirques have eaten into a mountain from more than two sides around a central area. Example: Matterhorn of the Swiss Alps.

Hot Spring—A spring that brings hot water to the surface. A thermal spring. Water temperature usually 15° F or more above mean air temperature.

Ice Age—The glacial period or Pleistocene Epoch of the Quaternary Period.

Icecap—A cap of ice usually over a large area. *See also* Continental Glacier.

Igneous Rocks—Rocks formed by solidification of magma.

Impression—The form or shape left on a soft surface by objects which have come in contact with it and which may have later hardened into rock. A type of fossilization consisting of the imprint of a plant or animal structure.

Intrusive Igneous Rock—Molten rock which did not reach the surface of the earth but hardened in cracks and openings in other rock layers.

Invertebrates—Animals without backbones.

Jaspar—Granular cryptocrystalline silica usually colored red by hematite inclusions.

Joint—A break in a rock mass where there has been no relative movement of rock on opposite sides of the break.

Joint System—A series of two or more sets of joints passing through a rock mass so as to separate it into blocks of more or less regular pattern.

Jurassic—The second, or middle, period of the Mesozoic Era.

Karst Topography—A type of landscape characteristic of some limestone regions, in which drainage is mostly by means of underground streams in caverns.

Kettle—A depression remaining after the melting of large blocks of ice buried in glacial drift.

Kipuka—An "island" of old land left within a lava flow.

Laccolith—Lens-shaped body of intrusive igneous rock that has domed up the overlying rocks.

Lacustrine—Pertaining to a lake, sediments on a lake bottom, or sedimentary rocks composed of such material.

Landslide—The downward, rather sudden movement of a large section of land that has been loosened from a hill or mountainside.

Lateral Moraine—A ridge of till along the edge of a valley glacier. Composed primarily of material that fell to the glacier from valley walls.

Lava—Hot liquid rock at or close to the earth's surface, and its solidified products.

Layer—A bed or stratum of rock.

Limestone—A sedimentary rock largely composed of calcium carbonate.

Lithification—The process whereby unconsolidated rock-forming materials are converted into a consolidated or coherent state.

Load—The amount of material that a transporting agency such as a stream, a glacier, or the wind, is actually carrying at a given time.

Magma—Molten rock deep in the earth's crust.

Mantle Rock—The layers of loose weathered rock lying over solid bedrock.

Marble—A metamorphosed, recrystallized limestone.

Marine—Belonging to, or originating in, the sea.

Mass-Wasting—Erosional processes caused chiefly by gravity. Example: a landslide.

Meanders—Wide curves typical of well-developed streams.

Mechanical Weathering—The process by which rock is broken down into smaller and smaller fragments as the result of energy developed by physical forces. Also called *disintegration*.

Mesa—A large, wide, flat-topped hill, usually a remnant of horizontal beds.

Mesozoic—The geologic era between the Paleozoic and Cenozoic Eras; the "Age of Reptiles"; contains the Triassic, Jurassic, and Cretaceous Periods.

Metamorphic Rocks—Rocks that have been changed from their original form by great heat and pressure.

Metamorphism—The process whereby rocks are changed bodily by heat, pressure, or chemical environment into different kinds.

Mineral—A natural, inorganic substance having distinct physical properties, and a composition expressed by a chemical formula.

Mineralogist—A geologist who specializes in studying minerals.

Mineralogy—The subdivision of geology which deals with the study of minerals.

Miocene—Fourth oldest epoch of the Tertiary Period of the Cenozoic Era.

Mississippian—The fifth period of the Paleozoic Era.

Monadnock—A residual hill or higher elevation left standing on a peneplain after erosion of the surrounding material.

Moraine—A ridge or mound of boulders, gravel, sand, and clay carried on or deposited by a glacier.

Mountain Glacier—Synonymous with alpine glacier.

Mud Cracks—Cracks caused by the shrinkage of a drying deposit of silt or clay under surface conditions.

National Monument—An area set aside by the President of the United States or by act of Congress because of its scientific or historical value, and administered by the National Park Service.

National Park—An area of greater importance, and usually of greater extent, than a National Monument, set aside by act of Congress, most commonly because of its scenic and geologic interest.

Névé—Compacted granular snow partly converted into ice; also called *firn*.

Nuée Ardente—Avalanche of fiery ash enveloped in compressed gas from a volcanic eruption.

Obsidian—A glassy rock formed from hardened lava, found just under the foamy top layer.

Oligocene—Third oldest epoch of the Tertiary Period of the Cenozoic Era.

Onyx—A translucent variety of quartz consisting of differently colored bands, often used as a decorative stone. Also, applied to similarly appearing varieties of calcite or limestone, such as dripstones and flowstones of caverns, and used for similar purposes.

Opal—Amorphous silica, with varying amounts of water. A mineral gel.

Ordovician—The second period of the Paleozoic Era.

Organism—Anything possessing life; a plant or animal body.

Orogeny—A major disturbance or mountain-building movement in the earth's crust.

Outcrop—An exposure of bedrock at the surface of the ground.

Outwash—Stratified sediments laid down by the meltwater of a glacier beyond the glacier itself.

Outwash Plains—Plains formed by the deposition of materials washed out from the edges of a glacier.

Overhang—The upper portion of a cliff which extends beyond the lower.

Oxidation—The chemical combination of substances with oxygen.

Pahoehoe Lava—Lava that has solidified with a smooth, ropy, or billowy appearance.

Paleobotany—The branch of paleontology which deals with the study of fossil plants.

Paleocene—Oldest epoch of the Tertiary Period of the Cenozoic Era.

Paleogeography—The study of ancient geography.

Paleontologist—A scientist who studies fossils.

Paleontology—The branch of geology which deals with the study of fossil plants and animals.

Paleozoic—The era of geologic time that contains the Cambrian, Ordovician, Silurian, Devonian, Mississippian, Pennsylvanian, and Permian Periods.

Parasitic Cones—Volcanic cones developed at openings some distance below the main vent.

Pass—A deep gap or passageway through a mountain range.

Peak—The topmost point or summit of a mountain.

Pebble—A smooth, rounded stone, larger than sand and smaller than a hen's egg.

Pediment—Broad, smooth erosional surface developed at the expense of a highland mass in an arid climate. Underlain by beveled rock, which is covered by a veneer of gravel and rock debris.

Pele's Hair—Volcanic glass spun out into hairlike form.

Peneplain—Extensive land surface eroded to a nearly flat plain.

Peneplanation—The process of erosion to base level over a vast area, which results in the production of a peneplain.

Pennsylvanian—The sixth period of the Paleozoic Era.

Period—A main division of a geologic era characterized primarily by its distinctive remains of life.

Permafrost—Permanently frozen subsoil.

Permeability—The degree to which water can penetrate and pass through rock.

Permian—The seventh and last period of the Paleozoic Era.

Petrifaction—A process in which the original substance of a fossil is replaced by mineral matter.

Petrology—The scientific study of rocks.

Piedmont—The area of land at the base of a mountain. That portion of the Appalachian Region which lies alongside the eastern side of the Blue Ridge.

Piedmont Glacier—A glacier formed by the coalescence of alpine glaciers and spreading over plains at the foot of mountains from which the alpine glaciers came.

Pillar—A column of rock in a cavern produced by the union of a stalactite and stalagmite. Also, any column of rock remaining after erosion of the surrounding rock.

Pillow Lava—A basaltic lava that develops a structure resembling a pile of pillows when it solidifies under water.

Pipe (volcanic)—The tube leading to a volcano, sometimes filled with solidified material.

Pit Crater—A crater formed by sinking in of the surface; not primarily a vent for lava.

Plain—A region of horizontal rock layers which has low relief due to a comparatively low elevation.

Plastic Deformation—The folding or flowing of solid rock under conditions of great heat and pressure.

Plateau—A region of horizontal rock layers which has high relief due to higher elevation.

Plateau Basalt—Basalt poured out from fissures in floods that tend to form great plateaus; also called flood basalt.

Pleistocene—The first of the two epochs of the Quaternary Period, and that which precedes modern time, known also as the Great Ice Age.

Pliocene—Last and youngest epoch of the Tertiary Period of the Cenozoic Era.

Plutonic—Applied to rocks which have formed at great depths below the surface.

Plutonic Rocks—*See* Intrusive Igneous Rocks.

Porphyry—A mineral texture of fairly large crystals set in a mass of very fine crystals.

Pothole—A rounded depression in the rock of a stream bed.

Precambrian—A collective name covering the Archeozoic and Proterozoic Eras and the rocks formed during those eras.

Precipitated Rocks—Sedimentary rocks formed by the precipitation of mineral matter out of solution, such as limestone or dolomite.

Proterozoic—The second of the geologic eras, also called Late Precambrian.

Pumice—A froth of volcanic glass.

Pyroclastic Rock—Fragmental rock blown out by volcanic explosion and deposited from the air; for example, bomb, cinder, ash, tuff, and pumice.

Quartz—One of the main rock-forming minerals, composed of pure silica.

Quartzite—A hard metamorphic rock composed essentially of quartz sand cemented by silica.

Quaternary—The second and last period of the Cenozoic Era. It includes the Pleistocene Epoch, or Ice Age, and all the time since.

Rapids—Stretches in a stream where the water drops over rock ledges or accumulations of loose rock, churning itself into foam and making navigation dangerous or impossible.

Recent—All time since the close of the Pleistocene Epoch or Ice Age.

Regional Metamorphism—The alteration of rocks over a very large area due to some major geological process.

Rejuvenation—Any action which tends to increase the gradient of a stream.

Relief—The difference in elevation between the high and low places of a land surface.

Replacement—The formation of mineral replicas of organic remains by the exchange of minerals for cell contents.

Residual Boulders—Large rock fragments formed in place by weathering of the solid bedrock.

Revolution—A time of major mountain building, bringing an end to a geologic period or era.

Rift—A large fracture in the earth's crust.

Rift Valley—A major topographical feature produced by the dropping down of a long segment of the earth's crust between two parallel faults.

Rift Zones—The highly fractured belts on flanks of volcanoes along which most of the eruptions take place.

Ripple Marks—Wavelike corrugations produced in unconsolidated materials by wind or water.

Roches Moutonnées—Bedrock that has been smoothed and "plucked" by the passage of glacial ice.

Rock—Any natural mass of mineral matter, usually consisting of a mixture of two or more minerals, and constituting an essential part of the earth's crust.

Rock Flour—Finely ground rock particles, chiefly silt size, resulting from glacial abrasion.

Rock Glacier—An accumulation of rocky material moving slowly down a valley in the manner of a glacier.

Rock Waste—Fragments of bedrock produced by weathering.

Runoff—The water which flows on the ground surface, tending to drain through streams toward the sea.

Sandstone—Sedimentary rock composed of largely cemented sand grains, usually quartz.

Schist—A finely layered metamorphic rock which splits easily.

Scoria—Slaglike fragment of lava explosively ejected from a volcanic vent.

Scarp—A steep rise in the ground produced either by the outcrop of a resistant rock or by the line of a fault.

Sea Cave—Cave formed as a result of erosion by sea waves.

Sea Cliff—Cliff formed by marine erosion.

Sediment—Solid material suspended in water, wind, or ice; such material transported from its place of origin and redeposited elsewhere.

Sedimentary Rocks—Rocks formed by the accumulation of sediment derived from the breakdown of earlier rocks, by chemical precipitation, or by organic activity.

Seismograph—An instrument which detects and records earthquake waves.

Seismologist—A person who studies and interprets the effects of earthquake activity.

Shale—A sedimentary rock formed by the hardening of mud and clay, and usually tending to split into thin sheets or layers.

Sheet Wash—A type of erosion in which water strips away exposed topsoil slowly and evenly on a slope.

Shield Volcano—A volcano having the shape of a very broad, gently sloping dome.

Silica—The chemical compound of oxygen and silicon which are the two commonest elements in the earth's crust.

Sill—A sheet of intrusive rock lying parallel to the bedding of the rock that is intruded.

Silt—Soil particles intermediate in size between clay particles and sand grains.

Silurian—The third period of the Paleozoic Era.

Sink—A depression in the earth's surface formed by the collapse of the roof of an underground cavern.

Slate—A metamorphosed clay rock with a pronounced cleavage along which it readily splits.

Slide-rock—*See* Talus.

Soil—Layers of decomposed rock and organic materials on the surface of the land areas of the earth.

Solfatara—A fumarole liberating sulfur-bearing gas.

Spatter Cones—Small cones that form in lava fields away from the main vent. Lava is spattered out of them through holes in a thin crust.

Speleology—The scientific study of caverns and related features.

Speleothem—A secondary mineral deposit formed in caves; for example, a stalagmite or a stalactite.

Spring—Water issuing from beneath the surface through a natural opening in sufficient quantity to make a distinct current.

Stack—An isolated column of rock left standing as waves erode a shoreline.

Stalactite—A stony projection from the roof of a cavern, formed of minerals deposited from dripping water.

Stalagmite—A raised deposit on the floor of a cavern, formed by minerals deposited from dripping water.

Strata—Rock layers or beds.

Stratification—The structure produced by the deposition of sediments in beds or layers.

Stratified Rocks—Rocks which occur in parallel layers.

Stratigraphy—The study of rock layers.

Stratum (pl. Strata)—A rock layer or bed.

Strato-volcano—A volcano having a cone of alternate layers of lava and solid fragments.

Striae—Scratches on the surface of rocks resulting from the movement of glacial ice.

Structural Geology—The study of rocks and their relationships.

Submergence—The flooding of land by the sea. Characteristic of most geologic periods.

Subsidence—Sinking of the earth's crust.

Syncline—A fold of layers of rock that dip inward from both sides toward the axis; opposite of anticline.

Synclinorium—A broad regional syncline on which are superimposed minor folds.

System—The rocks which accumulated during a period of geologic time.

Taiga—A type of vegetation characteristic of subarctic climates.

Talus—A mass of rock debris at the base of a steep mountain or cliff; also called *scree*.

Tectonics—The phenomena associated with rock deformation and rock structures generally; the study of these phenomena.

Temblor—An earthquake.

Tertiary—The first of the two periods of the Cenozoic Era; commonly called the "Age of Mammals."

Texture—The composite arrangement, shape, and size of the grains or crystal particles of a rock.

Till—Glacial deposits which have not been stratified or sorted by water action.

Tillite—A sedimentary rock composed of firmly consolidated till.

Topographic Map—A map showing surface features of a portion of the earth.

Topography—The relief and contour of the land surface.

Transport—The carrying by water, wind, or ice from one place to another.

Trap—Old name for a lava flow.

Travertine—A variety of limestone deposited by dripping or flowing water in caverns or by springs, such as stalactites and stalagmites.

Triassic—Oldest period of the Mesozoic Era.

Tributary—A stream which flows into a larger one.

Trilobites—An extinct group of arthropods possibly related to the crustaceans, with a trilobed dorsal skeleton.

Trough—A channel or long depression between two ridges of land.

Tundra—A type of climate in the zone of transition between the subarctic regions and the icecaps.

Unconformity—A break in the sequence of rock formations which separates younger groups from older ones; caused primarily by removal of some of the older rocks by erosion before those of a later sequence were laid down.

Uniformitarianism—The doctrine that the past geological record can be interpreted by reference to present-day phenomena and processes. "The present is the key to the past."

Uplift—The elevation of any extensive part of the earth's surface from a lower position by some geologic force.

Valley—A long depression on the earth's surface, usually bounded by hills or mountains, and typically traversed by a stream which receives the drainage from the adjacent heights.

Valley Train—The deposit of rock material carried down by a stream originating from a glacier confined in a narrow valley.

Vein—A thin and usually irregular igneous intrusion.

Vent—An opening where volcanic material reaches the surface.

Ventifact—A stone that has been smoothed by wind abrasion.

Vertebrates—Animals with backbones.

Vesicular—Having bubble-holes formed by gases.

Volcanic—Pertaining to volcanoes or any rocks associated with volcanic activity at or below the surface.

Volcanic Neck—A rock plug formed in the passageway of a volcano when magma slowly cools and solidifies there.

Volcanism—A general term including all types of activity due to movement of magma.

Volcano—The vent from which molten rock materials reach the surface, together with the accumulations of volcanic materials deposited around the vent.

Warping—The bending of sedimentary beds of rock into broad, low domes and shallow basins.

Waterfall—The dropping of a stream of water over a vertical or nearly vertical descent in its course.

Water Gap—A valley that cuts across a mountain ridge, through which the stream still flows.

Water Table—The upper boundary of the groundwater, below which all spaces within the rock are completely filled with water.

Wave-built Terrace—A seaward extension of a wave-cut terrace, produced by debris from wave action.

Wave-cut Terrace—A level surface of rock under the water along the shore, formed as waves cut back the shoreline.

Weathering—The natural disintegration and decomposition of rocks and minerals.

Selected Readings

Hopefully, this book has whetted the reader's appetite for additional information about planet Earth. The following list includes selected references of many types—any one of which will provide further reading on the various phases of geology. This list is by no means all-inclusive, and many other interesting and informative publications may be found in public, school, and college libraries. Wherever possible, the paperback edition of a book is listed and designated by an asterisk. Those not so marked are available only in hardcover.

TEXTBOOKS

General Geology

Bates, Robert L., and Sweet, Walter C., *Geology: An Introduction*. D. C. Heath & Company, Boston (1966).

Eardley, Armand J., *General College Geology*. Harper & Row, New York (1965).

Fagan, John J., *View of the Earth: An Introduction to Geology*. Holt, Rinehart & Winston, New York (1965).

Putnam, William C., *Geology*. Oxford University Press, New York (1964).

Shelton, John S., *Geology Illustrated*. W. H. Freeman and Co., San Francisco (1966).

Stokes, William L., and Judson, Sheldon, *Introduction to Geology: Physical and Historical*. Prentice-Hall, Englewood Cliffs, New Jersey (1968).

Physical Geology

Emmons, William H.; Allison, Ira S.; Stauffer, Clinton R., and Thiel, George A., *Geology: Principles and Processes* (5th ed.). McGraw-Hill Book Company, New York (1960).

Gilluly, James; Waters, Aaron C.; and Woodford, A. O., *Principles of Geology* (3rd ed.). W. H. Freeman and Co., San Francisco (1968).

Holmes, Doris L., *Elements of Physical Geology*. The Ronald Press Company, New York (1969).

Leet, L. Don, and Judson, Sheldon, *Physical Geology*. (4th ed.) Prentice-Hall, Englewood Cliffs, New Jersey (1971).

Longwell, Chester R.; Flint, Richard Foster; and Sanders, John E., *Physical Geology*. John Wiley & Sons, New York (1969).

Historical Geology

Clark, Thomas H., and Stearn, Colin W., *Geological Evolution of North America*. The Ronald Press Company, New York (1968).

Dunbar, Carl O., and Waage, Karl M., *Historical Geology* (3rd ed.). John Wiley & Sons, New York (1969).

Stokes, William L., *Essentials of Earth History: An Introduction to Historical Geology*. Prentice-Hall, Englewood Cliffs, New Jersey (1960).

INTRODUCTORY READINGS AND "OUTLINES"

* Foster, Robert J., *Geology*. Charles E. Merrill Books, Columbus, Ohio (1966).

* Leet, L. Don and Florence J., *The World of Geology*. McGraw-Hill Book Company, New York (1961).

* Matthews, William H., III, *Geology Made Simple*. Doubleday & Company, Garden City, New York (1967).

* ———, *Invitation to Geology*. The Natural History Press and Doubleday & Company, Garden City, New York (1971).

* Pearl, Richard M., *Geology: An Introduction to Principles of Physical and Historical Geology*. Barnes & Noble, New York (1960).

* ———, *Geology Simplified*. Barnes & Noble, New York (1967).

Scientific American. Resource Library. *Readings in the Earth Sciences*. W. H. Freeman and Co., San Francisco (1969).

* White, John F. (Ed.), *Study of the Earth: Readings in the Geological Sciences*. Prentice-Hall, Englewood Cliffs, New Jersey (1962).

General (Nontechnical)

Beiser, Arthur, and the Editors of *Life*, *The Earth*. Time-Life Books, New York (1962).

Clayton, Keith, *The Crust of the Earth: The Story of Geology*. The Natural History Press, Garden City, New York (1967).

Dunbar, Carl O., *The Earth*. World, New York (1966).

* Gamow, George, *A Planet Called Earth*. The Viking Press, New York (1963).

Mather, Kirtley F., *The Earth Beneath Us*. Random House, New York (1964).

Matthews, William H., III, *Introducing the Earth*. Dodd, Mead & Company, New York (1972).

Shimer, John A., *This Changing Earth: An Introduction to Geology*. Harper & Row, New York (1967).

* Viorst, Judith, *The Changing Earth*. Bantam Books, New York (1967).

EARTH MATERIALS AND PROCESSES

General

* Cailleux, André, *Anatomy of the Earth*. McGraw-Hill Book Company, New York (1968).

Fraser, Ronald G. J., *The Habitable Earth*, Basic Books, New York (1965).

* Spar, Jerome, *Earth, Sea, and Air: A Survey of the Geophysical Sciences*. Addison-Wesley, Palo Alto, California (1962).

Earth Materials

* Fenton, Carroll L., and Mildred A., *The Rock Book*. Doubleday, Garden City, New York (1940).

Gallant, Roy A., and Schuberth, Christopher J., *Discovering Rocks and Minerals*. The Natural History Press, New York (1967).

Hurlbut, Cornelius S., Jr., *Minerals and Man*. Random House, New York (1968).

* Hurlbut, Cornelius S., and Wenden, Henry E., *The Changing Science of Mineralogy*. D. C. Heath & Company, Boston (1967).

* Pearl, Richard M., *How to Know the Rocks and Minerals*. Signet, New York (1957).

————, *Rocks and Minerals*. Barnes & Noble, New York (1956).

Pough, Frederick H., *A Field Guide to the Rocks and Minerals* (3rd ed.). Houghton Mifflin, Boston (1960).

Geologic Processes

* Adams, William M., *Earthquakes: An Introduction to Observational Seismology*. D. C. Heath & Company, Boston (1964).

Bullard, Fred M., *Volcanoes: In History, in Theory, in Eruption*. University of Texas Press, Austin, Texas (1962).

Dyson, James L., *The World of Ice*. Alfred A. Knopf, New York (1963).

Engle, Eloise, *Earthquake! The Story of Alaska's Good Friday Disaster*. John Day, New York (1966).

Heintze, Carl, *The Circle of Fire: The Great Chain of Volcanoes and Earth Faults*. Meredith Press, New York (1968).

Herbert, Don, and Bardossi, Fulvio, *Kilauea: Case History of a Volcano*. Harper & Row, New York (1968).

* Hodgson, John H., *Earthquakes and Earth Structure*. Prentice-Hall, Englewood Cliffs, New Jersey (1964).

Iacopi, Robert, *Earthquake County: How, Why, and Where Earthquakes Strike California*. Lane Magazine & Book Co., Menlo Park, California (1964).

* Leet, L. Don and Florence, *Earthquake: Discoveries in Seismology*. Dell Publishing Co., New York (1964).

Macdonald, Gordon A., *Volcanoes*. Prentice-Hall, Inc., (1972).

* Macdonald, Gordon A., and Hubbard, D. H., *Volcanoes of the National Parks of Hawaii*. Hawaii Natural History Association, Hawaii Volcanoes National Park, Hawaii (1970).

Matthews, William H., III, *The Story of Glaciers and the Ice Age*. Harvey House, Inc., Irvington-on-Hudson, New York (1973).

————, *The Story of Volcanoes and Earthquakes*. Harvey House, Inc., Irvington-on-Hudson, New York (1969).

Roberts, Elliott, *Our Quaking Earth*. Little, Brown and Company, Boston (1963).

Vaughan-Jackson, Genevieve, *Mountains of Fire: An Introduction to the Science of Volcanoes*. Hastings House, New York (1962).

Wilcoxson, Kent H., *Chains of Fire: The Story of Volcanoes*. Chilton Books, Philadelphia (1966).

Landforms and the Earth's Surface

Baars, Donald L., *Red Rock Country: The Geologic History of the Colorado Plateau*. Doubleday/Natural History Press, Garden City, New York (1972).

* Bloom, Arthur L., *The Surface of the Earth*. Prentice-Hall, Englewood Cliffs, New Jersey (1969).

* Dury, G. H., *The Face of the Earth*. Penguin Books, Baltimore, Maryland (1966).

Shimer, John A., *This Sculptured Earth, The Landscape of America*. Columbia University Press, New York (1959).

EARTH HISTORY AND FOSSILS

* Barnett, Lincoln and the Editors of *Life*, *The World We Live In*. Golden Press, New York (1956).

Carrington, Richard, *The Story of Our Earth*. Harper & Row, New York (1956).

* Clark, David L., *Fossils, Paleontology and Evolution*. William C. Brown Company, Dubuque, Iowa (1968).

Fenton, Carroll L. and Mildred A., *The Fossil Book*. Doubleday, Garden City, New York (1965).

* Matthews, William H., III, *Fossils: An Introduction to Prehistoric Life*. Barnes & Noble, New York (1962).

* McAlester, A. Lee, *The History of Life*. Prentice-Hall, Englewood Cliffs, New Jersey (1968).

Moore, Ruth, *Man, Time, and Fossils*. Alfred A. Knopf, New York (1965).

Simak, Clifford D., *Trilobite, Dinosaur and Man: The Earth's Story*. St. Martin's Press, New York (1966).

* Simpson, George G., *Life of the Past: An Introduction to Paleontology*. Yale University Press, New Haven, Connecticut (1953).

Shimer, John A., *This Sculptured Earth, The Landscape of America*. Columbia University Press, New York (1959).

ACTIVITIES IN THE NATIONAL PARKS

Many areas in the National Park System provide special visitor activities as a part of their interpretive and conservation programs, and, in most cases, for the sheer pleasure and enjoyment of the parks. There are boating, fishing, camping, back-country travel, skiing, sledding, snowmobiling, and living history demonstrations.

The National Park Service has published six guides to these special activities, each offering a park-by-park listing and complete descriptions of the programs.

The publications listed below can be ordered from: Manager, Public Documents Distribution Center, 5801 Tabor Avenue, Philadelphia, Pa., 19120.

* CAMPING IN THE NATIONAL PARK SYSTEM. A guide to camping facilities and accommodations in the National Parks, including camping season, limit of stay, number of sites, fees, food and sanitary services, and recreational opportunities; chart lists park addresses and campsite locations; reference map.

I 29.71:971   S/N 2405–0266   25¢

* BOATING REGULATIONS IN THE NATIONAL PARK SYSTEM. Federal regulations covering permits, lifesaving equipment, commercial operations, lighting and signaling devices, and classifications; includes color-coded channel buoy guide and a guide to the uniform State Waterway Marking System.

I 29.9:B 63/970   S/N 2405–0243   40¢

* FISHING IN THE NATIONAL PARK SYSTEM. A guide to fresh-water and salt-water fishing in 61 parks, seashores, and recreation areas, including information on regulations, licenses, and special programs.
1 29.2:F 53/3/969   S/N 2405–0004   30¢

* WINTER ACTIVITIES IN THE NATIONAL PARK SYSTEM. A complete guide to downhill and crosscountry skiing, snowmobiling, and other winter sports in the parks, including information on accommodations, supplies, medical assistance, ski instruction, and winter access to parks by highway, bus, train, and plane; quick reference chart on facilities, activities, and accommodations; park and concessioner addresses.
1 29.2:W 73   S/N 2405–0262   35¢

* BACK-COUNTRY TRAVEL IN THE NATIONAL PARK SYSTEM. A comprehensive guide to more than 40 parks that permit travel in back-country areas on foot, on horseback, by canoe, and by other means; information on planning, conservation, safety, trail use, and regulations; and park topography and wildlife. Helpful hints for the backpacker; reference map.
1 29.9/2:B 12   S/N 2405–0267   35¢

* LIVING HISTORY IN THE NATIONAL PARK SYSTEM. An index of national parks offering live demonstrations, including living farms, arts, crafts and skills, outdoor life, military life, and Indian cultural activities; park-by-park listing with addresses.
1 29.2:L 76   S/N 2405–0216   30¢

All six National Park Service Activities guides, $1.95.

SPECIFIC PARKS

In addition to be the selected publications listed above, the National Park Service issues descriptive brochures and other informational material about each of the Parks. These may be obtained by writing the Superintendents of the respective Parks at the addresses listed in the appropriate chapter. In addition, many of the Parks are affiliated with museums and/or natural history associations that distribute specialized publications and maps about their respective areas.

The United States Geological Survey has also issued a number of publications about the geology of certain of the National Parks. For information about these, write Public Information Office, U. S. Geological Survey, Washington, D.C. 20242.

Index

Aa, 208
Abbe Museum of Stone Age Antiquities, 70
Acadia National Park, 62–72; address, 71; at a glance, 71; map, 62; plants and animals, 69–70; tips for tourists, 70–71
Alaska Range, 272, 279
Algal "reefs," 140
Alpine Visitor Center, 372
Alta Peak, 382
Alum Cave Bluffs, 188
American Museum of Natural History, 389
Amygdaloid Channel, 226
Amygdaloid Island, 224, 226
Andrews Glacier, 367
Anemone Cave, 68, 70
Angel Arch, 95, 98
Angel's Landing, 485
Annaberg Point, 407
Annaberg Ruins, 410
Antelope Springs, 348, 350
Anticlines, 26, 183
Antiquities Act, 8
Appalachian Geosyncline, 397–98
Appalachian Highlands, 181
Appalachian Mountains, 181, 393–99
Appalachian Revolution, 183, 397–98
Appekunny Argillite, 138
Appekunny Mountain, 138
Appetite Hill, 111
Araucarioxylon arizonicum, 333
Arbuckle Mountains, 347
Archeologists, 261, 265
Archeozoic Era, 36
Arches National Monument, 98
Arctic Circle, 271
Arête, 147, 457
Artist Point, 446

Avalanche chutes, 382
Avalanche Lake, 147

Baby Hippo, 110
Back-Country Travel in the National Park System, 512
Badger Pass Ski Area, 472
Badlands, 340
Baird, Donald M., 364
Baker Island, 71
Balcony House, 267
Balds, grass, 188
Balds, heath, 188
Barrier bars, 323
Basin, 75
Basketmaker Period, 262
Bat Flight, 113–14
Bathhouse Row, 216–17
Baths (Hot Springs N. P.), 216–17
Bats, 113–14
Bauer, C. Max, 430
Bear Lake, 367
Bearpaw Meadow, 382
Bears, 447–48
Beehives, 485
Beetle Rock, 382
Big Bend National Park, 74–83; address, 82; at a glance, 82; map, 74; plants and animals, 80; tips for tourists; 81
Big Boiler, 236
Big Break, 250
Big Meadow (Yosemite N. P.), 460
Big Meadows (Shenandoah N. P.), 400
Big Meadows-Stony Man area, 395
Big Room, 109, 111–12
Big Room Tour (Carlsbad Caverns), 111–12
Big Stump Basin, 389
Big Stump Trail, 386, 389

Biology, 12, 39–40
Bird Woman Falls, 148
Black Dragon's Caldron, 442
Black Forest, 332
Black Growler, 443
Black Hills, 421–22
Blake Point, 226
Blue Glacier, 320
Blue Mesa, 332, 342
Blue Ridge Mountains, 393–99
Blue Ridge Parkway, 401–2
Boating, 56, 61. *See also* specific Parks
Boating Regulations in the National Park System, 511
Bobby's Hole, 95
Boiling Springs Lake, 236
Boneyard, 104, 111
Books, 47–48, 507–12
Boone's Avenue, 252
Booth's Amphitheater, 252
Boquillas Canyon, 76, 79, 81
Bordeaux Mountain, 405–6, 410
Bottomless Pit, 112
Boulder Pass, 140
Box Canyon, 298
Boxwork, 422
Braided stream, 274
Bridalveil Fall, 467
Bridger, Jim, 1, 429
Bright Angel Shale, 157
Bright Angel Trail, 164
Broadway Avenue (Mammoth Cave), 252
Brokeoff Mountain, 235, 293
Bromide Hill, 349
Bromide Springs, 349
Bryce Canyon National Park, 84–89; address, 89; at a glance, 89; map, 84; plants and animals, 88; tips for tourists, 88–89
Bubbles, 66
Buffalo Springs, 348, 350
Bumpass, Kendall V., 239
Bumpass Hell, 235–37, 239
Bumpass Hell Trail, 238
Bumpass Mountain, 236, 239
Burnt Monarch, 389
Byrd Visitor Center, 401

Cable Mountain, 485
Cades Cove, 189–90
Cadillac Mountain, 63
Calderas, 120–21, 200, 205–7, 235
California Tree, 469
Cambrian Period, 37
Camp Curry, 472
Camp Denali, 279–80, 282
Campfire programs, 51. *See also* specific Parks
Campgrounds, 54–55. *See also* specific Parks
Camping, 54–55, 60. *See also* specific Parks
Camping in the National Park System, 55, 511
Canadian Shield, 222
Candlelight tours (Wind Cave), 424
Caneel, 407
Caneel Bay Plantation, 411
Canyonlands National Park, 90–91; address, 101; at a glance, 101; map, 90; plants and animals, 99; tips for tourists, 100–1
Canyon Village, 446, 449
Capitan Limestone, 103, 195
Capitan reef, 193–95
Capitol Hill, 431
Capulin Mountain, 120
Carbon Glacier, 297–98, 300
Caribbean Sea, 405
Carlsbad Caverns National Park, 102–15; address, 114; at a glance, 114; map, 102; plants and animals, 113–14; tips for tourists, 114; 195–96
Carmel Formation, 479
Carmichael Entrance (Mammoth Cave), 252
Cary, A. S., 309
Casa Grande, 77–78
Cascade Range, 118, 285–89, 307
Castle Crest Nature Trail, 123
Castle Geyser, 439
Cataract Canyon, 92
Cathedral Mountain, 486
Cathedral Rocks, 462
Catoctin Formation, 395
Cave boxwork, 422

Cave columns, 105, 249
Cave coral, 108, 112
Cave draperies, 107, 111–12, 249
Cave frostwork, 423
Cave pearls, 108
Cave pillars. *See* Cave columns
Cave popcorn, 108, 112
Cave "snowballs," 249
Caves, 103–12, 243–53, 382–83, 420–24
Cedar Mesa Sandstone, 95
Cedar Sink, 248
Cenozoic Era, 38
Centerline, 406, 410
Chaos Crags, 236
Chapel of the Transfiguration, 176
Chasm Lake, 367
Checkerboard Mesa, 478, 484
Cherokee Indians, 188, 354
Chesler Park, 95, 100
Chief Mountain, 142–44, 363
Chief Sequoyah, 354
Chief Sequoyah Tree, 388
Chilhowee Group, 182, 398
Chinese Temple, 112
Chinle Formation, 335–36, 476
Chisos Mountains, 75–76
Chocolate Hole, 406
Christine Falls, 299
Churchill, Sir Winston, 271
Cinder Cone (Lassen Volcanic N. P.), 236–37
Cinder Cones, 19, 27, 119–21, 236–37
Cinnamon Bay, 407
Cinnamon Bay Campground Commissary, 410
Circle Meadow, 388
Cirque Lakes, 147, 299
Cirques, 145, 147, 274, 298, 367, 457
Clagget, William, 3
Classic Pueblo Period, 263
Cleavers, 299
Clements Mountain, 148
Cleopatra Terrace, 441
Cleveland Avenue, 249, 252
Cliff House Sandstone, 259, 261
Cliff Palace, 264, 267
Clingman's Dome, 181

Clough Cave, 382
Coconino Plateau, 159–60
Coconino Sandstone, 159
Col, 147
Collapse blocks, 104
Colonnade, 302
Colorado Plateau, 92, 160, 339
Colorado River, 160–61
Colter, John, 1, 429
Colter Bay Visitor Center Museum, 175
Columbia Crest, 292
Columbia Plateau, 17, 288
Columbia River, 288
Columbia Tree, 469
Columnar jointing, 18, 395
Comet Falls, 299
Composite cones, 27, 118. *See also* Strato-volcanoes
Conducted tours, 51. *See also* specific Parks
Confluence Overlook, 100
Conglomerate Bay, 223
Congress Trail, 386, 388
"Consumptive huts," 244
Continental Divide, 372
Continental Drift, 363, 365
Contraction Theory, 363
Convection Theory, 365
Copper mines, 224
Coral Bay, 410
Correlation, 33
Cowlitz Chimneys, 302
Crabtree Meadows, 402
Cracker Lake, 147
Crandell, D. R., 293
Crater, pit, 208
Crater Lake Lodge, Inc., 126
Crater Lake National Park, 116–26; address, 125; at a glance, 125; map, 116; plants and animals, 123–24; tips for tourists, 124
Crater Observatory, 202
Crater Rim Drive, 208–9
Craters, 120. *See also* Volcanism
Craters of the Moon National Monument, 120
Crescent Formation, 318

Cretaceous Period, 38
Crevasses, 145–46, 297–98
Cross-bedding, 20, 478
Cruz Bay, 407, 411
Cruz Bay Visitor center, 410
Cryptovolcanic structure, 93–94
Crystal Cave, 382–83, 387
Crystal Forest, 331, 337
Crystal Lake, 251
Crystal Springs Dome, 108, 112
Cuthbert Rookery, 133
Cutler Formation, 95
Cycads, 334

Dana Glacier, 461
Deer Mountain, 366
Deer Park, 326
Del Norte State Park, California, 357
Denali, 271
De Soto, Hernando, 213
Devastated Area, 234
Developmental Pueblo Period, 263
Devil's Kitchen, 236
Devils Lane, 95, 100
Devils Postpile, 18
Devils Tower, 18
Devonian Period, 37
Diamond Grotto, 252
Diastrophism, 85
Dickey Ridge Visitor Center, 401
Dikes, igneous, 64, 140, 170
Dinosaur National Monument, 8
Discovery Point, 124
Display Springs, 216–17
Doane, Lieutenant, 3
Doll Theater, 112
Dome-pit, 248
Domes, granite, 463–66
Domes (Mammoth Cave), 248
Dorf, Erling, 434
Doughton Park, 402
Dripstone, 105–8. See also Caves
Druid Arch, 95, 98
Dyson, James L., 145

Eagle Peak, 236
Earth history, 28–38
Earthquake Lake, 443
Earthquakes, 443–46
East Temple, 484–85

Echo River, 248
Ecology, 40–42, 355
Ecosystems, 40–41, 356–57
Eielson Visitor Center, 279
El Capitan, 193, 462
Elastic rebound theory, 444–45
Elks Room, 424
Ellen Wilson Lake, 147
Emerald Ridge, 302
Emmons Glacier, 294, 297, 300
Emmons Vista, 294, 300
Entrada Sandstone, 98
Epochs, 33, 36–37
Eras, 33, 36–37
Erosion, differential, 162
Estes Park, 371
Evaporites, 195
Everglades National Park, 128–35; address, 134; map, 128; plants and animals, 131–33; tips for tourists, 133
Everglades National Park Company, Inc., 134
Excelsior Geyser Crater, 439–40
Exfoliation, 24, 382, 463–65

Fairgrounds, 424
Fairview Dome, 459–60
Fairy Ceiling, 250
Fairyland, 108, 112
Fall River Pass, 370, 372
Fantastic Lava Beds, 236
Fault-block mountains, 25–26, 169, 172, 378–79
Fault line, 26
Fault plane, 142
Faults, 24–26, 142, 183, 444–45
Fault scarp, 26, 207
Fault trough, 171
Fault zone, 237
Feldtman Ridge, 223, 226
Fenster, 184
Fin dome, 382
Finger Lakes, 415
Fins, 97
Fiord, 66
Firearms, 61
Fires, 44, 60
Firn, 150, 296
Fischer, William A., 444

Fishing, 55, 61. *See also* specific Parks

Fishing Bridge, 449

Fishing Cone, 441

Fishing in the National Park System, 512

Fishing licenses, 56, 61. *See also* specific Parks

Fissure eruptions, 121, 206. *See also* Volcanism

Flamingo Visitor Center, 133

Flattop Mountain, 366

Flattops, 342

Float trip (Grand Teton N. P.), 177

Florence Falls, 148

Floridian Plateau, 129

Florist's Garden, 249, 252

Flowstone, 105–8

"Fog belt" (Redwood N. P.), 355

Folded mountains, 182–84, 361–66

Folds, 24–26

Forest Canyon Overlook, 372

Forest fires, 44

Forest Service, 44

Fossil forest (Petrified Forest N. P.), 331–39

Fossil forest (Yellowstone N. P.), 433–35

Fossil trees, 31, 433–35

Fossils, 20, 29–31

Fountain Paint Pots, 442

Francis Bay, 407

Franklin Cliffs, 395

Fred Harvey, 167

Frost wedging, 162, 298

Frostwork, 423

Frozen Niagara Trip (Mammoth Cave), 248, 250–51

Frozen Waterfall 107–8, 110

Fumaroles, 236, 442–43

Fur Trade Museum, 175

Garden of Eden, 424

Garden Wall, 140, 147

Garfield Peak Trail, 124

General Grant National Park, 377

General Grant Tree, 386

General Lee Tree, 386

General Sherman Tree, 354, 386, 389

Geologic formations, 38–39. *See also* specific Parks

Geologic processes, 22–28

Geologic time scale, 33–39

Geological Survey of Canada, 364

Geology, 12–39. *See also* specific Parks

Geomorphologists, 160

Geosyncline, 138

Geyser basins, 436

Geyserite, 438

Geysers, 435–40

Giant Dome, 112

Giant Forest, 386, 388

Giant's Coffin, 252

Gibbon Canyon, 432

Glacial abrasion, 66. *See also* Glaciation

Glacial deposition. *See* Glaciation

Glacial drift, 148

Glacial erosion. *See* Glaciation

Glacial erratics, 65, 239, 275, 300, 319, 367, 460

Glacial grooves, 66, 148, 274

Glacial horn, 148

Glacial lakes, 67, 415. *See also* Glaciation

Glacial polish, 66, 239, 275, 298, 460

Glacial quarrying (plucking), 66, 147, 298, 456–58

Glacial stairway, 457–58

Glacial striations, 66, 148, 274, 298

Glacial trough, 148, 173, 299, 457

Glacial valley. *See* Glacial trough

Glaciation, 22–23, 64–66, 144–50, 226–27, 293–301, 319–21, 366–68, 381–82, 415, 455–67

Glacier, movement of, 144–46, 296–98

Glacier Basin, 370

Glacier Bay National Monument, 9

Glacier ice, 150, 296. *See also* Glaciers

"Glacier milk," 149

Glacier National Park, 136–53;

address, 152; at a glance, 152–
53; map, 136; plants and animals, 150–51; tips for tourists,
151
Glacier Natural History Associa-
tion, 151
Glacier Park, Inc., 152
Glacier Peak, 286, 308
Glacier Point, 461, 462, 471
Glaciers, 22–23, 64–66, 144–50,
366–68, 381–82, 455–67
Glaciofluvial sediments, See Out-
wash
Glaciologist, 296
Glen Canyon Dam, 161
Going-to-the-Sun Road, 138, 148–
49, 152
Grace Harbor, 223
Grand Canyon (Mammoth Cave),
252
Grand Canyon National Park,
154–67; address, 166; at a
glance, 166; map, 154; plants
and animals, 162–63; tips for
tourists, 163–66
Grand Canyon of the Yellowstone,
432, 446–47
Grand Canyon Series, 156–57
Grand Canyon Village, 165
Grand Central Station (Mammoth
Cave), 250, 253
Grand Lake, 371, 373
Grand Loop Road, 449
Grand Promenade, 217
Grand Teton, 170
Grand Teton Lodge Co., 178
Grand Teton National Park, 168–
78; address, 178; at a glance,
178; map, 168; plants and an-
imals, 174–75; tips for tourists,
175–77
Grandview Point, 93, 100
Granite, 14–16. See also Igneous
Rocks; Volcanism
Granite Park, 140
Granitization, 16
Grant Village, 449
Grant's Grove, 386
Grapevine Hills, 78
Great Arch, 484

Great Head, 68–70
Great Lakes, 227, 415
Great Smoky Mountains, 181–86,
393–99
Great Smoky Mountains National
Park, 180–91; address, 190; at a
glance, 190; map, 180; plants
and animals, 186–89; tips for
tourists, 189–90
Great Smoky Overthrust, 184
Great Valley of California, 378,
454
Great Wall of China (Mammoth
Cave), 251
Great West Wall, 484
Great White Throne, 478, 485
Green Gulch, 75
Green Lake Room, 110
Green River, 254
Green River Bottoms, 100
Greenstone Ridge, 226
Grinnell Falls, 148
Grinnell Glacier, 146, 149
Grinnell Point, 138
Grizzly Giant, 469
Gros Ventre Slide, 176
Grotto Geyser, 439
Groundwater, 237, 245–49, 347–
48, 436–41
Guadalupe Mountains, 193
Guadalupe Mountains National
Park, 192–96; address, 196; at a
glance, 196; map, 192; plants
and animals, 195–96
Guadalupe Peak, 193
Guano, 244
Guided Tour (Wind Cave), 423–
24. See also specific Parks
Gunsight Pass, 147
Gypsum deposits, 249

Haleakala "crater," 199–200
Haleakala National Park, 198–203;
address, 203; at a glance, 203;
map, 198; plants and animals,
201–2; tips for tourists, 202–3
Halemaumau vent, 207
Halemauu Trail, 202
Half Dome, 382, 462–65
Hall of Giants, 107, 111

Hammocks, 131
Hanging Gardens of Zion, 486
Hanging valleys, 148, 299, 367, 457
Harry W. Cole Tree, 354
Hart Tree, 386
Hat Creek, 234
Hayden, Dr. Ferdinand V., 3
Havasupai Indian Reservation, 165–66
Hawaii Natural History Association, 208
Hawaii Volcanoes National Park, 204–11; address, 210; at a glance, 210; map, 204; plants and animals, 208–9; tips for tourists, 209–10
Hawaiian Archipelago, 200
Hawaiian Islands, 199, 205, 295
Hawaiian Volcano Observatory, 210
Hawksnest Bay, 410
Hawksnest Beach, 411
Heat Exchange Building, 217
Hebgen Lake Earthquake, 443–46
Hedges, Judge Cornelius, 2
Helictites, 108, 249
Hermit Shale, 158
Herron Glacier, 273
Hidden Lake, 140, 147
Hidden Valley, 373
High Sierra Camps, 472
High Sierra Loop, 471
Highway Pass, 279
Hiking, 53, 60–61, 164. See also specific Parks
Historic Entrance (Mammoth Cave), 251
Historic Trip (Mammoth Cave), 244, 251–52
Hoh Glacier, 320
Hopi House, 165
Hopi Indians, 165
Horn, 299, 457
Horseback riding, 55. See also specific Parks
Horseshoe Falls, 367
Horseshoe Park, 370
Hosmer Grove, 202
Hot Springs, 214–15, 440–41

Hot Springs Mountain, 213–15
Hot Springs National Park, 212–18; address, 218; at a glance, 218; map, 212; plants and animals, 215–16; tips for tourists, 216–17
Houchins, Robert, 243
Houghton Point, 223
Howard Libbey Tree, 354
Hubert Glacier, 320
Hume Glacier, 320
Hunting, 61
Hurricane Ridge, 326
Hurricane Vent, 443
Hydraulic action, 67, 161, 322

Ice Age, 64, 150, 172, 185–86, 274, 289, 319, 454
Iceberg, 105, 110
Iceberg Lake, 146–47, 366–67, 372
Icebergs, 146
Ice Caves, 294
Igneous intrusions, 16. See also Volcanism
Igneous rocks, 14–19, 28, 432. See also Volcanism
Igneous rocks, extrusive, 16–19. See also Volcanism
Igneous rocks, intrusive, 14–16. See also Volcanism
Igneous rocks, plutonic, 14–16. See also Volcanism
Illilouette Fall, 467
Indian Circle, 470
Indian Mountain, 213
"Indian pits," 224
Indians. See specific tribes
Indians, Mesa Verde, 261–66
Ironbound Island, 70
Island in the Sky, 93, 100
Isle au Haut, 63
Isle Royale National Park, 220–31; address, 230; at a glance, 230; map, 220; plants and animals, 227–28; tips for tourists, 228–30
Isle Royale Natural History Association, 230

Jackson Hole, 171, 173–74

Jackson Hole Wildlife Range, 176
Jackson Lake, 177
Jasper Forest, 331
Jedediah Smith State Park, California, 357
Jeffers Glacier, 320
Jenny Lake, 172, 177
Jenny Lake Museum, 175
John Muir Trail, 472
Joints, 26, 456–57, 462–66
Jordan Pond, 65, 67
Jupiter Terrace, 441
Jurassic Period, 38
Juvenile water, 214–15

Kabetogama Lake, 416
Kabetogama Peninsula, 416
Kachina Point, 343
Kaibab Limestone, 159
Kaibab Plateau, 159–60
Kaibab Trail, 164
Karst topography, 246
Katmai National Monument, 9
Kau Desert, 208–9
Kaupo Gap, 200
Kayenta Formation, 477
Kentucky Avenue, 252
Keoneoio Flow, 201
Kern Canyon, 382
Kettle lakes, 275
Kettles, 173, 275
Keweenaw Peninsula, 223, 225
Keweenawan Series, 223
Kilauea Iki, 208
Kilauea Visitor Center, 209
Kilauea Volcano, 206–8
Kilauea Volcano House, 211
King's Bell Cord, 111
Kings Canyon National Park. See Sequoia-Kings Canyon National Parks
King's Draperies, 111
King's Palace, 107, 110
Kinnerly Peak, 148
Klamath Indians, 122
Klapatche, 302
Kolb's Studio, 165
Koolau Gap, 200

La Perouse Bay, 201

La Plata Mountains, 259–60
Labyrinthodont, 334
Lake Chelan National Recreation Area, 310–11
Lake Crescent, 299, 320
Lake George, 299
Lake Granby, 373
Lake Helen, 239
Lake Josephine, 149
Lake McDonald, 149
Lake Superior, 221–25
Lake Superior Basin, 223–25
Lake Sutherland, 320
Land of Standing Rocks, 93, 95
Landslide, Gros Ventre, 176
Landslide, Madison Canyon, 443
Langford, Nathaniel P., 3
Lantern Tour (Mammoth Cave), 253
Laramide Revolution, 143, 259, 363, 365
Lassen, Peter, 233
Lassen National Park Company, 240
Lassen Peak, 118, 233–35, 286
Lassen Volcanic National Park, 232–40; address, 239; at a glance, 239; map, 232; plants and animals, 237–38; tips for tourists, 238–39
Lava, 16–19, 118, 121. See also Volcanism
Lava, aa, 208
Lava, pahoehoe, 207–8
Lava, pillow, 18, 140, 315
Lava Beds National Monument, 120
Lava domes, 118, 205
Lava flows, 205–8, 288, 395. See also Volcanism
Lava flows, submarine. See Lava, pillow
Lava plateaus, 17, 288, 395
Leinster Point, 407
Lembert Dome, 460
Lewis Mountain, 400
Lewis Overthrust, 142
Liberty Cap (Mount Rainier), 292, 298

Liberty Cap (Yellowstone N. P.), 441

Liberty Cap (Yosemite N. P.), 462, 465

Lily-pads, 108, 112, 249

Lincoln Tree, 386

Litter, 60

Little Chief Mountain, 140, 143

Little Cranberry Island, 71

Little Hot Springs Valley, 236

Little Matterhorn, 148

Little Paradise Avenue, 249

Little Tahoma Cleaver, 299

Living History in the National Park System, 512

Logan Pass, 147

Lone Star Geyser, 439

Long Island (Isle Royale N. P.), 226

Long Logs, 331

Longmire, 295

Long's Peak, 367

Lookout Louise, 229

Loomis Museum, 238

Loomis Museum Association, 238

Loop (Glacier N. P.), 148

Lost Creek, 234

Lost John, 243, 252

Lost Mine Peak, 78

Lost River, 248

Louisenhoj Formation, 407

Lower Falls of the Yellowstone, 446

Lower Yosemite Fall, 467

Lyell Glacier, 461

Macaroni Factory, 251

McCargo Cove, 224

McKee, Bates, 309

Mackin, J. H., 309

McKinley Tree, 388

Maclure Glacier, 461

Madison Canyon, 443

Madison Junction, 449

Magma, 14–16. *See also* Volcanism

Magmatic water. *See* Juvenile water

Maho Bay, 407, 410

Majestic Mountain, 485

Mammoth Cave National Park, 242–55; address, 255; at a glance, 255; map, 242; plants and animals, 253–54; tips for tourists, 254–55

Mammoth Domes, 248

Mammoth Hot Springs, 431, 440, 449

Mancos Shale, 258

Many Parks Curve, 371

Manzanita Lake Visitor Center, 238

Maps, 47–48

Marias Pass, 143

Marine erosion. *See* Wave erosion

Mariposa Grove, 469

Mariposa Grove Museum, 470

Mariscal Canyon, 79

Markagunt fault block, 480

Markagunt Plateau, 480

Mark Twain Stump, 389

Marler, George D., 438

Martha Washington's Statue, 252

Marys Rock Tunnel, 394, 401

Mass wasting, 162

Mather, Stephen T., 4

Matterhorn, 148, 299

Maui, 199

Mauna Loa, 205–7

Mauna Loa Strip Road, 207–8

Mauna Loa Trail, 207

Maze, 95

Medical care, 58

Menefee Coal, 259, 261

Menor's Ferry, 176

Merced Canyon, 456

Merced Glacier, 466

Merced Grove, 469

Mesa Verde Company, 268

Mesa Verde Indians, 261–66

Mesa Verde National Park, 256–69; address, 268; at a glance, 268; map, 256; plants and animals, 266; tips for tourists, 266–68

Mesaverde Formation, 258

Mesozoic Era, 37–38

Metamorphic rocks, 20–22, 28

Metamorphism, 20–21

Metchosin Formation, 318
Meteoric water, 214
Miami-Oölite, 130
Middle Cascade Range, 286
Middle Island Passage, 223
Middle Teton, 170
Milner Pass, 370, 372
Mineral water, 348
Minerals, 13–15
Minerva Terrace, 441
Minong Ridge, 224
Mississippian Period, 37
Modified Basketmaker Period, 262
Moenave Formation, 477
Moenkopi Formation, 476
Mokuaweoweo Crater, 205–7
Monadnocks, 398
Mono Lake, 379
Montezuma Valley, 257
Monument Basin, 97, 100
Monument Rock, 227
Moonlight Dome, 251
Moose Visitor Center, 175
Moraine Park, 367
Moraines, 149, 173, 275, 300, 367–68, 458–59
Morning Glory Pool, 441
Moro Rock, 382
Mott Island, 229
Mountain building, 24–27, 141–43, 397–98
Mountain climbing, 56, 177, 303
Mountain of the Sun, 485
Mountaineering. See Mountain climbing
Mountaineering, Exum School of, 177
Mount Adams, 286
Mount Anderson, 320
Mount Angeles, 326
Mount Baker, 286, 308
Mount Broderick, 462, 465
Mount Carrie, 320
Mount Christie, 320
Mount Cleveland, 140
Mount Conrad, 235
Mount Crosson, 272
Mount Desert Island, 63
Mount Desert Range, 64
Mount Diller, 235

Mount Ferry, 320
Mount Foraker, 272–73
Mount Franklin, 229
Mount Harkness, 236
Mount Hayes, 272
Mount Hood, 118, 286
Mount Jefferson, 286
Mount McKinley National Park, 270–82; address, 281; at a glance, 281; map, 270; plants and animals, 275–78; tips for tourists, 278–81
Mount McKinley National Park Co., 281
Mount Mather, 272
Mount Mazama, 117–19, 235, 286
Mount Moran, 170
Mount Olympus, 315, 321
Mount Queets, 320
Mount Rainier National Park, 284–304; address, 304; at a glance, 304; map, 284; plants and animals, 301–3; tips for tourists, 303
Mount St. Helens, 286
Mount Shasta, 118, 286
Mount Silverthorne, 272
Mount Spry, 485
Mount Tehama, 235–36
Mount Tom, 320
Mount Whitney, 381–82, 472
Mount Wilbur, 140, 148
Mowich Lake, 299
Muav Limestone, 157
Mud cracks, 20, 138–39
Mud Geyser, 442
Mud pots, 236, 441–42
Mud volcanoes, 237, 442
Muir Woods National Monument, 8
Muldrow Glacier, 273
Mule trips (Grand Canyon N. P.), 165
Mummy Range, 372
Museums, 50–51. See also specific Parks
Mushpots, 442
Mystic Lake, 300

Namakan Lake, 416

Narrows, 486
National Cemeteries, 5
National Forests, 44
National Geographic Tree, 354
National Historical Parks, 5
National Lakeshore, 10
National Memorials, 5
National Monuments, 5–10
National Park Concessions, Inc. (Big Bend N. P.), 82
National Park Concessions, Inc. (Isle Royale N. P.), 231
National Park Concessions, Inc. (Mammoth Cave), 255
National Park Service, 4–6
National Park Service, uniformed personnel, 5
National Parks, history of, 1–5
National Parks, map of the, 7. See also specific Parks
National Parks Act, 4
National Parkway, 10
National Recreation Areas, 10
National Recreation Demonstration Area, 10
National Riverway, 10
National Seashore, 10
National Scenic Riverway, 10
National Waterway, 10
Natural features, preservation of, 61
Naturalist Program, 50–52. See also specific Parks
Naturalist Sea Cruise (Acadia N. P.), 70
Navajo Sandstone, 477–78
Needles, 93, 100
Nevada Fall, 457, 467
Névé, 150, 296
Never Summer Range, 366, 372
Nisqually Glacier, 294
Nisqually Valley, 295
Norris, P. W., 4
Norris Geyser Basin, 443, 449
North Cascades National Park 306–12; address, 311; at a glance, 311; map, 306; plants and animals, 310–11; tips for tourists, 310
North Kaibab Trail, 164

North Mountain, 213
North Rim (Grand Canyon), 160, 163
Northern Cascade Range, 286
Nuées ardentes, 234
Nunataks, 456

Oberhansley, Frank R., 387
Obsidian Cliff, 432
Ocoee Series, 182
Oconaluftee Visitor Center, 189
Ohanapecosh Formation, 287
Old Faithful Geyser, 438–39, 449
Old Faithful Museum and Visitor Center, 439
Old Rag Mountain, 394
Olympic Mountains, 315–18
Olympic National Park, 314–29; address, 328; at a glance, 328; map, 314; plants and animals, 324; tips for tourists, 327–28
Olympic Peninsula, 295, 317
Olympic rain forest, 324–25
Onyx Chamber, 251
Onyx Colonnade, 251
Opal Terrace, 441
Ordovician Period, 37
Orogenic movements. See Mountain building
Otter Cliffs, 68
Otter Cove, 68
Otter Point, 68
Ouachita Mountains, 215
Outer Brass Limestone, 407
Outlier, 142
Outside Range, 279–80
Outwash, 149, 300
Outwash plains, 173, 275, 300
Overthrust, 26, 142, 184
Owens Valley, 378, 455

Pacific Coast Area (Olympic N. P.), 321–24
Pacific Coast Ranges, 315
Pahasapa Limestone, 422
Pahoehoe lava, 207–8
Paint pots, 442
Painted Desert, 331, 340–43
Painted Grotto, 112
Paleobotanists, 333

Paleogeographic maps, 32
Paleontologists, 29–31
Paleontology, 29. *See also* Fossils
Paleozoic Era, 37
Palmer Cave, 382
Panther Junction, 81
Panther Pass, 75
Papoose Draperies, 107
Papoose Room, 107, 111
Paradise area (Mount Rainier N. P.), 303
Paradise Cave, 382
Paradise Glacier, 294
Paradise Valley, 295
Paradox Formation, 94
Park Archeologists, 5
Park Historians, 6
Park Naturalists, 5–6
Park Rangers, 5–6
Park roads, 52–53. *See also* specific Parks
Park trails, 53, 60–61. *See also* specific Parks
Pass, 147
Passage Island, 226
Paunsaugunt Plateau, 87
Peaks of Otter, 402
Pearly Gates, 424
Pediment, 260
Pele, 207
Pennsylvanian Period, 37
Periods, 33, 36–37
Permafrost, 276
Permian Period, 37
Peters Glacier, 273
Petrified Forest National Park, 330–44; address, 344; at a glance, 344; map, 330; plants and animals, 342–43; tips for tourists, 343–44
Pets, 61
Phantom Ranch, 165
Phantom Ship, 125
Photography, 57. *See also* specific Parks
Phreatic explosion, 293
Phytosaurs, 334
Picnicking, 52–53
Piegan Pass, 147
Pillow lava, 18, 140, 315

Pilot Pinnacle, 235
Pink Cliffs, 85
Pinnacle Wall, 140–41
Pinnacles (Crater Lake N. P.), 124
Pioneer Farmstead, 190
Pioneer Yosemite History Center, 470
Pioneer Yosemite Transportation Center, 471
Pithouses, 263
Pits (Mammoth Cave), 248
Plateau Point, 165
Platt National Park, 346–50; address, 350; at a glance, 350; map, 346; plants and animals, 349; Tips for tourists, 349–50
Platt National Park Museum, 349
Plug Dome, 235
Plutons, 16. *See also* Volcanism
Point Lookout Sandstone, 258
Point Success, 292
Porcelain Basin, 443
Porcupine Hills, 431
Porcupine Islands, 70
Post Office Room, 424
Pothole Dome, 460–61
"Potholes." *See* Kettles, 173
Prairie Creek State Park, California, 357
Precambrian, 36.37, 156–57, 182–83
President Tree, 388
Price Park, 402
Promenade Nature Trail, 217
Prospect Peak, 236
Proterozoic, 36
Ptarmigan Lake, 147
Ptarmigan Wall, 147
Pueblo Indians, 265
Pueblos, 263
Puget Sound Basin, 315
Pulpit Terrace, 441
Pummel Peak, 77
Punch Bowl Spring, 441
Puu O Maui, 200
Puu Ulaula, 202
Puyallup Cleaver, 299
Puyallup Glacier, 298

Quaternary Period, 38
Queen's Chamber, 107–11

Radioactive "clock," 35
Radiolarians, 407
Rain forest, Olympic, 324–25
Rainbow Curve, 372
Rainbow Dome, 251
Rainbow Forest, 332
Rainbow Forest Visitor Center, 332, 343
Rainier, Rear Admiral Peter, 289
Rainier National Park Co., 304
Rainy Lake, 416
Rankin Ridge, 425
Raspberry Island, 226
Red Arch Mountain, 485
Red Canyon Fault, 444
Red Mountain, 236
Redwall Limestone, 158
Redwood Mountain Grove, 386
Redwood National Park, 352–57; address, 357; at a glance, 357; map, 352; plants and animals, 355–57; tips for tourists, 357
Redwood trees, 353–55, 384–89, 469
Reef Bay, 406, 410
Reef, Capitan, 193–95
Religious services, 58
Rendezvous Bay, 407
Rental services, 58
Reynolds Mountain, 148
Ribbon Fall, 462, 467
Rim Drive, Bryce Canyon N. P., 88
Rimstone, 251
Rim Village, 124
Ripple marks, 20, 139
Riverside Geyser, 439
River Trail, 164
Roches moutonnés, 466
Rock Cut, 370, 372
Rock flour, 148, 298
Rock Harbor, 226, 229
Rock Harbor Lighthouse, 223, 229
Rock of Ages, 107, 112
Rock Reef Pass, 130
Rocks, 13–22. See also specific Parks

Rocky Mountain Geosyncline, 143, 362–64
Rocky Mountain National Park, 360–74; address, 374; at a glance, 374; map, 360; plants and animals, 368–71; tips for tourists, 371–74
Rocky Mountains, 142–43, 259, 361–66
Rocky Mountains (Mammoth Cave), 252
Room Tree, 388
Roosevelt, President Theodore, 332
Roosevelt Dome, 248, 250
Ross Lake National Recreation Area, 311
Rotunda, 244, 246, 248, 252
Royal Arches, 466
Royal Palm Visitor Center, 133
Ruhle, George C., 151
Ruins Road, 267

Sable Pass, 277
St. Croix Island, 405
St. Elias Mountains, 272
St. John Island, 405
St. Mary Valley, 148
St. Thomas Island, 405
Salt dome, 94
Saltpeter, 244
Sand Beach, 69
San Juan Mountains, 259–60
Santa Elena Canyon, 76, 78, 81
Scenic Trip (Mammoth Cave), 252–53
Schilderia adamania, 333
Schoodic Point, 63
Schooner Head, 68
Schulz, Paul E., 238
Sea Arch, 323
Sea caves, 227, 323
Sea stacks, 323
Seal Cove, 68
Sedimentary rocks, 19–20, 28
Sediments, 19
Seiche, 444
Seismic waves, 444
Seismograph, 210, 238
Seismologists, 444

Self-guiding trails, 51–52. *See also* specific Parks
Sentinel Dome, 463–65
Sentinel Mountain, 484
Sentinel Rock, 462
Sequoia and Kings Canyon National Parks Company, 391
Sequoia gigantea, 353, 384
Sequoia-Kings Canyon National Parks, 376–91; address, 391; at a glance, 391; map, 376; plants and animals, 382–86; tips for tourists, 386–87
Sequoia Natural History Society, 387
Sequoia sempervirens, 353, 356
Sequoia trees, 353–55, 384–89, 469
Service stations, 58
Shadow Mountain National Recreation Area, 371, 373
Sheep Lake, 370
Sheep Mountain, 176
Shenandoah National Park, 392–403; address, 402; at a glance, 402; map, 392; plants and animals, 399–401; tips for tourists, 400–2
Shenandoan Natural History Association, 400
Shenandoah Valley, 398
Shield volcano, 27, 118, 205
Shields, Precambrian, 222
Shinarump Conglomerate, 476
Ship Harbor, 68
Sierra Block, 378–81, 455
Sierra del Carmen, 77
Sierra Nevada, 377–81, 453–55
Sieur de Monts Spring, 70
Signal Mountain Road, 175
Silent City, 87
Siliceous sinter, 438
Sills, 140. *See also* Volcanism
Silo Pit, 248, 250
Silurian Period, 37
Silverswords, 201
Singleshot Mountain, 138
Sinkholes. *See* Sinks
Sinking Creek, 248
Sinks, 246–48
Sinnott Memorial Overlook, 124

Sinopah Mountain, 148
Sioux Indians, 421
Siyeh Limestone, 140
Skyland, 400
Skyline Drive, 400–2
Skyscraper Mountain, 302
Sled Dog Demonstration, 280
Slicks, 188
Slide Lake, 176
Sliding Sands Trail, 202
Snake River (float trip), 177
Snorkel trips (Virgin Islands N. P.), 411
Snow Lake, 299
Snowball Room, 249, 252
Soleduck Formation, 316–17
Solfataras, 237
Solution. *See* Caves
Somes Sound, 66
South Bubble, 65
South Kaibab Trail, 164
South Rim (Big Bend N. P.), 78
South Rim (Grand Canyon), 160, 163
Southern Cascade Range, 285
Specimen Avenue, 249, 252
Specimen Mountain, 366, 372
Specimen Ridge, 435
Speleothems, 105. *See also* Caves
"Spelunkers," 103
"Spelunking" tours (Wind Cave), 424
Sperry Glacier, 146, 149
Split Mountain, 148
Spruce Canyon, 367
Spruce Tree House, 264
Square Tower House, 264
Squaw Flat, 100
Stalactites, 106–7, 249. *See also* Caves
Stalagmites, 106–7, 249. *See also* Caves
Star Chamber, 252
Steam vents. *See* Fumaroles
Stegocephalian, 334
Stone Mountain, Georgia, 463
Strait of Juan de Fuca, 319
Strato-volcanoes, 27, 118, 235, 290–91

Stream erosion, 23, 92–93, 160–62, 480–81
Success Cleaver, 299
Sugar Loaf Mountain, 213
Sugarlands Visitor Center, 189
Sulfur water, 348
Sulfur Works, 235–37
Summit Mountain, 142
Sunset Amphitheater, 298
Sunset Crater National Monument, 120
Supai Formation, 158
Superintendent of Documents, 44
Superposition, Law of, 35
Suzy's Cave, 229
Swimming, 56. See also specific Parks
Synclines, 26, 183
Synclinorium, 272

Taiga, 275
Tansill Formation, 103
Tapeats Sandstone, 157
Tectonism, 85
Tehipite Dome, 382
Temple Butte Limestone, 158
Temple of Sinawava, 486
Temple of the Sun, 112
Tenaya Canyon, 456, 460
Tenaya Glacier, 465
Tenaya Lake, 460
Terlingua, 81
Tertiary Period, 38
Teton Fault, 171–72
Teton fault block, 172
Teton Range, 169–72, 378
Theater Curtain, 251
Thermal areas, 236–37, 435–43
Thomas A. Jaggar Memorial Museum, 209
Three Brothers, 463
Three Patriarchs, 485
Three Sisters, 286
Thunder Hole, 68
Till, 148–49, 300
Tiltmeter, 210
Tips for Tourists, 59. See also specific Parks
Tipsoo Lake, 299
Tonto Group, 157

Tonto Platform, 157
Topographic maps, 48
Topography, 28, 246
Toroweap Formation, 159
Totem Pole, 112
Tours, conducted, 57. See also specific Parks
Tower Fall, 449
Tower Ruin, 95
Towers of the Virgin, 485
Trail Ridge Road, 371–73
Trails, 51–53. See also specific Parks
Trapping, 61
Travertine, 105, 249, 348, 440–41
Travertine Creek, 350
Travertine Falls, 348
Travertine Island, 348
Triassic Period, 38
Trunk Bay, 410–11
Tufa, 216–17, 348
Tule Mountain, 77
Tundra, 276, 369–70
Tundra Trail, 370, 372–73
Tuolumne Grove, 469
Tuolumne Meadows, 459
Turtleback Dome, 465
Tutu Formation, 407
Twin Brothers, 485
Twin Buttes, 431
Twin Domes, 112
Two Medicine Lake, 147
Tyndall Glacier, 366

Udall, Stewart L., 91
Unconformities, 33, 141, 157
Underground Lunchroom (Carlsbad Caverns), 109, 111
Uniformitarianism, Principle of, 31
U. S. Geological Survey, 48, 210
U. S. Government Printing Office, 44
U. S. Office of Map Information, 48
Upheaval Dome, 93–94, 100
Upper Falls of the Yellowstone, 446
Upper Geyser Basin, 439
Upper Two Medicine Lake, 147
Upper Yosemite Fall, 462, 467

Utah Parks Company, 167, 487

Valley train, 300
Van Trump Park, 302
Vancouver, Captain George, 289
Vandalism, 61, 435, 441
Vegetation, native, 43–45. See also specific Parks
Veiled Statue, 110
Vernal Fall, 457, 467
Virgin Islands National Park, 404–12; address, 411; at a glance, 411–12; map, 404; plants and animals, 409–10; tips for tourists, 410–11
Virgin River, 480–81
Virginia Falls, 148
Virginia Park, 95
Virginia Sky-Line Company, Inc., 403
Vishnu Schist, 156
Volcanic bomb, 18
Volcanic cones, 16–17, 27. See also Volcanic mountains; Volcanism
Volcanic mountains, 25, 27. See also Volcanism
Volcanic rock, 16–19
Volcanism, 14–19, 23–24, 233–37, 431–33
Volcano, active, 205–8
Volcano, composite. See Stratovolcano
Volcano, dormant, 201, 233–35
Volcano, lava dome, 205
Volcano, plug dome, 235
Volcano, shield, 205
Volcano, strato-volcano, 27, 118, 235, 290–91
Voyageurs, Les, 415–16
Voyageurs National Park, 414–18; address, 417; at a glance, 417; map, 414; plants and animals, 416–17; tips for tourists, 417
Vulcanism. See Volcanism
Vulcan's Castle, 236

Wapowety Cleaver, 299
Wasatch Formation, 86
"Washboard" topography, 226
Washburn, Henry D., 2

Washburn Expedition, 3
Washington Column, 462
Washington Harbor, 226
Water Island Formation, 406
Waterfalls. See specific falls, Parks
Wave-cut cliffs, 68, 227, 322
Wave-cut terrace, 322
Wave erosion, 24, 67–69, 321–24
Wawona Tunnel, 465
Wawona Tunnel Tree, 469
Weathering, 24, 87–88
Weeping Rock, 485
West Mountain, 213–14
West Rim Trail, 163
West Temple, 478
West Thumb, 441, 449
Whale's Mouth, 107, 110
White, Jim, 109
White Glacier, 320
White Hill, 202
White Rim, 93, 100
Wild Cave Tour (Mammoth Cave), 253
Wildlife, 42–43, 61. See also specific Parks
Williams, Howel, 119
Willis Wall, 298
Wilson Dome, 248
Wind Cave National Park, 420–26; address, 426; at a glance, 426; map, 420; plants and animals, 425; tips for tourists, 425–26
Windigo, 229
Window, 78
Winter Activities in the National Park System, 512
Winter sports, 56. See also specific Parks
Wizard Island, 121, 125
Wonder Lake, 277
Woodworthia arizonica, 333

Yaki Point, 164
Yakima Park, 300
Yavapai Museum, 158
Yellowstone Lake, 441
Yellowstone Library and Museum Association, 430, 438, 444, 449
Yellowstone National Park, 428–

50; address, 450; map, 428; plants and animals, 447–48; tips for tourists, 448–49
Yellowstone Park Bill, 3–4
Yellowstone Park Co., 450
Yosemite Glacier, 466
Yosemite National Park, 452–73; address, 473; at a glance, 473; map, 452; plants and animals, 468–69; tips for tourists, 469–73
Yosemite Park and Curry Company, 472

Yosemite Valley, 453–67
Yosemite Village, 470
Yosemite Visitor Center, 470

Zion Canyon Junction, 485
Zion Canyon Road, 485
Zion-Mt. Carmel Tunnel, 484
Zion National Park, 474–87; address, 486; at a glance, 486; map, 474; animals and plants, 482–83; tips for tourists, 483–86